Energy Audits and Improvements for Commercial Buildings

Energy Audits and Improvements for Commercial Buildings

Ian M. Shapiro

Cover image: Cover drawings by Florence Baveye; Cover photos by Ian M. Shapiro
Cover design: Wiley

This book is printed on acid-free paper. ∞

Published by John Wiley & Sons, Inc., Hoboken, New Jersey.
Published simultaneously in Canada.

For general information about our other products and services, please contact our Customer Care Department within the United States at (800) 762-2974, outside the United States at (317) 572-3993 or fax (317) 572-4002.

Wiley publishes in a variety of print and electronic formats and by print-on-demand. Some material included with standard print versions of this book may not be included in e-books or in print-on-demand. If this book refers to media such as a CD or DVD that is not included in the version you purchased, you may download this material at http://booksupport.wiley.com. For more information about Wiley products, visit www.wiley.com.

Library of Congress Cataloging-in-Publication Data:

ISBN 978-1-119-08416-7 (hardback) ISBN 978-1-119-08422-8 (ePDF)
ISBN 978-1-119-08421-1 (epub) ISBN 978-1-119-17485-1 (oBook)

Names: Shapiro, Ian M., author.
Title: Energy audits and improvements for commercial buildings : a guide for
    energy managers and energy auditors / Ian M. Shapiro.
Description: Hoboken, New Jersey : John Wiley & Sons, 2016. | Includes
    bibliographical references and index.
Identifiers: LCCN 2015044563 (print) | LCCN 2015050091 (ebook) | ISBN
    9781119084167 (cloth) | ISBN 9781119084211 (ePub) | ISBN 9781119084228
    (Adobe PDF)
Subjects: LCSH: Commercial buildings—Energy conservation—Handbooks,
    manuals, etc. | Energy auditing—Handbooks, manuals, etc.
Classification: LCC TJ163.5.B84 S53 2016 (print) | LCC TJ163.5.B84 (ebook) |
    DDC 696—dc23
LC record available at http://lccn.loc.gov/2015044563

Printed in the United States of America

10 9 8 7 6 5 4 3 2 1

Contents

Acknowledgments   ix

Chapter 1   Introduction   1

Chapter 2   Overview   5

Chapter 3   Field Guide   17

Chapter 4   Envelope   27

Chapter 5   Lighting   71

Chapter 6   Heating   99

Chapter 7   Cooling and Integrated Heating/Cooling Systems   117

Chapter 8   Heating and Cooling Distribution   137

Chapter 9   Ventilation   151

Chapter 10   Identifying Heating and Cooling Equipment   167

Chapter 11   Controls   171

Chapter 12   Water   189

Chapter 13   Electric Loads (Other than Lighting)   217

Chapter 14   Gas Loads (Other than Heating and Domestic
             Hot Water)   237

Chapter 15   Advanced Energy Improvements   239

Chapter 16   Estimating Savings   247

Chapter 17   Financial Aspects of Energy Improvements   257

Chapter 18   Reporting   **265**

Chapter 19   Sector-Specific Needs and Improvements   **271**

Chapter 20   Project Management   **281**

Chapter 21   Operation, Maintenance, and Energy Management   **295**

Chapter 22   Portfolio Programs   **305**

Chapter 23   Resources   **317**

Appendix A   Building Material R-Values   **319**

Appendix B   Window Ratings   **321**

Appendix C   Air-Mixing Method of Airflow Measurement   **323**

Appendix D   Recommended Illuminance   **325**

Appendix E   Lighting Power Allowances—Space-by-Space   **327**

Appendix F   HID Lighting Designations   **329**

Appendix G   Lighting Software   **331**

Appendix H   Lighting Reflectances   **333**

Appendix I   Room Air Conditioner Efficiency Requirements   **335**

Appendix J   Chiller Efficiency Requirements   **337**

Appendix K   Existing Exhaust Schedule   **339**

Appendix L   Existing Outdoor Air Schedule   **341**

Appendix M   Proposed Outside Air Schedule   **343**

**Appendix N   Simplified Model of a Building Entering or Recovering
from Setback**   **345**

Appendix O   Gas Pilot Sizes and Gas Use   **347**

Appendix P   **Estimated Existing Motor Efficiencies, Pre-1992**   **349**

Appendix Q   **Equipment Expected Useful Life**   **351**

Appendix R   **Request for Proposal for Energy Audits**   **359**

Appendix S   **Energy Audit Review Checklist**   **363**

Appendix T   **Energy Preventive Maintenance Schedule**   **367**

Index   **369**

Acknowledgments

Big thanks go to Florence Baveye, of Taitem Engineering, who prepared all the drawings in the book and annotated the photographs. And thanks to Susan Galbraith, who did background research on a variety of topics.

Thanks also go to the staff at Wiley who answered my many publishing-related questions and made the book happen, including Margaret Cummins, Mike New, Paul Drougas, Justin Mayhew, Sharon Kucyk, and Kerstin Nasdeo.

On the technical side, thanks to the energy audit staff at Taitem Engineering, including Rob Rosen, Vaibhavi Tambe, Umit Sirt, Sara Culotta, Jim Holahan, and Mahbud Burton. Thanks to Tim Allen for a very helpful discussion about benchmarking and billing analysis. Thanks to Anna Legard, Bill King, Mahbud Burton, Susan Galbraith, Rob Rosen, Nate Goodell, Betsy Parrington, and Don Wells for reviewing and editing portions of the writing. Thanks to Melissa Conant-Gergely for assistance with organizing the images. Thanks to Taitem interns Colin Diamond, Korin Carpenter, Ihotu Onah, and Lucas Sass for review/edits, research, and other assistance. And to all the other wonderful Taitem staff, from whom I learn so much every day and have for so many years. Thanks to Mark Lorentzen, TRC, for review and input on portfolio programs. Thanks to Don Fisher, Fisher Nickel Inc., for insights into commercial cooking equipment energy efficiency. Thanks to Timothy Sutherland, Navigant Consulting, Inc., for information on commercial clothes washer standards and interpretation of ratings. Thanks to my brother, Phil Shapiro, for putting his library skills to work in finding hard-to-find resources.

And, last but not least, for their patience and support, my family: Dalya, Shoshana, Tamar, and Noa. Shoshana also helped with edits, and took the photo of an old suspended gas heater, something I had been looking for, at a drive-in movie theater. I take responsibility for all other photos, taken during my work in the field. Noa also helped with review and with organizing the images.

The book is dedicated to the loving memory of my mother and father.

Ian M. Shapiro
Ithaca, New York
March 2016

Chapter 1

Introduction

Buildings account for 40 percent of U.S. energy use. A strong interest in energy conservation is motivated by concerns over climate change, pollution, energy costs, and reliance on fossil fuels.

Energy audits and improvements are being driven by an increasing number of bold goals to reduce energy use and carbon emissions. The widely recognized Architecture 2030 program has set a goal of reducing energy use in existing buildings by 50 percent. The federal government has had a goal of reducing energy intensity in federal buildings by 3 percent per year. States are also setting energy reduction goals. For example, New York State has set a goal to reduce greenhouse gas emissions by 80 percent by the year 2050.

An *energy audit*, also called an *energy assessment*, is an evaluation of a building's energy systems in order to identify opportunities for reducing energy. *Energy improvements* form the implementation of the work, actually reducing energy by making changes in buildings. By addressing both energy audits and improvements in one book, we seek to take the recommendations of energy audits and translate them into reality. We seek to transform our buildings.

This Book's Focus

The book's focus is how to reduce energy use in existing commercial buildings, including such buildings as offices, hospitals, multifamily dwellings, schools, universities, hotels, retail, religious, not-for-profit, institutional buildings, and more.

The book covers a broad variety of commercial building types, including both larger and smaller buildings (Figure 1.1), for all climates. This scope is intentional. Historically, energy audits for larger buildings have primarily been performed by engineers, and the focus has been on heating, cooling, lighting, and controls, while the building envelope (insulation, windows and doors, infiltration) has not received sufficient attention. However, audits for smaller buildings have largely been done by non-engineer energy auditors, with a strong focus on building envelope, and less attention directed to heating, cooling, lighting, and controls. This book is intended to bring the strengths of larger-building energy audits to smaller building energy audits, and vice versa.

The book goes beyond *energy audits* to also cover *energy improvements*—in other words, the installation of energy conservation improvements in order to deliver substantial and persistent energy savings—addressing such topics as project management, quality control, financing, and operation and maintenance. It also covers *portfolio programs*, in which government agencies, utilities, or owners of building portfolios seek to plan and implement energy improvements across multiple buildings.

Throughout, we maintain an interest in what we might call *transformational energy improvements*. We are interested in energy improvements that transform our buildings, which measurably reduce energy use, and which bring our aging building stock up to current standards of energy use, comfort, health, and safety.

Figure 1.1 Commercial buildings comprise a wide variety of building types.

Figure 1.2 Effective energy work requires not only good energy audits but effective energy improvements—the actual changes that can transform a building.

The book is intended for energy auditors, energy managers, energy engineers, building performance contractors, and students in these fields. The book may also be of use to energy policymakers and utility demand-side management professionals.

We frequently address both the energy auditor and the energy manager in the book. These two individuals can form a powerful team to address energy problems. The energy auditor brings a knowledge of many buildings, of best practices, of specialty topics like energy modeling, and more. The energy manager brings in-depth knowledge of their own building. The energy auditor brings specialty tools for diagnosing energy problems. The energy manager has the ability to take measurements in their buildings over time. The energy auditor might not include an inventory of every refrigerator in a 500-unit apartment complex, but such an inventory may already be maintained by an energy manager, or may be of interest to an energy manager, and will lead to a far better assessment of the potential for refrigerator replacements. The complementary skills and abilities of the energy auditor and the energy manager can be most effectively put to use through collaboration. This book, therefore, does not stop only at the information sought in a typical energy audit, but rather suggests deeper investigations to understand and reduce energy use in buildings.

The book is improvement-centric. We focus our attention on energy improvements. In much energy work, the excitement about new products and what we are proposing to evaluate and recommend for a building leads us to focus our attention extensively on the "thing that we want to install." In the process, we can make the mistake of paying too little attention to "the thing that is already installed," in other words, the baseline against which energy savings will be measured. If our evaluation of the baseline is inadequate, we run the risk of overestimating savings, if we assume, for example, that the baseline is worse than it really is. Conversely, if we assume that the baseline is better than it really is, we may underestimate potential savings, and prematurely rule out a good energy improvement. The baseline is fully as important, in estimating savings, as the new product about which we might be so enthusiastic. In this book, we try to direct equal attention to the baseline, to establishing what a building already is and has in it, and how each component of the building uses energy. We try to provide authoritative sources that will support energy calculations for these existing components, whether it is the average spray duration of a prerinse spray valve in a commercial kitchen, or how to estimate the efficiency of a 40-year-old chiller.

The book seeks to provide a broad set of solutions to the problem of building-related climate emissions, by offering technical and programmatic guidance to reduce energy use in buildings to as low as net-zero. Climate change has become the most pressing environmental challenge facing society. The book has as its goal to be an evidence-based reference for energy conservation and associated climate emissions reductions in the building sector.

The book is based on fundamentals of building science, along with practical discussions of energy improvements, to broadly cover the emerging field of building performance as it relates to existing commercial buildings. It seeks to help energy auditors and energy managers deliver measurable savings for purposes of mitigating climate change impacts of buildings, reducing energy costs, and achieving related goals such as improved indoor comfort, human health, and air quality.

In seeking to support high-quality fieldwork to establish baseline energy conditions, we try to provide a comprehensive field guide for examining buildings and their energy components. An early chapter serves as a general field guide, and then detailed field guidance is integrated throughout the book on how to examine buildings and energy components of buildings. There is value to good fieldwork for energy audits. Extra effort in the field, rolling up our sleeves in buildings, will pay off with more improvements, better quality assumptions and measurements, and deeper and more accurate savings.

Seeing in Buildings

Energy work starts with knowing what we are seeing in buildings. We must learn to identify building components and energy-consuming equipment. We then need to move expeditiously beyond just recognizing these elements to understanding how much energy they might be using, and then, further, to identifying possible energy savings opportunities. As such, our end goal is not only to know what we are seeing, but a new kind of seeing, one that sees deeply, and sees the potential for transformational energy savings in buildings.

This book has photographs of real buildings and real energy components in real buildings: lighting, heating, insulation, windows, appliances, and more. We believe that it is important to recognize what we are seeing in a building, in the building's real state. It is important to be able to see through the rust, dirt, and deterioration, in order to assess the energy use of what is already there (Figure 1.4). We do not attempt to provide shiny photographs of buildings and new equipment. To the contrary, we seek to highlight the deficiencies that contribute to energy inefficiency, and that contain the potential for saving energy. We need to be able to distinguish between real energy inefficiencies and perceived energy inefficiencies. Sometimes, dirt and rust are covering perfectly good energy systems. Saving energy is our goal, and we seek to do it by understanding existing buildings and real energy components, along with what might replace deficient components, and so deliver energy savings.

Goals of Energy Improvements

What are the goals of energy improvements? There are many. Environmental goals include reduced carbon emissions, reduced reliance on depletable fossil fuels, and reduced pollution due to extraction and use of these fuels. Financial and economic goals include reduced energy costs, increased property value, and economic development. Political goals include reduced dependence on foreign fuels, reduced conflict over fuels, and reduced strain on electric power grids. Health and comfort goals include improved human health due to reduced pollution, improved indoor air quality, and improved comfort due to reduced drafts, glare, cold and hot indoor spaces, and light pollution.

Goals are often harmonized—when we save energy, we achieve multiple goals. But sometimes priorities are different for different individuals and organizations, or

Figure 1.3 Understanding what we see leads to better energy improvements.

Figure 1.4 Real energy systems of real buildings.

even individuals within one organization. In one building, the owner may first want to save energy, but the chief financial officer (CFO) may first want to save costs, while the facility manager may first want to reduce comfort complaints, and the occupants may first want a bright and healthy work environment. Recognizing different priorities, and sometimes different goals, and jointly navigating through these differences is an important part of energy work.

There are many potential pitfalls along the way, risks that cause energy savings to not be delivered, to be eroded before they are delivered. Knowing these risks helps us to prevent them. Erosion of potential energy savings occurs when energy auditors miss identifying or evaluating good energy improvements; when energy auditors overestimate savings; when owners decide not to implement good energy improvements; when contractors install inferior products or do not complete a job properly; when commissioning of improvements does not happen, for example, when controls are not properly set; and when equipment is not operated correctly or maintained.

The main goal of the energy auditor is to evaluate and prioritize energy improvements. The main goal of the energy manager is to deliver energy savings. Between the two, the potential arises to entirely transform our buildings.

Chapter 2
Overview

Principles

Principles that guide energy improvements include:

Comprehensiveness. We want to look at a building as a whole, treat it as a whole, and improve it as a whole.

Transparency. We want our work to be clear and our assumptions to be clear so that they can be checked and confirmed.

Do-no-harm. We want energy improvements to not damage a building or place the occupants at risk.

Evidence-based. We want energy predictions to be based on evidence, not based on wishfulness or sales-based promotions. Savings estimates should be based on proposed physical changes to the building, not on assumed percent savings or other rule-of-thumb estimates, with savings based either on first principles of physics or on published and peer-reviewed empirical studies.

Truthfulness. We want our energy savings predictions to be truthful. We want to avoid the temptation to overestimate savings.

Objectivity. We want to avoid bias in improvements, to avoid natural tendencies to favor specific improvements, either because they were promoted by vendors or for other reasons.

Cost-effectiveness. We want to maximize energy savings by evaluating, choosing, and recommending improvements that are the most cost-effective, which allow the most savings to be delivered for a given budget.

Choice. We want to give building owners and property managers choices; we want to avoid second-guessing their priorities or making decisions for them.

Robustness and persistence. We want to evaluate, choose, and recommend improvements that are more likely to deliver energy savings, that are proven, and that are more likely to last over time, to deliver persistent savings.

Throughout, we err on the conservative side in estimating savings and in promising savings. Throughout, we try to maintain the glorious goal of delivering measurable savings, of exceeding the owner's expectations.

One focus of past energy work has been the *limited*, or *targeted*, or *no-cost/ low-cost* energy audit, in which simple changes to a building are made in an effort to see modest energy savings at little up-front cost. The effectiveness of a limited, targeted, or no-cost/low-cost energy audit has come to be questioned. We get what we contract for. If we contract for a limited, targeted, or no-cost/low-cost energy audit, the results are likely to be limited, targeted, and low cost/no cost. Rather than obtaining results that can barely be seen in the noise of utility bills, a more successful approach has been repeatedly proven to be in-depth, comprehensive energy audits,

Figure 2.1 We seek to deliver measurable energy savings, to exceed the owner's expectations.

5

Figure 2.2   Unsealed attic floor in a school.

Figure 2.3   Compact fluorescent downlights.

followed by in-depth and comprehensive implementation of energy improvements. The results are measurable and enduring and transform buildings.

There is merit to focusing on larger improvements, the improvements that can deliver substantial energy savings. By doing so, something magical happens, and suddenly the energy savings are no longer 5 to 10 percent, but rather 50 percent, or 60 percent or more, and then within reach of supplying a measurable portion of the balance of usage with renewable energy. The number of improvements might not be many more than a conventional audit. And the energy audit effort does not necessarily even need to be much more than a conventional audit. By using a few strategies, and a methodical approach, deep energy reductions are possible. A primary strategy with transformational energy work is to focus on substantive improvements, rather than symbolic improvements.

For example, a symbolic improvement might be to weather-strip a front door, where visible cracks are a symbol of energy loss. Instead, a substantive improvement is to air seal an attic floor, which is not as visible, but where there are typically far more and larger infiltration sites (Figure 2.2).

As another example, a symbolic improvement might be to place an outdoor light, which is observed to be sometimes left on during the day, on a photocell control. The photocell will prevent the light from coming on during the day on those occasions when the light was inadvertently left on. Instead, a substantive improvement would be to put the light on a combined photocell and motion sensor control. The new control will not only keep the light off during the day, it will keep it off for most of the night as well, and provide light when it is needed, with no sacrifice in functionality or safety.

As another lighting example, a symbolic improvement would be to replace a compact fluorescent (CFL) recessed downlight (Figure 2.3) with a light-emitting diode (LED) downlight, and see perhaps 30 percent savings. Instead, a substantive improvement would be to replace the downlight fixtures with linear fixtures and save well over 50 percent.

As a heating example, a symbolic improvement might be to replace steam traps and save 10 percent for perhaps five years. A substantive improvement would be to convert the steam system to hot water, and save 50 percent for the remaining life of the building.

In many ways, energy audits call for thinking outside the box. A traditional inside-the-box energy audit considers changing one piece of equipment for a higher-efficiency version of the same equipment, without considering more broadly how the equipment is being used, or if it is even needed. For example, an inside-the-box energy audit recommends replacing an outdoor wall-mounted high-intensity discharge (HID) light fixture with an LED fixture. Thinking outside the box would engage the energy manager in a discussion about whether the fixture is even needed at all. As another example, an inside-the-box energy audit would look at replacing a low-efficiency oil boiler with a higher-efficiency boiler. Thinking outside the box would look at converting the boiler to gas and converting from steam to hot water, or even converting the entire system to a heat pump. This is a theme that will run throughout this book—the option of going beyond "like-for-like" replacements and so only marginally reducing energy usage, and instead to transform or eliminate entire systems, dramatically reducing energy use without sacrificing comfort, health, or safety.

We might consider the traditional approach to energy audits and improvements as *incremental*. We have for too long tried to improve buildings and building components incrementally. We have changed one lightbulb out for another as an incremental improvement. We have changed one boiler out for another as an incremental improvement. We have changed one chiller out for another as an incremental improvement. We have changed one commercial dishwasher out for another as an incremental improvement.

In recent decades, a newer approach, termed loosely *holistic* energy work, moved the industry forward by treating the building as a whole, exploring a wider variety of improvements, including envelope improvements, and also directing attention to health and safety issues.

We would suggest that a third evolutionary step in energy work is happening, beyond incremental and holistic, which might be called *transformational energy improvements*. With transformational energy improvements, we seek to transform the building and transform its energy components. Instead of just changing a lightbulb, we change the light fixture, reduce the number of fixtures to "right-light" the space, and add controls, with a transformational change in lighting energy use and savings of 80 to 90 percent, without sacrificing any lighting quality, and in fact improving lighting in underlit areas. Instead of just changing a boiler, we change the entire heating system perhaps to an air-source heat pump, bring the distribution system within the thermal boundary, size the system correctly, and provide better zone temperature control, with transformational savings of 50 to 60 percent, instead of 10 percent. Instead of just changing a commercial dishwasher, we also change the prewash spray nozzle, add a gas-fired preheater, change the primary water heater to high efficiency, and transform energy use by the dishwashing system with savings of 70 to 80 percent. To reduce infiltration, we do not stop at adding weather-stripping to windows and doors, but we also caulk the window and door frames, seek and eliminate infiltration at the attic floor, seal penetrations into vertical shafts to slow stack effect airflow, and more, to reduce infiltration by over 50 percent, rather than 10 percent or less. At every step of the way, we seek measurable changes, rather than token or cosmetic changes, to transform our old buildings into new buildings that work for us, that use little energy, of which we can be proud, and in which we can be comfortable and safe.

Trends

There is a trend towards deeper energy savings in comprehensive energy improvements. While it might seem that there are endless possible energy improvements, options can be summarized in a few groups. Lighting improvements are three-fold: efficient replacements, reduced overlighting, and controls. Lighting controls themselves have four main options: improved manual control, occupancy controls, photocontrols, and timers. Heating and cooling improvements may be grouped as plant improvements (high-efficiency replacements), distribution improvements, and controls. Envelope improvements primarily involve insulation and air sealing. Ventilation improvements are reduced duration and magnitude. Motor improvements are high-efficiency replacements and variable speed drives. Plug-loads are high-efficiency replacements and reduced duration. Water improvements are efficient end uses (reduced flow and/or reduced duration of flow), increased heater efficiency, and reduced leaks. These improvements form the technical core of energy conservation and are supplemented with less technical strategies, such as behavioral approaches and training. Renewable energy forms an important complementary set of improvements, delivering clean energy, rather than reducing energy use.

Commercial energy audits are required or offered by a rapidly increasing number of laws or state and utility programs, and independent private initiatives. In 2009, New York City passed Local Laws 84 and 87. Local Law 84 requires annual utility benchmarking data to be submitted by owners of buildings with more than 50,000 square feet. Local Law 87 mandates that buildings 50,000 square feet or larger undergo energy audits and retrocommissioning every 10 years (Figure 2.4). Philadelphia and Boston passed similar laws in 2013, and other large cities are following. Many states also operate energy audit and improvement programs.

Figure 2.4   Commercial energy audits are required by an increasing number of local laws and portfolio programs.

These programs provide energy audits at no charge or at subsidized cost, as well as incentives for energy improvements. Electric and gas utilities in many areas operate commercial energy audit programs, typically at the direction of public service commissions, and often also at no charge to the building owner. The federal government requires energy and water audits of its buildings every four years, and has established a goal to reduce energy use by 3 percent per year. Some commercial energy audits are provided by energy service companies (ESCOs), typically private firms that offer both energy audits and improvements, and who do so with their own financing. The turnkey offering of audits, improvements, and financing is known as *performance contracting*. Performance contracting has been active for over 20 years, primarily in larger facilities such as universities and hospitals.

Clients for commercial energy audits include not only building owners but also property managers, facilities managers, government entities, school districts, hospitals, and more. In many ways, states and utilities operating energy conservation programs are clients, providing market demand for energy audits. Federal and state tax incentives also drive energy improvements. For example, the federal government has offered a 30 percent tax credit for renewable energy improvements. State incentives have become so numerous that an active web site has started to track and update these incentives.

However, whereas *residential energy auditing* has matured considerably in the 40 years since the 1973 energy crisis, *commercial energy auditing* has not seen similar advances. While residential energy work today benefits widely from on-site measurements such as blower door testing for infiltration, infrared thermography for heat transfer problems, and duct leakage testing, it is far less frequent that such instrumentation is used in commercial energy audits. Residential energy work also more often includes evaluation of health and safety issues, to ensure that energy conservation work does not inadvertently result in negative health and safety impacts, for example, in indoor air quality or in the improper operation of combustion appliances. Residential energy programs, such as the federal Weatherization Assistance Program, have sophisticated approaches to life-cycle costing of energy improvements and associated prioritization, and account for interactive effects between energy improvements. We are beginning to see some of these advanced strategies in commercial building energy work, and expect this trend to continue.

Finding the balance between too much and too little detail is the great challenge of energy audit work. Some energy program administrators fear the cost of comprehensive energy audits, resulting in shortcuts and limited-scope energy audits. We seek to balance the effort of energy audits with the potential rewards. But there are risks to limited energy audits. There is evidence that low-cost/no-cost energy audits prevent owners from considering deeper energy audits or improvements. Once an energy audit is complete, even if limited in scope, owners frequently decline further energy audits on the basis that "an energy audit was already done."

Over the last 30 to 40 years, energy auditing and energy management have evolved. At the same time, energy work has in some spheres become a commodity. Energy audits have become thinner, energy auditors have trimmed their fees, and for many there has been a race to the bottom. But something funny happened along the way: Owners became interested in energy use in their buildings. Whether because of the reality of climate change, or because of a newfound interest in buildings, many owners now seek deep reductions in energy use, and are no longer interested in "low-cost/no-cost" energy audits or improvements. This book is built on the premise that low-cost/no-cost energy work has possibly been a failure, does not lead to further and deeper energy work, and essentially commodifies energy work. This book is built on the shoulders of those who have shown not only that deep energy savings are possible but that they are worthwhile and important, that many tangential benefits accrue from significant energy improvements, that understanding our buildings is a rich and fulfilling endeavor, and that every building is unique and is deserving of attention. Those interested in low-cost/no-cost energy work should probably look elsewhere.

The energy field is also changing, as we find ourselves examining a wide mix of building types and building efficiencies. Many old buildings still exist, with many energy deficiencies and significant potential energy savings. However, many buildings already have made energy improvements, and we need to be sure not to assume that all components in an old building are inefficient. We also must not assume that components that once were efficient are still efficient. The federal ENERGY STAR program, for example, is already over two decades old. Much equipment that earned the ENERGY STAR label from the earlier years of the program is already inefficient by today's standards. The ENERGY STAR program, and other above-code efficiency programs, have substantially contributed to the advancement of energy conservation and will continue to do so. But if we assume that any equipment we find that once was rated as efficient is not worth examining, we may be making a mistake and losing potential energy savings. We are faced with the paradox that above-code energy programs could possibly retard energy work if we make such mistakes. We cannot assume that Leadership in Energy and Environmental Design (LEED)-certified buildings, ENERGY STAR equipment, or other above-code buildings and components cannot be improved.

We are also increasingly facing the challenge of evaluating buildings that already have renewable energy, such as photovoltaic systems. For these buildings, we can no longer simply examine the utility bills and assume that the usage in the buildings is equal to the usage as shown on the utility bills, because some of the building energy usage is now coming from the renewable energy source. The task of analyzing usage and predicting savings becomes more complicated.

Definitions

Energy improvements are changes to a building that reduce energy use. Other terms used include *energy conservation measures* (also called *measures* or ECMs), and *energy conservation opportunities* (ECOs). Sometimes, ECO refers to a recommended improvement, and ECM refers to the improvement after it is installed. In this book, we will use the term *energy improvement* to generally cover all of these terms.

Interacting improvements, or *interactive improvements*, refer to two or more improvements that affect each other's savings. For example, installing a high-efficiency boiler at the same time as installing attic insulation will result in total savings that are less than the sum of the savings if each improvement were made on its own, without the other. Other examples of interacting improvements include lighting and cooling improvements, heating/cooling and ventilation/infiltration improvements, and heating/cooling and control improvements such as indoor temperature setback. Any improvements that separately reduce the *load* on a building (e.g., heat loss or heat gain) and increase the efficiency of the heating or cooling *plant* (equipment) are interacting improvements.

Packages of improvements, or *packages*, refer to a simultaneous set of improvements that may or may not interact. A package also has its own set of metrics such as total installed cost, total annual energy savings, and total payback. A package of recommended improvements may need to have its individual improvements prioritized in case a building owner chooses not to implement the entire package.

Deep energy retrofits are packages of improvements that substantially and measurably reduce energy use in a building.

Net-zero buildings are buildings that have had deep energy retrofits, in combination with the installation of renewable energy, and so produce as much or more energy than they use, on an annual-total basis.

Units of Measure

The U.S. building energy field uses a mixture of U.S. units and international (SI) units. Electrical measurements are in metric units. Thermal measurements are mostly in U.S. units. Some measurements use a mix of both units, such as lighting power density in watts per square foot.

Unit conversions are provided in Table 2.1.

The Role of the Energy Auditor

Commercial building *energy auditors* include many engineers, but also contractors who have moved from energy auditing of smaller buildings into larger commercial work, as well as energy professionals with non-engineering backgrounds. Energy auditors work for a variety of types of companies, including engineering firms; ESCOs; building performance contracting firms; heating, ventilating, and air conditioning (HVAC) equipment and controls manufacturers; state energy agencies; gas and electric utilities; and more.

Energy auditors must be independent and objective in order for energy audit reports to avoid bias in favor of specific products or services. They must also avoid telling the owner or property manager only what they want to hear.

TABLE 2.1

Common Energy Unit Conversions

U.S. Units	Multiply By	To Obtain
Btu/ft^2-F-hr	5.6783	W/m^2-K
Btu/hr	0.29	Watts
Btu/lb	2.326	kJ/kg
Btu/lb-F	4.1868	kJ/kg-K
Cubic feet per minute (CFM)	0.472	Liters/second
CFM/ft^2	5.08	Liters/second per m^2
CFM/watt	0.472	Liters/second per watt
Cubic feet (CF)	28.3168	Liters
Energy-efficiency ratio (EER)	0.293	Coefficient of performance (COP)
Feet	0.3048	Meters
Foot-candles	10.764	Lux
Ft^2-F-hr/Btu	0.1761	m^2-K/W
Gallons	3.7854	Liters
Gallons per minute (GPM)	3.7854	Liters/minute
Horsepower (HP)	0.7457	Kilowatts (kW)
Inches	25.4	Millimeters
Inches water gauge	249.09	Pascals (Pa)
kbtu	0.00106	Gigajoules (GJ)
kbtu/SF/year	3.1548	kWh/m^2/year
Mbtu	1,000	Btu
Square feet	0.0929	Square meters
Square inches	645.16	Square millimeters
Therms	100,000	Btu
Tons of cooling	3.517	kW
W/SF	10.764	W/m^2

Even while being independent and objective, the energy auditor can and should serve as an advocate for energy conservation. An advocate explains the goals of energy improvements and the resulting benefits, even while explaining risks. An advocate speaks not only for the business case but also for the societal case for reducing energy use. An advocate does not say only what the owner might want to hear. An advocate is an educator. An advocate does not make presumptions about what the owner can and cannot afford to do immediately, but provides information to allow energy planning into the future. An advocate presents the case for maximizing energy improvements, with all the information necessary for the building owner to make decisions that fits their goals and needs. An advocate is aware of what is on the cutting edge. An advocate makes proposals. An advocate leads. Over time and with experience, the energy professional will develop comfort with the role of being an advocate. It is an important role, and good advocacy is the work of good energy professionals, even while remaining objective and taking a scientific approach to building analysis.

While informed by the constraints of real buildings, such as which improvements can be fit into a specific building or which improvements are allowable by building code, the energy auditor does not need to deliver a final design as part of the energy audit. The energy auditor's goal is to deliver an energy audit, which is a feasibility study, not a final design.

The energy auditor is not responsible for the decision to implement, or not to implement, specific improvements. In other words, the energy auditor does not need to second-guess the building owner or property manager and make *a priori* decisions to exclude a specific improvement. The role of the energy auditor is to provide information on which the owner or manager can make decisions. The energy auditor should not rule out improvements by making assumptions for the owner or manager. For example, the energy auditor should not assume that the owner is "not interested in any payback over two years."

The energy auditor must recognize that there are multiple stakeholders for any one building. There might be an owner, a property manager, a board of directors, a facilities manager, building occupants who may or may not be tenants, and more. Each stakeholder group has a different set of interests. The energy auditor should not assume that the person with whom she is interacting during the energy audit speaks for all stakeholders. The energy auditor needs to take a neutral, professional stance.

The energy auditor has a responsibility to report observed health and safety deficiencies in a building, even if it means that correction of these deficiencies increases energy use. Examples include underventilation, inadequate lighting, and insufficient cooling or heating capacity. This can present challenges in estimating energy savings because increased energy use from correcting these deficiencies will offset other energy savings. One option is to correct the baseline (existing) energy use accordingly and to evaluate improvement savings with the presumption that the deficiency was not there. Government and utility energy portfolio programs should allow for and encourage such corrections without penalizing meeting goals.

The main role of the energy auditor is to conduct and deliver energy audits. This involves visiting buildings, identifying energy inefficiencies, developing improvement recommendations, evaluating the installed cost and energy cost savings for these improvements, and describing the building and recommendations in an energy audit report. The main purpose of the energy audit is to allow prioritizing energy improvements.

Energy auditors are point people in the field at the front lines of energy work in buildings. They are the ones in contact with building owners, property managers, and tenants. They are the ones who need to get utility bills and to understand them. They are the ones who need to identify energy savings. As such, they can and do identify unmet needs in the field of energy work, and they can and should convey these needs to government officials, policymakers, program administrators, educators, nongovernmental organizations with interest in energy conservation, and others.

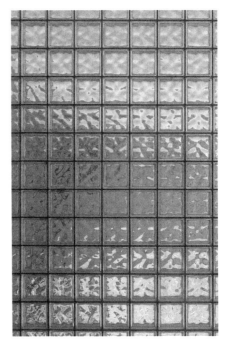

Figure 2.5 While serving as an independent expert, the energy auditor can and should serve as an advocate for energy conservation.

For example, there is an urgent need for databases of efficiency ratings of existing equipment, the products that energy auditors and energy managers find in buildings, which may no longer have ratings labels attached to them. There is an urgent need for standardization of utility billing data, and for ease of access to these data. There is an urgent need for hard data on the performance of new energy-efficiency products. All these needs must be continuously identified and articulated back to those who can do something about it.

Energy audit work is physically demanding. To access and inspect areas such as attics and crawlspaces requires the ability to climb and to crawl. Accessing roofs requires the ability to climb ladders, which may be 30 feet high or higher. The work can include spending periods of time in extreme temperatures, as boiler rooms can be over 100 degrees, and work outdoors can be below 0 degrees. The work is often dirty and can include exposure to dust and other irritants to which people suffering from asthma might be allergic. At least one member of any energy audit inspection team must be physically able to perform these tasks. If not, important areas of a building will not be inspected, and important potential energy improvements will be missed.

The Role of the Energy Manager

An *energy manager* is responsible for reducing energy use for a building or portfolio of buildings. In large complexes, like universities, there may well be a dedicated energy manager. In smaller organizations, the job of energy manager may be performed by the owner, facility manager, or even others, such as the person responsible for finances. Energy manager broadly refers to the individual responsible for energy at a facility, who is affiliated with the facility, and who represents the facility.

The energy manager has many roles and responsibilities. These include keeping and maintaining records, such as drawings, equipment user manuals, R-values of original and added insulation, and records of other energy improvements.

The energy manager keeps utility billing records, including water. The energy manager uses these records to track and benchmark energy usage, comparing a building's energy use to the use of similar buildings and to the building's own use as it changes over time.

The energy manager may also obtain and maintain further information about the building, such as occupancy information (how many people are in the building, and when), schedules (lighting schedules, ventilation schedules), and similar energy-related information that can help in programming controls and in supporting energy audits.

The energy manager may possibly even do energy estimates for some improvements.

The energy manager may want to keep up to date on developing technologies, but without being too quickly sold on these by vendors who come calling to promote products or services.

The energy manager should check for and eliminate leaks, such as gas leaks, water leaks, or compressed air leaks. These can be checked using leak dials on gas or water meters, when there is no other gas or water use. Or leaks can be sought with leak detectors or visual observation.

In addition to checking for leaks, the energy manager can perform other responsibilities that are less feasible for energy auditors, because energy managers are on-site more often and longer than energy auditors. An example is to check that photocell-controlled lights turn on and off at the right time (dawn and dusk), rather than turning off too late in the morning or turning on too early in the evening. This kind of check is typically not done by energy auditors because they are not on-site at dawn or dusk.

The energy manager can periodically procure energy audits or develop capability to do these in-house. The energy manager should be able to issue requests for proposals for energy audits, select a qualified energy auditor, review and check the energy audit, and interpret it for building management.

The energy manager also implements energy improvements, whether supervising improvements that are done in-house, or contracting out larger improvements. Energy improvement projects are often construction projects and require the same high attention that construction projects demand. Project management responsibilities include scope control, schedule control, cost control, quality control, and risk management. The energy manager may well do all these tasks or may contract out some or all of these.

Finally, the energy manager should report on energy issues to building management, with recommendations and results. Like the energy auditor, the energy manager can and should become comfortable with the role of being an advocate for energy conservation.

Types of Energy Audits

ASHRAE has defined three levels of energy audits: Levels 1, 2, and 3, and a necessary background step called Preliminary Energy-Use Analysis (PEA).[1]

The PEA entails billing analysis, to establish an Energy Utilization Index, in units of kBtu/SF/year, by adding up all energy use in the building over a year's period, converting all energy into units of kBtu, and dividing by the building floor area (square feet).

A Level 1 energy audit is called a Walk-Through Survey, and is limited to low-cost/no-cost improvements, along with a list of possible capital improvements for future consideration.

A Level 2 energy audit is called an Energy Survey and Analysis, and includes evaluation of energy by end use, as well as peak demand analysis. Energy savings and installed cost estimates are provided for "all practical energy efficiency measures", as well as operating and maintenance (O&M) improvements, and, again, a list of possible measures recommended for future in-depth analysis.

A Level 3 energy audit is a Detailed Analysis of Capital-Intensive Modifications. It calls for more detailed field data, more rigorous analysis, and possibly building energy modeling. Installed cost estimates are conducted with a higher degree of detail and confidence, and life cycle analysis is commonly performed.

In addition to the PEA and audit levels 1-3, ASHRAE suggests an optional separate energy audit, focusing on a single energy use or load, and called a Targeted Audit.

ASHRAE allows that there are not strict boundaries between the different levels of energy audits.

In this book, in addition to ASHRAE's energy audit options, we suggest another possible approach. Instead of a sequential multi-level approach, we suggest the option of single energy audit, including in-depth analysis of energy conservation and renewable energy options, with the goal to transform energy use in a building, and to serve as a long-term planning tool. In some ways, such a transformational energy audit would be similar to moving immediately to a Level 3 audit, and including the PEA, Level 1 and 2 audits. Such a transformational energy audit approach, and its pros and cons, is explored as a theme through the book.

Traps and Pitfalls

While doing our fieldwork, we want to be vigilant for signs of energy inefficiency—open windows in winter, a front door without weather-stripping, high indoor air temperatures in winter, low indoor air temperatures in summer, exterior lights that are on during the day, and the like. However, we must avoid overreacting to such signs when they are few—the single open window, the single door lacking weather-stripping, the single hot or cold room, the single light found on during the day. As dissatisfying as these token signs are, and as important as it is to fix these deficiencies, they may not be significant energy problems and should not detract from more serious evaluation. We must not paper over an energy investigation with minor, token signs of inefficiency.

We are constantly faced with questions of how much detail to seek in energy evaluations. The effort and cost of energy audits, and energy improvements in general, must always be weighed against the potential energy savings. In this book, we present and advocate for more detail, rather than less. We believe that every extra hour spent in buildings, assessing what is there, will save multiple hours in subsequent effort, from energy analysis to implementing improvements to avoiding poor work that resulted from inadequate up-front work. We caution against underbudgeting energy evaluation work. The quality and benefit of the end goal, measurable energy savings, directly depends on the quality of the work up front.

Every high-efficiency energy product brings its own set of advantages and disadvantages. Many of these characteristics require adaptation by building owners and occupants. High-efficiency dishwashers tend to run longer in order to deliver the same cleanliness, using less energy. Temperature setbacks in buildings may surprise people who arrive after hours and who have not been told how to override the setback. Energy auditors and energy managers need to be aware of these pros and cons, to discuss and describe them openly, and to actively educate people about them. It is important that building owners and occupants are not surprised by changes and can reap the energy savings benefits without being disturbed by these changes or removing the energy improvements.

Sometimes new products simply do not perform well. In the rush to release new products, manufacturers may not have adequately tested them, or a single underperforming product slips through due to poor quality control, or products fail prematurely. These occurrences are very bad for our industry. They give detractors an opportunity to criticize the energy field as a whole, and they give building owners and occupants the type of experience that prevents them from proceeding with further energy conservation work. We energy auditors and energy managers need to honestly call out failures as failures. We are in a unique position to tell good from bad. We cannot be caught on our heels, defending a bad product, or continuing to recommend it, or dismissing poor performance as in any way acceptable. We need to give honest feedback to manufacturers. We need to demand quality.

Likewise, we need to stand up for quality when inspecting the work of installing contractors. This means identifying when a job was done poorly and insisting that deficiencies be corrected. It takes character to do so. But not standing up for quality means that energy savings may not be delivered, and that is the worst result of all.

There has been a consistent and widespread problem with overestimating energy savings in energy audits. We recognize that estimating savings is difficult under any circumstances. We cannot expect savings estimates to be perfectly accurate. It is to be expected that some estimates will be high and some will be low. However, there has still been a trend to overestimate savings. During energy audits, there is a natural tendency to be overly optimistic. We want our proposed improvements to do well. This can lead to a form of wishfulness, in which we almost wish our proposed improvements to do well. In some government and utility programs, there can be built-in incentives that promote overestimation of savings.

However, overestimating savings, in the long run, serves no one. It leads to poor decisions in which investments are made where they could better have been made elsewhere. It leads to disappointed owners. It does not help the environment. It gives the energy field a bad name. It develops bad habits that lead to further bad work. The overestimation of savings should be regarded as the bane of energy auditors and energy managers.

There is both a blessing and a curse to energy cost savings, the reduction in operating costs when we use less energy. The blessing is that energy cost savings can help to justify, and in some ways to pay for, energy conservation work. The curse is that we have come to expect these savings, and we compare investments in energy conservation to other investments. Energy consumption hurts the environment and is a major contributor to climate change. If it were a form of pollution for which mitigating the pollution did not reduce cost, we would be taking on this mitigation without being distracted by whether it is paying for itself, just as we do with asbestos or with toxic waste dumps. The fact that there are cost savings associated with energy conservation may well have slowed the effort down, as we look at it through a lens of investment. For some important improvements that deliver significant energy savings, there may not be an attractive return on investment. On occasion, energy prices are such that there is no attractive return on investment for any improvement. In these instances, conservation efforts slow and even stop. We cannot allow this to happen.

We cannot rely on markets alone to achieve our energy conservation goals. If our efforts to reduce energy use succeed, the drop in demand for energy will likely result in lower energy prices, and the attractiveness as an investment may evaporate. However, the urgency to continue to reduce energy will still be there. What happens if fuel and electricity prices drop as demand drops? We should most prudently regard energy cost savings as a side bonus, not as a prime driver, to reduce energy use.

The Energy Efficiency Field

The field of energy efficiency is new and exciting. The energy professional is part detective, part scientist, part environmentalist, part advocate, part builder, part engineer, and part architect. There are roles for everyone, and we need the best and brightest of our generation: those who are interested in understanding buildings, those who are interested in understanding equipment, those who interested in understanding sales, those who are interested in understanding government, those who are interested in understanding organizations and how organizations work and make decisions and change. And most of all, we need people who are interested in understanding people and what motivates them and how they can be inspired to improve their buildings and their environments and their whole world.

Some people would claim that the field is somewhat mystical, and that only those with deep experience can identify worthwhile improvements. Nothing could be farther from the truth. Buildings are physical objects, and energy use in buildings follows the laws of physics. The behavioral component of how people interact with buildings makes things a little more interesting. But there is nothing mystical about it. And the more we share our approaches to understanding building science, the more progress we will make. There is no room for mystics or mysticism in the energy conservation field. Mysticism is a way for consultants to argue that the energy field should be left to them, rather than be made accessible to buildings owners and energy managers. Mysticism can and should be replaced with shared approaches, shared resources, shared stories of success, and a shared vision to transform our buildings.

Figure 2.6 We should most prudently regard energy cost savings as a side bonus, not as a prime driver, to reduce energy use.

Many people who are new to the field assume that it is well established, and that much of what needs to be learned has already been learned. Just the opposite is true. The field is so young and so new that, if you see a possibly new opportunity to save energy, it is often safe to assume that nobody has tried it, and it is worth trying. The possibilities are endless for saving energy in new ways or in reducing the cost of saving energy and transforming our buildings.

Reference

1. ASHRAE, *Procedures for Commercial Building Energy Audits*. Second Edition. 2011.

Chapter 3

Field Guide

Owner Records

Frequently, the building owner or manager has various records that can be of help in evaluating energy systems in a building. These records include original drawings, as-built drawings, vendor records, purchasing records, test results, and owners' manuals. The energy manager should keep copies of these accessible for energy work.

It is helpful to learn to interpret energy aspects of building drawings and specifications. The most useful drawings are the mechanical drawings (usually labeled M or H), some of the plumbing drawings (labeled P—ones of interest include domestic hot water and drawings that show plumbing fixtures), electrical drawings (labeled E) that show lighting and major electrical loads such as large motors, and some of the architectural drawings (labeled A). *Schedules* refer to tables on drawings, which frequently contain energy-related information (Figure 3.1). *Details* on drawings and *Floor plans* also can be useful for energy work (Figure 3.2 and Figure 3.3). The architectural drawings of interest include floor plans (for room sizes and types), elevations (for building height and window/door count/sizes), and insulation details (wall, roof, and foundation). Occasionally, there is some energy-related information in the project specifications, typically provided as a soft-cover book. *As-built* drawings can be useful if available. These are marked-up drawings that show changes made to a building's design during construction.

Evacuation route drawings (Figure 3.4), frequently posted in buildings, can be copied, to provide a convenient set of reduced floor plans on which to take notes.

Tools

The following lists of tools for energy work are somewhat subjectively grouped as required, recommended, optional, or advanced.

REQUIRED

Required tools for energy work include a camera, tape measure, infrared thermometer, lighting ballast checker (Figure 3.5), screwdriver, adjustable wrench, stopwatch, stepladder, and flashlight. A clipboard or tablet computer is also helpful, to take notes.

RECOMMENDED

Recommended tools for energy work include a light meter, a smoke generator for infiltration inspection, a digital clamp-on electric meter, and a calibrated container for measuring fixture water flow rates (or commercially available measuring bag). A tool belt and/or tool bag can be useful for conveniently carrying tools.

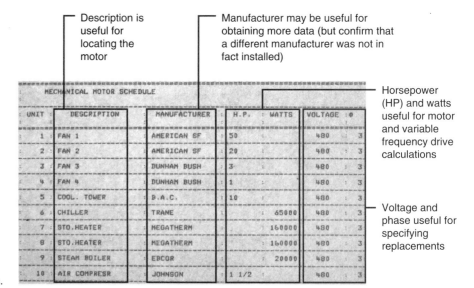

Figure 3.1 Example of a drawing schedule.

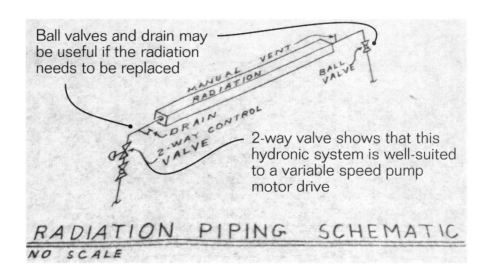

Figure 3.2 Example drawing detail.

OPTIONAL

Optional tools for energy work include a feeler gauge or gap gauge to measure the size of small infiltration gaps, walkie-talkies, plug-in watt-meter, drill bit sizes 71 through 80 to measure gas pilot orifice sizes, and a digital voice recorder.

ADVANCED

Advanced tools for energy work include a blower door (Figure 3.6), combustion analyzer, fiberscope (borescope), balometer (flow hood), low-airflow measurement device, data-loggers and associated sensors (temperature, humidity, current, power, etc.), low-e window tester, carbon dioxide meter, anemometer, digital pressure gauge, tachometer (to measure motor speed), duct leakage measurement device, laser distance meter, and a window gas fill analyzer.

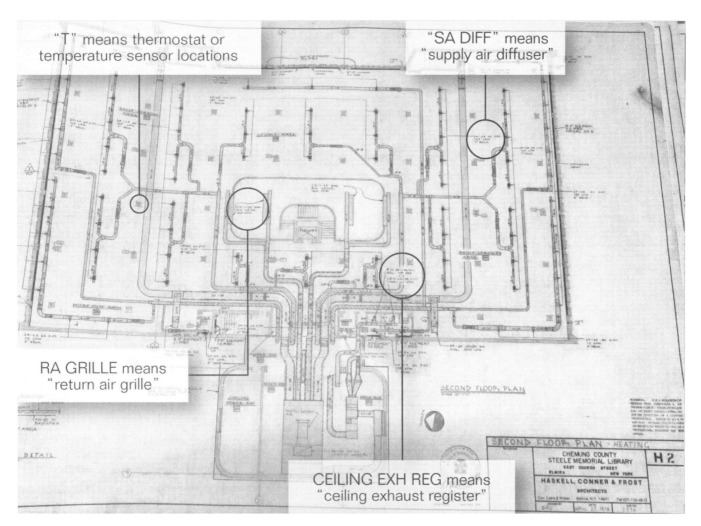

Figure 3.3 Mechanical floor plan.

Types of Measurements

There are three types of measurements: direct measurements, indirect measurements, and reported measurements. A direct measurement might be the temperature of water measured at a faucet with a thermometer. An indirect measurement might be the temperature setting on a commercial water heater. And a reported measurement is when someone else reports a measurement to an energy auditor. It is helpful to identify in our reports the type of measurement that was taken. Our confidence is greatest in direct measurements, lower in indirect measurements, and lowest in reported measurements.

PHOTOGRAPHS AND VIDEO

A camera is an essential tool for energy work. Photographs are a rapid way to obtain equipment nameplate data. Photographs can also document energy problems, such as air leakage sites, which can be helpful when corrective action is undertaken. Cameras can also be used to obtain information in tight locations where the human eye cannot

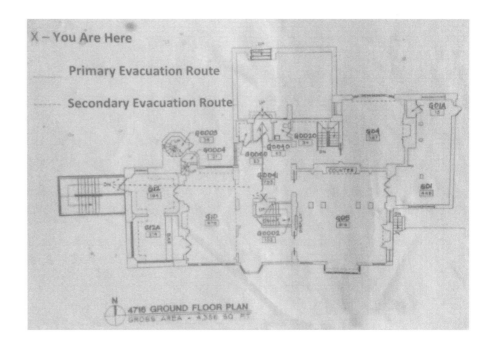

Figure 3.4 Evacuation drawing.

Figure 3.5 Ballast checker.

reach, for example, to obtain nameplate data on a piece of equipment that is within a few inches of a wall, or the lightbulb type in cove lighting, and the like.

Frequently, it is helpful to take photos at a distance. For example, motors and other equipment are often suspended from a tall ceiling, and a camera can sometimes be used to obtain nameplate data from the ground, without use of a ladder. Also, photos can possibly be taken of rooftop equipment without having to access the roof (Figure 3.7). For example, a small point-and-shoot digital camera with 30× optical zoom and 120× digital zoom provides sufficient resolution to photograph an equipment nameplate from as far as 60 feet or more. A more typical point-and-shoot digital camera with a 4× to 5× zoom can take photos of nameplates not much more than 10 feet away. A smartphone camera can typically only take photos of nameplates up to approximately 5 feet.

Photographs also add a visual dimension to an energy audit report. A nice touch is to include a photograph of the building on the front cover of a report, for example, showing the name or address of the facility, highlighting an entrance sign if possible.

Photographs are also useful for elevation views of buildings. From these, it is often possible to obtain rough measurements of window/door and wall areas, for example, to estimate insulation quantities or window/door sizes and quantities. Elevation views also allow identification of some outdoor equipment and associated possible energy improvements (Figure 3.8).

Videos can be helpful to document deficient conditions for building owners, such as infiltration as evidenced by smoke movement, or to allow recorded voice descriptions of existing conditions.

TEMPERATURE MEASUREMENTS

Infrared (IR) thermometers are an essential energy audit and evaluation tool. In this book, we describe a wide variety of applications for IR thermometers, including air temperature measurement (outdoors, indoors, forced hot and cold air, and more), pipe and duct temperatures, window and window frame temperatures to indirectly estimate energy efficiency, door and wall temperatures to indirectly estimate insulation levels, aspects of lighting, and more. The speed of measuring temperatures with IR thermometers allows them to be used productively to understand many aspects of a building's energy losses and associated potential savings. The main caution with

IR thermometers is to avoid taking measurements on highly reflective surfaces. On reflective surfaces, place masking tape or electrical tape, and allow the tape time to come to temperature, before taking measurements. High-temperature tape is available to allow measurements on hot reflective surfaces.

AREA MEASUREMENTS

Floor Area

Floor area measurements are required for benchmarking and for such space-specific measurements as lighting power density (watts per square foot). Benchmarking area measurements, for example, with EPA Portfolio Manager, are based on *gross floor area*, the measurements to the outside of exterior walls. Gross floor area does not include exterior spaces, such as balconies or loading docks. For tall spaces, such as atriums, the floor area includes just the base floor area. Lighting power density and other interior measurements are based on *net floor area*, the measurements to the interior of walls. Volume measurements, such as those used for blower door testing, are also based on net area.

Major vertical penetrations are such spaces as elevator shafts, stairwells, pipe and duct chases, and the enclosing walls of these spaces. Major vertical penetrations are subtracted from gross floor area, for each floor, in order to calculate net rentable or leasable floor area. These and other more subtle calculations of floor area are provided by a variety of standards.[1]

Window and Door Area

Window and door areas are based on the clear opening of the window or doorframe.

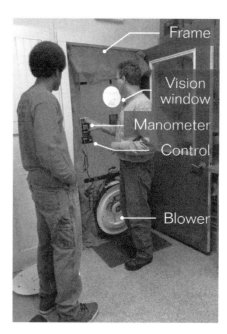

Figure 3.6   Blower door.

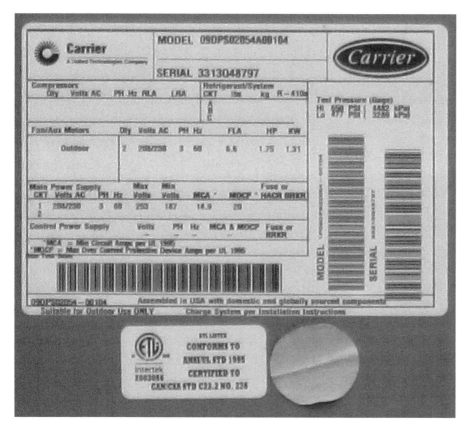

Figure 3.7   Photograph of air conditioning nameplate on a rooftop, taken standing at ground level, 66 feet away, 30x optical zoom with 120x digital zoom.

Figure 3.8 Building elevation view showing equipment and energy improvement opportunities.

FOLK MEASUREMENTS

We can take some measurements by eye or by hand, and we might refer to these as *folk measurements*, or *folk quantification*.

For example, water temperature may be estimated from the observations given in Table 3.1.

We note that the range of temperatures that is not painful is only 60 to 100 F.

With experience, water flow rates may be roughly estimated by eye. A typical faucet flow is approximately 1 gallon per minute (GPM), a typical shower flow is 2 GPM, and a typical bathtub spout flow is 5 GPM.

Distances can be measured by paces. The pace of a person of average height is 2.2′. A typical pace is a person's height × 0.4, so the pace of a person 5′ tall is 2′, and for a person 6′ tall is 2.4′. Length measurements can also be made by counting ceiling tiles (typically 2′ × 2′ or 2′ × 4′), or floor tiles (typically 1′ × 1′, although older tiles are 9″ × 9″). Building height, or floor-to-floor height, can be estimated visually (or on photographs) against door heights. A door itself is typically 6′8″, but adding the frame brings it very close to 7′. Heights can also be estimated relative to the height of a person—the average height of adults is 5′7″.

TABLE 3.1

Water Temperature Folk Quantification

Temperature Range		Perception/Sensation
Greater than 110 F	—	Steam appears
105–110 F	—	Approximate limit of too hot to touch
100–105 F	—	Hot
95–100 F	—	Warm
90–95 F	—	Neutral on skin
85–90 F	—	Slightly cool
80–85 F	—	Cool
60–80 F	—	Cold
40–60 F	—	Very cold (painful within 10 seconds)
Less than 40 F	—	Feels ice cold (painful immediately)

The dimensions of concrete masonry units (CMUs) can be used to estimate wall dimensions. The most common size is 8″ high by 16″ long, and 8″ wide. These dimensions include the width of the mortar, so these dimensions can be used directly when counting lengths or heights of multiple concrete blocks.

Standard brick dimensions are 7-$^5/_8$″ long, with widths either 3-$^5/_8$″ or 2-$^1/_4$″ high depending on how the brick is laid, with mortar joints $^3/_8$″ thick. Therefore, they can be used for estimating dimensions by assuming a total length of 8″ per brick, and either 2-$^5/_8$″ or 4″ in the width dimension, depending on how the brick is laid. Note also that there are a variety of non-standard dimensions, so measuring a single brick is worthwhile, to calibrate the dimension, if using bricks for measurement.

Pipe sizes may be estimated using one's fingers: An adult thumb width is approximately 1″, a typical ring finger is $^3/_4$″, a little finger is $^1/_2$″, and these can be adjusted for an individual's sizes or combined for larger pipes. It is important to note that most pipes (iron pipes and most copper pipes) are nominally designated by interior diameter, except for air conditioning and refrigeration (ACR) copper piping, which is nominally designated by exterior diameter.

Light pole heights, or building heights, may be estimated by comparing to the size of a typical car, 14′ long, 6′ wide, and 5′ high. The width of a typical street (curb to curb) is 36′, also helpful for estimating the height of a parking/area light pole.

A test to estimate low airflow into or out of grilles is referred to as the *garbage bag test*.[2] A 26 × 36-inch garbage bag is taped to an opened wire hanger. The time for it to fill (supply airflow) or empty (exhaust or return airflow) can be used to estimate the flow rate. See Table 3.2.

TABLE 3.2

Low Airflow Measurement

Time in Seconds	Supply Airflow (CFM)	Exhaust/Return Airflow (CFM)
12	11	13
10	15	16
8	21	20
6	29	25
4	46	35
2	NA	66

Paper taped on one edge to small ceiling exhaust/return grilles can be used to estimate airflow.[2] If a single-ply tissue is held in place to the grille, airflow is over 30 cubic feet per minute (CFM). If a single-ply paper towel is held in place, airflow is over 60 CFM. If 20-pound copier paper is held in place, airflow is over 110 CFM.

Involving Energy Managers in Field Measurements

There is often merit to enlisting an energy manager's help with field measurements. For example, for areas in which access is difficult, such as tenant spaces, the energy manager can inventory lighting and appliances. As another example, for multifamily buildings, the energy manager can inventory the make and model of all refrigerators. This ensures a more accurate assessment of savings and recommended replacements than one that is based on a small sample of refrigerators. Another example might be for the energy manager to inventory as-found lighting and occupancy in each office of a large office building. Or a restaurant energy manager might measure dishwasher water use by running the dishwasher, without any other water loads, and reading the water meter before and after. Involving the energy manager is not intended to reduce work for the energy auditor, but rather to increase accuracy of data in the audit. Enlisting energy managers in field measurements also engages them in the process. For example, an energy manager who has found that a high percentage of offices have lights that are left on but are unoccupied is more likely to advocate for a program to encourage occupants to turn lights off, or to advocate for motion sensors to prevent lights being left on.

Reading Nameplates

Equipment *nameplates* are a useful source of information for energy work. They are typically located on an exterior side of large air conditioners and boilers, inside the control section of furnaces when the cover is removed, inside refrigerators, visible through the supply grille of room air conditioners, and on the bottom or back of small plug-load appliances.

Nameplate ratings are frequently referred to as *nominal* ratings, where nominal means *by name*. For example, the nominal full load current of an air conditioner may be 10 amps, but this does not mean that the air conditioner draws 10 amps under all conditions. A nominal value is still useful, and can help with a variety of calculations, but we recognize that it is not the actual value, under all conditions, for any piece of equipment.

Manufacturer (*make*) and model are important if seeking more information on existing equipment. For equipment such as air conditioners, the model can also provide a clue to the rated capacity of the equipment. For example, 048 in an air conditioner model number typically means 48,000 Btu/hr nominal capacity. Sometimes the serial number can also be useful, as it can identify the date of manufacture.

The electrical characteristics can be helpful in seeking replacement equipment. Typically, we abbreviate the electrical characteristics as "voltage-phases-frequency." For example, 240-3-60 represents 240 volts, 3-phase, 60 hertz. The voltage may be 110 volts, 208 volts, 230 volts, dual voltage such as 208/230, 240 volts, 460 volts, and higher. The phase, often abbreviated PH, is either 1 or 3 (single-phase or three-phase). The frequency (HZ) is 60 in the United States, Canada, much of the Caribbean and Central America, about half of South America, Taiwan, the Philippines, and a few other countries. It is 50 in most other countries around the world. Some equipment can work on either 50- or 60-Hz frequency. There are often a variety of currents (amps) shown on a nameplate, some of which relate to energy work, and some of which do not. These include:

Maximum fuse or circuit breaker amps. The maximum rating of the fuse or circuit breaker. Generally not of interest for energy work.

Full load amps (FLAs). Typically used for motors like fans.

Rated load amps (RLAs). Typically used for hermetic motors like cooling compressors.

Locked rotor amps (LRAs). The locked rotor amp is the (high) current that a motor will draw if its shaft is not allowed to turn. This is not used for energy work, but rather is intended to help in sizing the overcurrent protection.

Minimum circuit ampacity (MCA). The minimum circuit ampacity is the current for which wire sizes are calculated. Typically equal to $1.25 \times$ the total FLA. Maybe easiest to remember as synonymous with "wire sizing amps."

Chillers and air conditioner nameplates also typically list the refrigerant (R22, R410a, R134, R123, etc.).

Anomalies and Catastrophes

In doing energy work, we are particularly interested to find abnormal situations, where significant energy is lost. An example is ductwork that is broken open in an attic, where heated and cooled air is literally being blown away. Another example is a heat pump that has lost most of its refrigerant, with a compressor running continuously, trying to keep up with heating and cooling loads that it cannot meet, and drawing power continuously, 24 hours a day. Yet another example is an attic in a complex of multiple townhouse buildings in which the attic insulation was inadvertently never installed, somehow overlooked in a fast-moving attic insulation project. Yet another example is ice dams along the length of a building, indicating catastrophic roof heat loss, as seen in Figure 3.9. These are all real examples. We might call these situations anomalies or catastrophes. However, despite each being abnormal, anomalies and catastrophes are remarkably common.

Figure 3.9 Ice dams as an indicator of catastrophic heat loss.

Figure 3.10 Catastrophic room air conditioner fin damage.

Anomalies and catastrophes can be found only through field inspection. They cannot be found through billing analysis, remote energy audits, or phone surveys. Finding anomalies and catastrophes requires rolling up one's sleeves and inspecting a building thoroughly, with a vigilant eye for the unusual and the abnormal. It is very much detective work. However, the work can be methodical and does not necessarily require either luck or unusual skill. Most anomalies and catastrophes are very discoverable (see Figure 3.10). In this book, we seek to provide some tools, strategies, and signs to find anomalies and catastrophes.

Field Guide Summary

Every hour of time spent in a building teaches us more about it and sheds light on opportunities to save energy. Every building is different. The energy auditor and energy manager bring different skills, lenses, tools, and opportunities to see a building. Their collaboration can result in a greater building transformation than if they work in isolation.

This field guide is purposefully intended to serve as an introduction to energy field work in buildings. We will continue exploring specific techniques for field measurements within each technical chapter, and, in this way, the full field guide is integrated throughout the book. Fieldwork is a critical aspect of energy work. With good fieldwork, we can transform our buildings.

References

1. Michael Deru, "Summary of Building Area Definitions." *National Renewable Energy Laboratory*, November 8, 2004.
2. ASHRAE Guideline 24–2008, "Ventilation and Indoor Air Quality in Low-Rise Residential Buildings."

Chapter 4
Envelope

A building *envelope* is composed of its walls (and associated components such as windows and doors), roof, and foundation. The term *enclosure* is also seeing some use as an alternative to the term *envelope*. Energy losses at the envelope affect heating and cooling energy use.

Insulation

PRINCIPLES

Insulation retards heat flow, keeping heat in buildings in winter, and keeping heat out in summer. The energy code mandates required levels of insulation. A first principle of effective insulation is to provide continuity of the *thermal boundary*, defined as the path that insulation takes around a building. We want insulation to be unbroken as it covers the surface of a building. Common discontinuities or sites of insulation weakness include uninsulated or partially insulated wall cavities, attic floors, and where floors meet walls.

APPROACH

A structured approach to identifying insulation deficiencies and energy-saving opportunities includes:

- Locating the *thermal boundary*, all the way around a building, from drawings and by examining the building.
- Identifying existing insulation, following the thermal boundary.
- Identifying *discontinuities* (breaks) in the thermal boundary.
- Taking infrared images of the thermal boundary to identify the location and quantity of *voids* (missing insulation) and *thermal bridging* (structural interruptions of insulation).

IDENTIFYING EXISTING INSULATION

As-built drawings can be examined to see what existing insulation was reportedly installed. However, this information is best field-verified, because insulation can degrade in various ways, or as-built drawings may themselves not be up-to-date. Blown insulation can settle, insulation might have been missed during installation, suspended insulation batts can detach (e.g., on a basement ceiling), and insulation is often disturbed (Figure 4.1) or removed.

There are a variety of best practices for examining insulation in buildings. Inspecting insulation in unconditioned spaces is productive, such as above dropped ceilings, in mechanical rooms, in attics, in basements, and in crawlspaces. Insulation is frequently not concealed in these locations. Beyond these more readily inspected

Figure 4.1   Disturbed attic insulation.

locations, it is important to open wall and ceiling cavities to inspect insulation. This can be done in inconspicuous locations, such as closets, or at electrical boxes, or behind baseboard radiation, through small holes cut with a shop knife, or drilled. A non-conductive probe, such as a plastic knitting needle, can be used to assess the depth of the cavity and depth of the insulation. For a more detailed nondestructive investigation of the type and condition of insulation, a *fiber-optic* camera or video image may be taken. Also referred to as a *borescope* or *boroscope*, fiber-optic imagery is obtained by inserting a flexible tube with an eyepiece into the wall or ceiling cavity. *Thermal imaging*, also called *infrared thermography*, is regarded as essential in order identify the location and quantity of insulation voids (missing insulation). For flat roofs, inspecting insulation can be difficult, but the option is always available to take a core sample. In all measurements, a best practice is to direct attention to measuring the area of missing, compressed, or otherwise degraded insulation.

Common types of insulation in current use include:

- *Fiberglass*. Most commonly used in the form of batts, fiberglass can also be blown and is available in board form. In batt or blown form, it is spongy in feel and is irritating to the skin. Fiberglass is an effective insulator, but in batts or blown, it does not resist the flow of air.
- *Cellulose*. Generally blown, it is also spongy in feel and resists the flow of air when dense-packed.
- *Spray foam*. Fairly rigid and resists airflow (Figure 4.2).
- A variety of *rigid insulations*, typically available in board form. These can resist the flow of air when their seams are taped.
  - *Polyisocyanurate*, often referred to by its nickname, *polyiso*. Typically foil-faced, polyisocyanurate is a higher-cost insulation. Polyisocyanurate has one of the highest-rated R-values per inch. However, there are reports that its thermal resistance degrades at lower outdoor temperatures.[1]
  - *Extruded polystyrene*, often referred to as *XPS*, or by the trade name Styrofoam. Note that the name Styrofoam is widely (and incorrectly) used to refer to *expanded* polystyrene (not extruded polystyrene, where its use is correct). XPS is largely impervious to moisture. XPS is also referred to as blue board or pink board.
  - *Expanded polystyrene*, often referred to as *EPS*, or informally as *beadboard*. EPS is low in cost. EPS is widely (and incorrectly) referred to as Styrofoam. EPS is typically white.

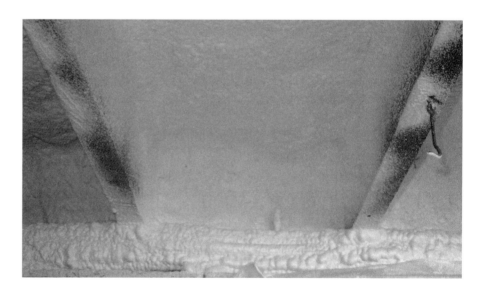

Figure 4.2   Spray foam insulation.

■ *Polyurethane.* Like polyisocyanurate, polyurethane foam board is a closed-cell insulation, with a high thermal resistance.

■ *Mineral wool*, also referred to as rock wool.

Forms of insulation found in existing buildings but not widely in current use include vermiculite, urea formaldehyde, balsam wool, and cotton batt insulation.

Vermiculite insulation is typically found in attics. It is a mined mineral. When heated to high temperatures, vermiculite expands into lightweight pieces between $1/4''$ and $1''$ in diameter, which can be easily poured. It has a speckled appearance and is gold or light brown in color. Vermiculite is also referred to by the brand name Zonolite. One of the largest sources of vermiculite was a mine in Libby, Montana, which was closed in the early 1990s. This mine also contained asbestos, a known carcinogen and lung hazard, so much of the vermiculite insulation that was installed contains asbestos. The Environmental Protection Agency (EPA) recommends that vermiculite be presumed to contain asbestos and should either be left undisturbed or should be removed by trained and certified professionals.[2]

Balsam wool insulation was made from wood fiber (not from wool), and was reportedly sold in the early part of the twentieth century, with its height of use in the 1940s.[3] It typically had a black paper backing. The fibers are tan/brown, and the fibers look like sawdust. It does not contain asbestos.[4]

Another insulation no longer commonly installed is *urea formaldehyde*, also known by the acronym *UF* or as *UFFI* (pronounced "you-fee"). It is found in buildings dating to the early 1980s but has seen little use since that time because of concerns over formaldehyde off-gassing. Injected as a foam, urea formaldehyde can be identified by its dull appearance, its yellow or gold color (sometimes gray with dirt), being prone to crumble, softness, fragility, and shrinkage (Figure 4.3).

Cotton batt insulation was manufactured in the 1930s and 1940s, most prominently by the Lockport Cotton Batting Company, under the Lo-K® brand. It was originally white but grays over the years with dust accumulation. It was installed with paper facing. Cotton insulation is seeing a comeback as a natural insulating material.

Once the type and thickness of existing insulation is known, the thermal properties can be established. A common measure of resistance to heat flow is the *R-value*, in units of ft^2-F-hr/Btu. Common insulation types in buildings, and their R-values per inch, are provided in Table 4.1.[5, 6] Frequently, the R-value of insulation is printed on its face, as shown in Figure 4.4.

R-values can diminish over time, and diminish significantly if insulation becomes wet. This can be of particular concern with flat roofs, where water

Figure 4.3   Urea formaldehyde insulation.

Figure 4.4   Printed thermal resistance rating (R-value).

TABLE 4.1

Common Insulation R-values

Insulation	R-Value per Inch
Cellulose	3.6–3.8
Cementitious foam	3.9
Fiberglass	3.1–4.3
Mineral wool	3.7
Perlite	3.1
Phenolic foam	4.8
Polyisocyanurate, with foil	7.1–8.7
Polystyrene board (EPS, XPS)	3.8–5.0
Polyurethane	5.5–6.5
Urea formaldehyde foam	4.4
Vermiculite	2.4

penetration is an issue. It may be identified with thermal imaging or may warrant taking a core sample to evaluate moisture content if water penetration is suspected.

We are also interested in the thermal resistance provided by other materials in a wall, ceiling, floor, or other assembly, such as gypsum board, masonry, and siding or cladding. Even though these materials have lower thermal resistance than insulation, they still contribute to the overall thermal resistance. The R-values of common building materials are provided in Appendix A.

INFRARED DIAGNOSTICS

Infrared cameras are used to diagnose envelope problems in a process called *thermal imaging*. Thermal imaging requires a minimum 18 F temperature difference between indoors and outdoors.[7] Early morning is a time when there is the best chance of obtaining the required temperature difference. Other conditions to avoid are sun exposure and high wind. Thermal imaging is used to identify thermal bridging and insulation voids. It should be used to identify both the location and the quantity of these deficiencies, in order to support estimating energy savings. For buildings with insulation voids, *all* walls and ceiling/roof elements should be documented with thermal images in order to guide the installation of insulation. One or two example images can only show that voids or thermal bridging are present, but are not sufficient to quantify the extent of the problems, to accurately calculate energy savings, or to guide the improvement work. Thermal imaging is also used in conjunction with blower door testing to locate the sites and impacts of infiltration. This is addressed in the subsequent discussion of infiltration.

ASSEMBLY R-VALUES, THERMAL BRIDGING

Once the R-value of insulation and other materials in an assembly are identified, the overall R-value may be calculated by adding the individual R-values of each heat transfer path and then weighting parallel heat transfer paths. For example, the overall R-value of a 2×6 wood frame wall with fiberglass batt insulation between the studs is calculated as shown in Table 4.2.

Thermal bridging is where a solid material, such as a structural member of a building, penetrates the thermal boundary and allows excessive heat to be transferred. Thermal bridging is undesirable. In the above example, wood studs in a frame building reduce the effectiveness of the insulation by 12 percent. If we had accounted for studs at the top and bottom plates, and around windows and doors, the

TABLE 4.2

R-Value Impact of Thermal Bridging

	R-Value through Stud	R-Value through Cavity	Overall R-Value
Exterior air film	0.17	0.17	
4″ face brick	0.44	0.44	
Air gap	1.00	1.00	
2×6 wood stud	6.88	—	
Cavity/insulation	—	21.00	
$^1\!/_2$″ gypsum board	0.45	0.45	
Interior air film	0.68	0.68	
Total R-value	9.62	23.74	
U-value = 1/R	0.10	0.04	
Percent of wall	9.4%	90.6%	
U-value × percent of wall	0.0097	0.0382	
Overall R-value =			
1 / (Ustud × % stud + Ucavity × % cavity) =			20.9

TABLE 4.3

Impact of Continuous Insulation

	R-Value through Stud	R-Value through Cavity
Exterior air film	0.17	0.17
4″ face brick	0.44	0.44
Air gap	1.00	1.00
Rigid insulation, 2″ polyiso.	13.00	13.00
2 × 6 wood stud	6.88	—
Cavity/insulation	—	21.00
½″ gypsum board	0.45	0.45
Interior air film	0.68	0.68
Total R-value	22.62	36.74
U-value = 1/R	0.04	0.03
Percent of wall	9.4%	90.6%
U-value × percent of wall	0.0041	0.0247
Overall R-value = 1 / (Ustud × % stud + Ucavity × % cavity) =		34.7

negative effect of thermal bridging would be seen to be even higher than 12 percent. There are many examples of thermal bridging in a building if continuous insulation is not provided to limit heat flow through such components: metal studs, wood studs, shelf angles, parapets, details where floors meet walls, balconies, porches, and more. We seek to reduce the effects of thermal bridging by providing a continuous layer of insulation across the components that are conducting heat. Insulation can be continuous on the exterior of such components, on the interior, or both. Interior insulation needs to be carefully applied, to avoid risk of condensation, which will occur if moist indoor air finds its way past the insulation to touch a cold building component in winter.

By adding continuous insulation over thermal bridging, its negative impact can be reduced, and the overall R-value of the assembly significantly increased. In the preceding example, if 2″ of rigid polyisocyanurate were added, the overall R-value of the assembly would increase by 66 percent. See Table 4.3.

The R-values of metal studs are slightly more complicated to calculate due to flanges. A simplified approach has been developed by the American Society of Heating, Refrigerating, and Air-Conditioning Engineers (ASHRAE), providing a correction factor to facilitate the calculation. See Table 4.4.

TABLE 4.4

Effective R-Value of Metal Studs

Nominal Stud Size	Spacing of Framing (inches)	Cavity Insulation	Correction Factor	Effective R-Value
2 × 4	16″	R-11	0.50	R-5.5
2 × 4	16″	R-13	0.46	R-6.0
2 × 4	16″	R-15	0.43	R-6.4
2 × 4	24″	R-11	0.60	R-6.6
2 × 4	24″	R-13	0.55	R-7.2
2 × 4	24″	R-15	0.52	R-7.8
2 × 6	16″	R-19	0.37	R-7.1
2 × 6	16″	R-21	0.35	R-7.4
2 × 6	24″	R-19	0.45	R-8.6
2 × 6	24″	R-21	0.43	R-9.0
2 × 8	16″	R-25	0.31	R-7.8
2 × 8	24″	R-25	0.38	R-9.6

Note how the degradation of overall R-value is even higher with metal studs than with wood studs, as we see a loss of overall thermal resistance of almost 70 percent.

As an alternative to calculating R-values, there are emerging equipment and approaches to measure R-values in the field.[8]

IDENTIFYING THE THERMAL BOUNDARY AND OPPORTUNITIES TO INSULATE

In seeking poorly insulated walls, ceilings, and floors, it is helpful to identify the thermal boundary around a building. We are looking for discontinuities or weak spots—in other words, breaks in the thermal boundary.

Common breaks in the thermal boundary occur around unconditioned spaces—spaces that do not require heating or cooling. For example, consider an attached parking garage. Typical insulation weaknesses might include an uninsulated overhead door, or the ceiling of the garage below a conditioned space above, or the wall between the garage and the conditioned space. The garage door may be kept open much of the day, so the garage interior may be at a temperature close to that of the outdoors. Or even if the garage door is kept closed, the garage may not be conditioned (heated or cooled), so either the exterior walls or interior walls should be insulated, whereas frequently neither are.

In identifying the thermal boundary, a sketch may help. Ideally, we want to be able to draw a line continuously around a building, to show the location of the thermal boundary, without lifting the pencil at any discontinuities of the thermal boundary. Many such discontinuities go back to the original building construction, where insulation was never installed.

Complex building details are a common site of missing insulation. Knee walls, walls around stairs up to attics (Figure 4.5) or down to basements, and cantilevered building spaces are examples of surfaces that frequently are lacking insulation.

Some breaks in the thermal boundary occur over time, when insulation is disturbed or removed or detached. An example is insulation on the attic floor of a pitched-roof building, very common in low-rise commercial buildings. The insulation is frequently removed during renovations, such as to route cable wiring, and is not re-installed after the renovation. Another example is insulation over a crawlspace or basement. Typically stapled to joists, such insulation frequently comes detached, as seen in Figure 4.6.

Another example of a break in the thermal boundary occurring over time is insulation in a wall cavity, which settles over time.

In evaluating the impact of missing insulation, we cannot simply average the thickness of insulation and perform a single calculation. The areas without insulation transfer more heat "from the average" than is saved in heat transfer by the areas that have more insulation than the average. For example, consider an attic floor with 6″ insulation rated at R-21, which has an overall assembly R-value of 19.8, accounting for thermal bridging at the joists, and accounting for the R-value of non-insulation materials such as the gypsum board ceiling. If insulation were entirely removed from a quarter of the attic, the overall assembly R-value plummets to below R-10. If we treated the 25 percent missing insulation as a 25 percent reduction in the insulation R-value, we would mistakenly estimate an overall assembly R-value of R15, far higher than the real overall assembly R-value, below R-10. Missing insulation dramatically increases heat transfer, even if the fraction of missing insulation is small.

The effect of voids on overall R-value can be calculated by adding a third column to the calculation above. This three-column approach is useful, in that most frame walls, whether wood frame or metal frame, have thermal bridging (one column), insulation between the framing (second column), and voids in the insulation (third column). Similarly, attic floors, basement ceilings, and other surfaces also have three adjacent heat flow paths: thermal bridge, insulation, and voids. See Table 4.5.

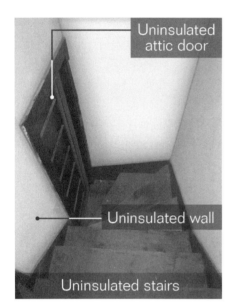

Figure 4.5   Attic walkup stairs—missing insulation

Figure 4.6   Detached insulation.

TABLE 4.5

Three-Column Approach to Estimate Overall R-value

	R-Value through Stud	R-Value through Cavity with Insulation	R-Value through Cavity without Insulation
Exterior air film	0.17	0.17	0.17
Siding and sheathing	1	1	1
2 × 4 wood stud	4.38	—	—
Cavity/insulation	—	11	1
1/2 gypsum board	0.45	0.45	0.45
Interior air film	0.68	0.68	0.68
Total R-value	6.68	13.30	3.30
U-factor	0.15	0.08	0.30
Percent of wall	0.09	0.87	0.04
U-factor × percent of wall	0.01	0.07	0.01

Overall R-value = 1/(Ustud × percent stud + Ucavity × percent cavity with insulation + Ucavity × percent cavity without insulation) = 11.0

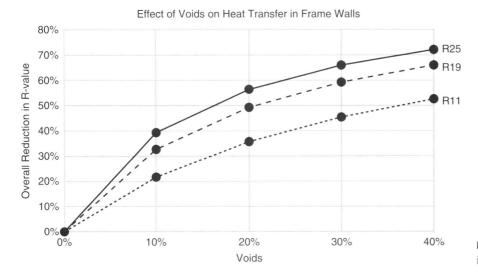

Figure 4.7   Effect of voids on heat transfer in frame walls.

Using this approach, we can examine the effect of different voids on a variety of insulation levels. See Figure 4.7. We note that the reduction in R-value is higher for more insulated walls, and the rate at which the R-value is reduced is greatest for small voids. The results are similar to those presented in a classic study by Vinieratos and Verschoor in testing performed for the U.S. Navy.[9] It is important and easy to make these corrections, rather than rely on rules of thumb, which underestimate the degradation and therefore underestimate savings from insulating voids.

Voids further degrade thermal resistance due to convective airflow within the voids in wall cavities, as air is heated on the warm side, rises, and falls as it is cooled on the cold side.[10]

In short, insulation voids seriously degrade the overall thermal resistance of building surfaces, in far greater proportion than the size of the voids. Voids should be actively sought and eliminated with insulation.

Insulation may also be compressed during installation (Figure 4.8), reducing its thermal resistance.

Figure 4.8   Compressed insulation.

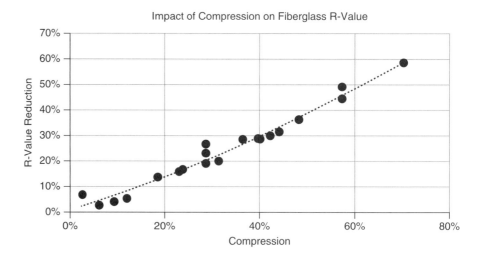

Figure 4.9 Impact of compression on fiberglass R-value. Graph based on data from Building Insulation Compressed R-value Chart, Technical Bulletin, Owens Corning, Publication Number 10017857, 2012.

Figure 4.10 Double-hung window sash counterweight.

Figure 4.11 Flexible outdoor room air conditioner cover.

Fiberglass that has been compressed by half its label thickness will deliver almost 40 percent less R-value. Fiberglass compressed by 70% its label thickness will deliver 60 percent less than its rated R-value. See Figure 4.9.

Another site where insulation is frequently missing is in the space around windows and doors. This can be evident in infrared imagery or if trim is removed.

We also look for opportunities to insulate in unexpected locations. For example, counterweight cavities for windows with ropes, pulleys, and counterweights (Figure 4.10) can be insulated if the counterweights are removed and either left without counterweights or replaced with spring-loaded counterbalances, available since the late 1800s. The width of a typical counterweight cavity is approximately 3 to 4″, the height is the height of the window, and the depth is typically the depth of the wall cavity. An additional benefit is reduced infiltration if spray foam or other airtight insulation approaches are used.

Another type of cavity is empty air conditioner sleeves. Apartment buildings frequently provide empty sleeves, in which air conditioners are installed in summer, or in which air conditioners are optionally rented. However, many tenants do not choose to rent air conditioners, and so the sleeves stand empty. Tight-fitting fiberglass batts are an effective insulation fill for air conditioner sleeves, as are foam boards, although foam boards tend to break at the edges if the insulation is removed and reinstalled seasonally. An empty sleeve has an equivalent R-value of approximately 1. A simple flexible outdoor cover, as shown in Figure 4.11, has been shown to provide little additional thermal resistance. Interior flexible covers, as shown in Figure 4.12, are equally ineffective.

A sleeve filled with insulation has been found to have an equivalent R-value of approximately 2.5. Ostensibly, the R-value of the insulation is reduced by heat conduction through the metal sleeve walls. If a flexible cover is used on the outdoor side of the sleeve, along with the insulation fill, the R-value increases only to approximately R-3. However, if a prefabricated rigid cover is used on the indoor side, in addition to the insulation fill, the R-value of the assembly is approximately R-6.[11]

Surface R-value can be estimated by measuring interior surface temperature. Measurements are better when cold outside. See Figure 4.13. Such measurements should be regarded as rough estimates only, as they depend on a variety of assumptions, such as surface and air temperatures being constant, absence of wind, absence of solar gain, and more. This approach is particularly effective for surfaces that have low R-value, in other words, surfaces that need insulation.

CEILINGS AND ROOFS

Pitched roofs and the attics they cover are a highly problematic area of the thermal boundary. We think of pitched roofs as being associated with single-family homes, but they are common in commercial buildings as well (Figure 4.14), especially in low-rise and medium-rise construction. Attics are typically vented, and so are at a temperature that can be close to the outdoors, for example, in winter at night, and hotter than outdoors in the summer sun. Insulation problems in attics are typically at the attic floor.

Attic floors not only have missing insulation but may be inadequately insulated to begin with (Figure 4.15). Additional attic insulation is relatively easy to install. A

Figure 4.12   Interior room air conditioner cover.

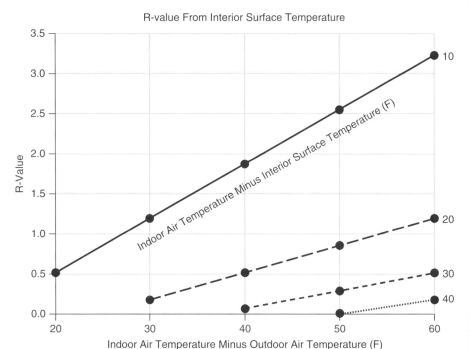

Figure 4.13   R-value estimated from interior surface temperature and air temperatures.

Figure 4.14   Pitched roofs and attics are common in commercial buildings.

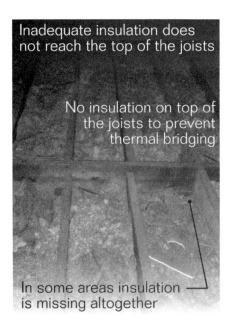

Figure 4.15 Attic insulation deficiencies.

Figure 4.16 Cellulose insulation in attic.

common approach to insulating attics is blown cellulose, as shown in Figure 4.16. Installed to a depth that covers the floor joists, cellulose can reduce the effects of thermal bridging at the joists, while also reducing airflow from the conditioned space up into the attic. Cellulose insulation requires the use of chutes at the edges of attics, to allow attic ventilation (Figure 4.17). It also requires dams around penetrations, such as attic hatches, to prevent cellulose from spilling into these areas (Figure 4.18). Another best practice is to require the use of depth markers (Figure 4.19), typically one for every 300 square feet, to ensure that adequate insulation depth has been reached and that the depth is uniform.

The floors of attics are not the only areas with inadequate insulation. Details such as knee walls frequently lack insulation, and therefore become breaks in the thermal boundary.

Dropped ceilings often offer an opportunity for added insulation. Frequently, loose batts are laid on top of dropped ceiling tiles. This is not regarded as effective, because air can move freely through the loose batts. More effective options include insulating the wall at the edge of the plenum formed by the dropped ceiling. This surface may be easier to insulate than the wall below the ceiling because it does not need to be finished and painted. The wall surface above the dropped ceiling typically has no windows or doors to work around, or electrical components such as light switches and receptacles. The savings may not be as high as insulating walls below the ceiling, because the space above the ceiling is essentially an unconditioned space. However, the space is still typically within or in line with the thermal boundary, so savings can accrue from insulating these walls. This space is frequently as much as several feet high in commercial buildings and so can offer a large area to insulate. If the dropped ceiling is on the top floor of a building, insulating below the roof above is another option. Insulation in these areas should be air sealed, to prevent moist interior air from reaching cold structural components on the cold side of the interior insulation.

Insulation can be added to flat roofs. A variety of rigid foams are commonly used. Insulation on flat roofs is typically tapered, to promote drainage of water to roof drains (Figure 4.20).

The pitch of this taper is generally required by code to be a minimum of $1/4''$ per foot (1:48), although it is not unusual to find slopes as low as $1/8''$ per foot (1:96) or as high as $1/2''$ per foot (1:24) or higher. The R-value of tapered roof insulation (Figure 4.21) has historically been slightly underestimated, so existing roofs may well have less insulating value than believed.[12] The underestimation has come about because an average thickness of the roof insulation has been used to calculate the R-value. However, thinner insulation transfers more heat than thicker insulation saves. The underestimation is more significant if the existing insulation is less. For example, a rectangular piece of insulation that is sloped in one direction from 5″ thickness to 0.5″ thickness has an actual R-value approximately 20 percent less than simply calculating the R-value from the average thickness.[13] The loss in R-value is more pronounced for roofs with tapered insulation forming a peak, draining to perimeter gutters. An online calculator is available that can be used to estimate the R-value more accurately.

Spray foam is another option, which can be used on top of existing flat or low-pitch roofs, covered in a weatherproof coating.

Ceiling/attic heat transfer can also be reduced with radiant barriers, as shown in Figure 4.22. Residential research on radiant barriers indicate modest savings in buildings with uninsulated ductwork in the attic in the deep South, and lower savings for other applications (insulated ductwork, no ductwork, buildings other than the deep South).[14] The insulating benefits of vegetated (green) roofs (Figure 4.23) have been found to be small.

WALLS

Uninsulated walls offer several options for insulation.

Cavity walls can have insulation blown in from either the interior or the exterior. Common types of insulation include cellulose, fiberglass, and spray foam. Costs can be reduced if insulation is done at a time when other renovations are happening, either at a time when interior walls are being painted or at a time when new cladding or siding is being installed on the exterior.

Another approach is to install rigid installation on the exterior, either by removing the siding/cladding or as an *exterior insulated finishing system* (EIFS, pronounced "ee-fis"). The benefit of this approach is to reduce thermal bridging. Rigid insulation installed below siding/cladding should be taped, to minimize air leakage and to maximize the effectiveness of the insulation. Attention is required to maintain a drainage plane, to ensure that water that penetrates the cladding can be shed and drained, with particular attention required around components such as windows and doors.[15]

Insulation on the interior of frame buildings is also possible. This may be desirable if the exterior is difficult or costly to insulate. Continuous insulation is preferable to furred-in insulation in order to counter any existing thermal bridging and to avoid introducing new thermal bridging. Rigid insulation should be taped to prevent moist indoor air from migrating between insulation seams and coming into contact with cold building components on the exterior of the insulation.

Masonry buildings can also be insulated. An old concern has been to avoid moisture-related issues if insulation is installed on the interior. Approaches to avoid such problems have been developed.[16] Insulation on the exterior of masonry buildings should also not be ruled out, and, again, insulation during larger renovations of a building exterior or façade are most cost-effective.

Retrofit insulation of basement walls is usually more easily done on the interior, due to the higher cost of below-grade exterior insulation, and the challenges of protecting above-grade exterior insulation. Interior insulation must avoid moisture-related problems, primarily avoiding moisture becoming trapped between the insulation and the wall. This means allowing moisture in the foundation wall to dry to the interior. A variety of approaches are recommended,[17] including, for example:

- For the lowest cost, insulate the top half of the interior foundation wall with foil-faced polyisocyanurate (Figure 4.24). This does not require a fire-rating cover, and it allows the lower half of the wall to dry to the interior.

- For additional energy savings, insulate the entire interior foundation wall with rigid unfaced extruded polystyrene (EPS), covered with $1/2''$ gypsum board. EPS allows the wall to dry to the interior. The EPS should be taped to prevent moist basement air from reaching the foundation wall. The gypsum board provides the required fire rating. Do not install a vapor barrier.

Note that in both these approaches, rigid insulation is installed directly on the interior of the foundation wall.

If exterior insulation is an option, XPS can be used, although it must be protected above grade, and damp-proofing is recommended between the insulation and the foundation wall.

Unusual opportunities to insulate should always be sought, such as abandoned doors and windows (Figure 4.25).

FLOORS

There may be occasions in which floors need to be insulated. Examples include cantilevered structures with exposed floors, and floors over crawlspaces and basements. Best practices include continuous rigid insulation with taped seams, whether insulating on top of the floor or below, to prevent air migration between insulation

Figure 4.17 Chutes to prevent cellulose insulation from blocking vents.

Figure 4.18 Attic hatch with dam to prevent cellulose spilling into hatch.

Figure 4.19 Insulation depth marker.

Figure 4.20 Roof drain on flat roof, to which roof insulation is used to provide pitch for drainage.

and to minimize thermal bridging. Spray foam is another option. Rigid insulation generally has good strength in compression and can bear the weight of people and furniture, so it can be used on top of existing floors and covered with subflooring and floor finishes, or even covered in a concrete topping slab. Attention must be directed to risks of condensation in cold climates, so insulation on the exterior is preferable.

Insulation can be used to reduce heat transfer through concrete floor slabs out to a building wall or to balconies, a common site for losses for buildings where thermal breaks were not originally provided. Ideally, insulation should be applied around the balcony itself. Interior insulation of the floor is possible but presents condensation risks.[18]

Configurations such as heat loss through balcony slabs, or even through floor slabs that are connected to an uninsulated exterior wall, present a more complex challenge to calculate energy savings. Most heat loss configurations are simpler, such as heat loss through a wall or roof, and require what we call one-dimensional heat loss calculations. If we draw arrows to show the direction of heat flow, all arrows point in one direction only, symbolizing one-dimensional heat transfer (Figure 4.26, top). However, for floor slabs exchanging heat with an exterior wall or balcony, a more complex two-dimensional heat flow calculation is required.

The flow of heat in two-dimensional heat transfer is shown graphically by arrows that are not all moving in the same direction (Figure 4.26, bottom). The heat transfer could even be three-dimensional, moving in directions perpendicular to the section. Complex heat transfer configurations such as these are best handled with a type of calculation known as *finite element analysis* (FEA), which is not provided by most energy audit firms (Figure 4.27). However, the calculation is readily done, and may well be worthwhile for larger buildings.

UNCONDITIONED SPACES

Unconditioned spaces are those spaces that have neither heating nor cooling.

Unconditioned spaces are perhaps the most common locations of insulation deficiencies in the thermal boundary. Unconditioned spaces frequently include stairwells, utility rooms, electrical rooms, mechanical rooms, janitor's closets, attics, basements, crawlspaces, vestibules, and more. Unconditioned spaces also include less recognized spaces such as plenums above ceilings, mechanical shafts, and elevator shafts.

Unconditioned spaces are often placed between conditioned spaces and the outdoors. As such, unconditioned spaces can define both an *inner envelope* (between the conditioned space and the unconditioned space) and an *outer envelope* (between the unconditioned space and the outdoors). Consider a parking garage that forms the first floor of a building. The walls of the garage and the overhead door form the outer envelope. The ceiling of the garage forms the inner envelope.

Two major problems exist with unconditioned spaces:

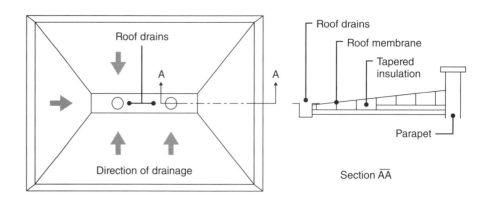

Figure 4.21 Tapered roof insulation.

- Frequently, there is no insulation of either the outer envelope or the inner envelope. Or, over time, insulation has been removed or detached.

- Unconditioned spaces can have high infiltration, and therefore are at a temperature close to the outdoors. Consider a parking garage with the overhead door left open, or a vented attic or crawlspace. Unconditioned spaces are frequently located on the lower floors of a building (basement, crawlspace, parking garage, mechanical rooms, etc.) or on the upper floors (attic, penthouse mechanical room, etc.), and therefore are subject to the highest positive or negative *stack effect* pressures, promoting infiltration. Stack effect pressures arise from air temperature differences between indoors and outdoors. If the inner envelope is poorly insulated or not insulated, the conditioned space loses and gains heat from the near-exterior temperature of the unconditioned space.

So, unconditioned spaces are a weak point of buildings, and can contribute to excessive heat loss and gain. In evaluating unconditioned spaces for energy improvements, we seek to make them work for us, instead of against us. In this way, several characteristics of unconditioned spaces can be put to good use:

Figure 4.22 Radiant barrier.

- Unconditioned spaces typically do not require the high cost of finishes (paint, carpeting, etc.), and so the cost of insulation can be lower.

- Unconditioned spaces (Figure 4.28) offer two surfaces that we can insulate: the outer envelope and the inner envelope. We can choose the more appropriate surface, or we can choose to insulate both surfaces for additional energy savings.

- If infiltration is reduced, the unconditioned space can serve as a buffer space, essentially allowing the space itself to serve as another layer of insulation.

- Unconditioned spaces can serve as *air locks*, reducing infiltration when people (or vehicles, in the case of garages) enter and leave a building.

The decision whether to insulate the inner envelope or outer envelope depends on each space and its uses. In general, if the space is to be actively used, insulate the outer envelope. However, if a space is in fact to be actively used, then it typically is conditioned. So a rule of thumb for unconditioned spaces is to first try to insulate the

Figure 4.23 Vegetated (green) roof.

Figure 4.24 Rigid insulation directly fastened to a basement wall, foil-faced to provide required fire resistance.

Figure 4.25   Abandoned window.

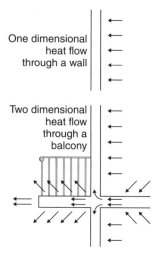

One dimensional heat flow through a wall

Two dimensional heat flow through a balcony

Figure 4.26   One-dimensional versus two-dimensional heat flow.

inner envelope. For example, give priority to insulating an attic floor, rather than the roofline. Or first try to insulate a basement ceiling, rather than the basement walls. Insulating the inner envelope minimizes the surface over which heat transfer occurs, and it minimizes the amount of insulation necessary. If a basement ceiling or an attic floor is insulated, it is important to continue the thermal boundary around the access to these spaces, such as the stairway and door.

As we take advantage of not needing to finish insulation in unconditioned spaces, we need to remain aware of fire protection requirements for combustible materials. For example, foil-faced polyisocyanurate does not need any protection. Most other types of insulation do need protection, which can be provided by $1/2''$ gypsum board or intumescent paints or coatings.

Crawlspaces are an unusual type of unconditioned space. Developed after World War II as an affordable alternative to full basements, crawlspaces are typically vented spaces, and are frequently built over exposed soil. Recurring problems with crawlspaces include high energy loads, high humidity, freezing pipes, a cold floor above, and distribution system energy losses. Exposed soil has been shown to release moisture to the air at the same rate as an exposed pool of water, per square foot of area. An effective solution has been to bring crawlspaces inside the thermal boundary, insulating the walls, eliminating the vents, and placing a vapor barrier on the floor.[19] Benefits of bringing crawlspaces into the thermal boundary include eliminating risks of freezing pipes, reduced humidity problems, a warmer floor above, and reduced energy usage.

In examining an unconditioned space, we frequently need to assess whether it is vented or not. Attics and crawlspaces are typically vented; basements are not. Mechanical rooms are typically vented because of combustion equipment. Crawlspace vents are typically wall-mounted. Attics are vented in several ways: ridge vents, soffit, vents, gable vents, and turbine vents. Look for daylight from inside a space to identify if it is vented. Measuring the temperature of an unconditioned space in winter is another way to assess if it is vented.

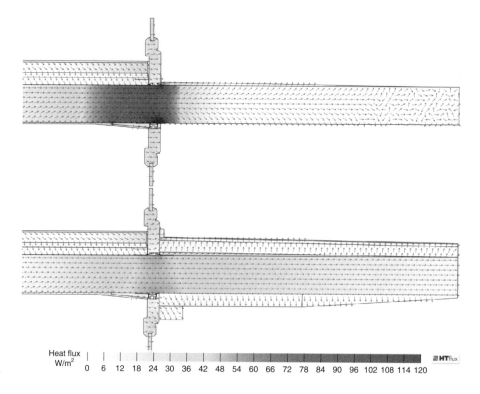

Figure 4.27   Analysis of energy savings from insulating balconies. Heat flow before insulation (top) and after insulation (bottom). *Source*: Daniel Rudisser, HTflux.

Heat flux W/m²   0 6 12 18 24 30 36 42 48 54 60 66 72 78 84 90 96 102 108 114 120

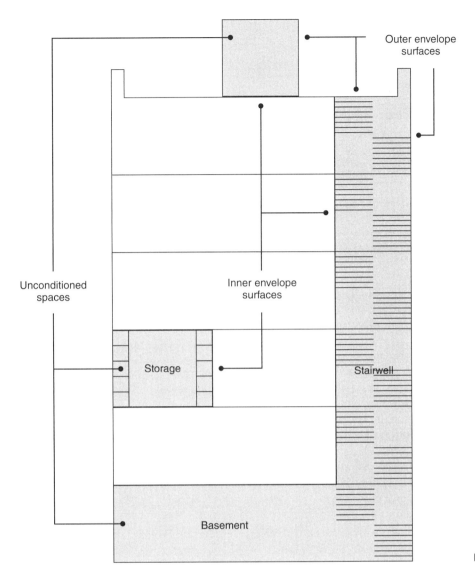

Figure 4.28 Unconditioned spaces.

WATER AND MOISTURE CONTROL

Water and moisture control are critical aspects of insulating a building. Added insulation can either introduce new water and moisture problems, or it can act to constructively prevent problems and even solve old problems.

The interactions between insulation and water/moisture are extensive. Insulation can serve multiple functions, to not only retard heat flow but to also serve as a drainage plane, or to act as a permeable material through which water can be dried out. Another interaction is that many types of insulation, when wet, have lower thermal resistance. If moist indoor air can move through insulation, in winter, it risks condensing on cold structures on the cold side of the insulation. If low-permeable insulation materials are in a location where water or moisture can be trapped, such as on the underside of a roof or between a basement wall and interior insulation, mold growth is a risk. Wherever possible, proven insulation/structure assemblies should be used.[20]

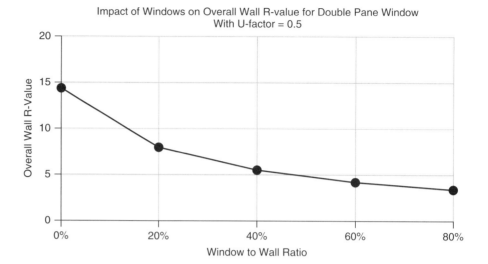

Figure 4.29 Windows reduce the overall thermal resistance of the thermal envelope.

Sliding

Single-hung or double-hung

Awning

Hopper

Casement

Fixed

Figure 4.30 Window types.

Windows

The word *window* comes from the Old Norse word *vindauga*, with "vindr" meaning wind, and "auga" meaning eye. In other words, a window is an eye to the wind.

Windows are valued for the natural light they provide, for views to the exterior, for solar gains in winter, and for natural ventilation. Windows also lose significant energy through heat losses in winter, heat gain in summer, solar gain in summer, and infiltration. The best window, with an R-value of approximately R-5, is typically far worse in allowing heat losses than the worst wall in any given building. See Figure 4.29. Recalling the strong negative impact of voids on an insulated wall's overall thermal resistance, we recognize that windows substantially reduce the thermal resistance of buildings. Windows can also cause other problems, including glare, temperature stratification, drafts due to convective currents (even if windows are not leaky), radiant heat loss and gain from occupants to the outdoors, and more. As much as we value them, windows truly are holes in our walls, from an energy perspective.

TYPES OF WINDOWS

Windows can either be operable, in other words, they can open and close, or they can be fixed. Different types of operable windows include sliding windows (in other words, slide laterally), hung windows (in other words, slide vertically, either single-hung or double-hung), and hinged windows, as shown in Figure 4.30. Hinged windows include casement windows (hinged on the side), hopper windows (hinged at the bottom), and awning windows (hinged at the top).

Hinged windows provide the best air seal of any operable windows.

Windows also appear in doors, where they are referred to as *lights*, or around exterior doors, above and/or on the sides. Windows can also form the entire structure of doors, such as fully glazed sliding porch or balcony doors.

Windows are also classified by the material of their frames. In low-rise buildings, wood and vinyl frames are common. In high-rise buildings, metal frames are necessary for structural reasons and to withstand wind forces. In the United States, over 90 percent of nonresidential window frames are reported to be aluminum.[21] Steel frames were common in the 1890–1950 time period and are considered to have historic characteristics.[22] Fiberglass frames, with lower thermal conductivity than metal windows, are used in low-rise buildings and have reportedly been used in medium-rise buildings and lower high-rise buildings, primarily up to 15 stories.

Impact of Windows on Overall Wall R-value for Double Pane Window With U-factor = 0.5

Metal frames have a high thermal conductivity and therefore allow high levels of heat to be transferred through them. To resist this heat flow, thermal breaks have been developed within the frame. However, it is not possible to visually assess if a metal frame is thermally broken. We might infer from the age of the building/window if a metal frame is thermally broken. Kawneer, a major manufacturer of aluminum windows, introduced its thermal break in 1988,[23] and thermal breaks for steel and bronze window frames were introduced to the United States in 2004. It is reported that 20 percent of aluminum windows were thermally broken in 2005, and this increased to 38 percent in 2012,[21] roughly equal to the fraction of nonresidential buildings that are certified as green (41 percent in 2012).[24] Accordingly, in our fieldwork, we might assume that pre-1990 metal windows do not have thermal breaks, and that thermal breaks through the present are primarily found in more energy efficient buildings, such as Leadership in Energy and Environmental Design (LEED)-certified buildings.

The thermal conductivity of different metals varies. Aluminum has a high thermal conductivity, and stainless steel a lower thermal conductivity, for example. However, the thermal conductivity of all metals is so high that the U-factor of metal frames is dominated by the thermal resistance of the air films on either side of the frame. So metal frames are not typically distinguished by the type of metal, for frame U-factors.

The area of windows used for energy calculations includes both the glazing and the frame (Figure 4.31). By definition, it is the rough opening of the window, minus clearances.[25]

Accurate window dimensions may not be possible, when interior or exterior trim cover the rough opening size. For more accurate measurements, consult building drawings. For approximate estimates, make a reasonable estimate of the size of the rough opening, adjusting approximately from the size of the window including trim.

In discussions of window sizes, we sometimes encounter the term *united inches*, sometimes referred to as *unified inches*, which is the sum of a window height and width.

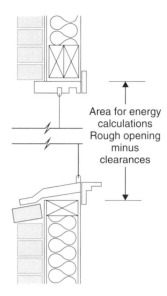

Figure 4.31 Window area for energy calculations.

PERFORMANCE

Windows are rated according to a system administered by the National Fenestration Research Council (NFRC). Ratings include:

- *U-factor*, in English units of Btu/(h-F-ft^2).
- *Solar heat gain coefficient*, or *SHGC*, a metric that is between 0 and 1, which indicates a window's ability to allow solar heat gain (higher SHGC) or resist solar heat gain (lower SHGC).
- *Visible transmittance*, or *VT*, a metric that is between 0 and 1, which indicates a window's ability to allow visible light (higher VT) or prevent it (lower VT).
- *Air leakage*, in units of cubic feet per minute per square foot of window area (CFM)/ft^2.

The units of U-factor are the inverse of the units of R-value. The U-factor accounts for the thermal resistance of the air films on the indoor and outdoor surfaces of the window, as well as for the frame of the window. It does not account for wall framing around the window.

In the early 1990s, NFRC ratings were introduced for U-factor and SHGC. In 1995, the CABO Model Energy Code recognized NFRC's U-factor, and in 1996 recognized SHGC. In 1998, the IECC took over from the Model Energy Code, and ENERGY STAR introduced window performance requirements. A summary of code requirements is shown in Appendix B. The table does not include requirements for fixed windows and skylights, and should not be taken as national codes, because different states adopted different versions at different times.

A few observations may be made about trends in window efficiency:

- Window energy requirements for residential buildings have generally been more stringent than requirements for commercial buildings.
- ENERGY STAR requirements apply only to residential windows.
- ENERGY STAR requirements have not always been more efficient than the model energy codes.

The NFRC rating method divides a window into three areas: Center of glass (COG), edge of glass (EOG, the area of glazing next to the frame, 2.5″ wide), and frame. Each area has an associated U-factor, and the area-weighted impact of the three U-factors is used to calculate an overall U-factor. For many windows, the center of glass U-factor is lower than the edge of glass or frame U-factors, and so the frame and edges drag down the overall U-factor. It also means that a smaller window typically has a higher U-factor (in other words, is less efficient) than a larger window of the same construction. For this reason, a large window that is divided into smaller panes is better treated as multiple windows, to account for the added frame area where the window is divided, particularly where the dividers (muntins; Figure 4.32) fully divide the glazing, in other words, where the frame at the divider is a full frame, rather than being a surface element for aesthetic purposes only.

Computer programs such as WINDOW, available from Lawrence Berkeley National Lab[26] may be used to calculate U-factors and other properties for window assemblies of known materials. WINDOW is relatively easy to learn, and is a powerful and flexible program. It comes with a library of glazing systems and frames. WINDOW is particularly strong for advanced glazing systems, although it does have glazing as simple as a single pane of glass. Highly complex frames may be modeled, in conjunction with a program called THERM, making the program appropriate for use by manufacturers, for the design of new window systems. But the program is simple enough for energy auditors and energy managers to assess the

Figure 4.32   Metal frame muntins.

TABLE 4.6

Calculating Whole-Window U-Factor

	Frame	Edge of Glass (EOG)	Center of Glass (COG)
U-factor	1.00	0.58	0.48
Area (Square Inches)	412	410	1,338
Percent area	19%	19%	62%
U-factor × percent area	0.19	0.11	0.30

Overall R-value = $1 / (U_{frame} \times Area_{frame} + U_{EOG} \times Area_{EOG} + U_{COG} \times Area_{COG})$ = 1.67
Overall U-factor (1/R) = 0.60

benefits of various window design strategies. The default frames provided with the program are limited, comprising thermally broken aluminum, wood, and vinyl, but not steel; aluminum that is not thermally broken; and some other frames that are found in existing/older buildings.

A simple whole-window U-factor calculation can also be performed with a spreadsheet, allowing U-factors to be calculated for a variety of existing window types (see Table 4.6). The results of this spreadsheet match WINDOW well.

ENERGY IMPROVEMENTS

Traditionally, windows had a single pane of glass, and many buildings still have these old single-pane windows. Double-pane windows have become standard for new windows, and are essentially a minimum required by energy codes. Triple-pane windows are also available, and advanced window assemblies, including features such as suspended films between the panes, offer the lowest U-factors, as low as 0.15. Double- and triple-pane window systems that are hermetically sealed are commonly referred to as *insulating glass* (IG), or *insulating glass units* (IGUs), or *sealed insulating glass units*.

Single-pane windows have the worst (highest) U-factors, estimated to be over 1. For windows without known ratings, a summary of presumed U-factors and solar heat gain coefficients (SHGCs) is shown in Table 4.7. It is worthwhile to check U-factors with WINDOW or calculation, because U-factors can vary significantly with window size and other factors.

Storm windows are effective in reducing heat transfer through windows. Storm windows can be installed on the interior or exterior of most windows, fixed or operable.[27] Interior storm windows are sometimes referred to as *interior window panels* because they do not offer protection from storms.

The performance of storm windows is actually fairly forgiving to a small amount of infiltration around the storm window. Many storm windows are purposefully vented to allow moisture to escape. This venting does not substantially reduce the effectiveness of the storm window.[28] In other words, storm windows are still effective, even if there is some air leakage.

Storm windows are vulnerable to being left open, and therefore typically require an active operation and maintenance program to ensure that they are closed in winter. Interestingly, a double-hung storm window, if left open (in place for the upper half of the window, open for the lower half), still delivers much of its intended insulating function. The air between the primary window and the upper half storm is warmer than the outdoor air below it, and so mostly does not move, despite being open to the outdoor air below. Therefore, we cannot assume that a half-open storm window is entirely ineffective; we can only assume it is not functioning for the area that is open.

TABLE 4.7

Estimated Properties of Existing Windows

Frame	Glazing	Notes	U-Factor	SHGC
Wood/vinyl	Single		0.95	0.80
Wood/vinyl	Single	Exterior storm	0.49	0.56
Wood/vinyl	Single	Interior DIY film	0.50	0.57
Wood/vinyl	Single	Exterior storm, interior film	0.32	0.51
Wood/vinyl	Single	Low-e storm	0.38	0.48
Wood/vinyl	Double		0.55	0.70
Wood/vinyl	Double	Storm	0.32	0.47
Metal	Single		1.20	0.80
Metal with thermal break	Single		1.10	0.80
Metal	Double		0.80	0.70
Metal with thermal break	Double		0.65	0.70
Glazed Block	NA		0.60	0.60

The actual performance of storm windows depends on where the storm window is mounted, for field-installed storms. For example, if mounted on the sash of a double hung window (as is the case with do-it-yourself interior film, for example), the storm window only reduces glazing heat transfer, but does not reduce frame heat transfer. For an existing single-pane window with an aluminum frame (thermally unbroken), adding a storm window only over the glass reduces the overall U-factor from 1.06 to 0.64, whereas adding a storm window over both the glass and frame further reduces the overall U-factor to 0.49. For a double-pane window with a thermally broken aluminum frame, adding a storm window over only the glass reduces the overall U-factor from 0.57 to 0.45, whereas adding a storm window over both the glass and frame reduces the overall U-factor to 0.35.

Various methods are available for fastening storm windows. They can be fixed (permanent), temporarily installed with magnets, temporarily installed with latches to allow removal, or operable such as with a triple track. Operable storm windows offer the benefit of allowing the window to be opened. However, they raise the risk of the storm window being inadvertently left open.

Interior storms are more affordable and are easier to install and remove seasonally. There can be a risk of condensation in winter if moist interior air gets past the interior storm to reach the cold prime window (which is not itself vented to the outdoors). However, in practice, this risk has been found to be small, even in cold climates, as long as the interior storm is fairly tight. The opposite is true in summer, as infiltration of warm, moist outdoor air past an exterior storm is at risk of condensing on the prime window, but, again, the risk is low. Most window condensation issues are believed to relate to heat transfer at the frames, rather than to infiltration. Interior storms (Figure 4.33) also allow a low-emissivity coating to be added to the assembly, on a preferable surface, the exterior surface of the interior storm. Likewise, the interior surface of an exterior storm is a good location for a low-emissivity coating.

The air gap between the storm window and the prime window does not impact its performance significantly, except for smaller gaps. Based on a study of varying window gaps, the optimum gap appears to be approximately 0.5″ to 0.6″.[29] Between 0.6″ and 1.2″, the center-of-glass U-factor is fairly constant. Many sources indicate that the U-factor for air gaps between 0.5″ and 4″ is a constant 1.[30] A simulation using WINDOW, with 1/8″ clear glass and a variable air gap between two panes of glass, confirms these reports. See Figure 4.34.

Figure 4.33 Interior storm on stained-glass window in 130-year-old building in a historic district.

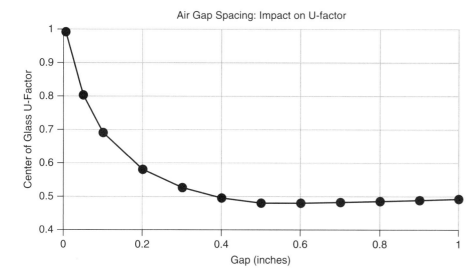

Figure 4.34   A small gap between a storm window and prime window reduces its thermal performance.

Most significantly, the U-factor increases steadily below a gap of 0.5″, and rapidly below a gap of 0.2″. So, we want to avoid small gaps between storm windows and prime windows.

For insulated glass systems, *spacers* were originally metal-only, typically aluminum, with high thermal conductance. Increasingly, *warm-edge* (or low-conductance) spacers have been developed, introducing a thermal break at the spacer. Warm edge spacers can reduce the overall U-factor of a $3' \times 4'$ window by 0.01 to 0.02.[31]

There are reports of emerging equipment and approaches to measure window U-factors in the field.[32] Such equipment and approaches might be useful for unknown assemblies, but cannot replicate ratings developed according to National Fenestration Rating Council (NFRC) standards.

Window energy performance can be improved through the use of low-emittance (low-e) coatings. For two-pane windows, the low-e coating is typically on the inside surface of the outer pane. We call this surface number 2, as the numbering starts with 1 for the outer surface of the outer pane, and proceeds inward. For triple-pane windows, the low-e coating is typically on surfaces 2 and 5 (Figure 4.35).

For a double-pane window, low-e coatings are preferable on either surface 2 or 3. A coating on surface 4 is unusual but would perform almost as well. A coating on surface 1 performs poorly, almost as poorly as having no coating at all. Performance of low-e coatings on different surfaces are shown in Figure 4.36.

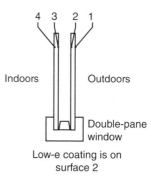

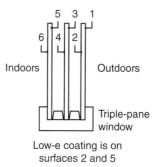

Figure 4.35   Window surface numbering and low-e coating.

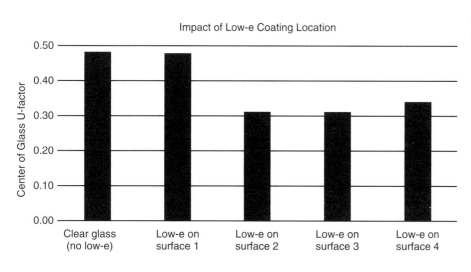

Figure 4.36   Low-e coating works best on surface 2, still works well on surfaces 3 and 4, but does not work on surface 1.

Figure 4.37 Single-pane window.

Figure 4.38 Double-pane window.

Figure 4.39 The indoor frame temperature can be used to estimate if an aluminum frame is thermally broken.

The number of panes of an existing window can be identified by holding a candle or flashlight close to the glazing. Each pane will provide two reflections, close to each other, one reflection for each surface of the pane. So a single-pane window will provide two reflections, as shown in Figure 4.37, while a double-pane window, as shown in Figure 4.38, will provide four reflections (in two distinct groupings of two each), and a triple-pane window will provide six reflections in three groups of two reflections.

The presence of low-e coatings on existing windows can be determined using low-e detectors. These devices not only detect the presence of low-e coatings, but can indicate which surface(s) the coatings are on. Handheld gas analyzers are also available to identify if a window has a gas fill, the type of gas, and its concentration.

Whether aluminum window frames are thermally broken may be assessed by measuring the surface temperature of the frame and the outdoor air temperature. Assuming the indoor air temperature is 70 F, whether or not the frame is thermally broken may be assessed from Figure 4.39. If the frame temperature is on or above the solid line, the frame likely has a thermal break. If the frame temperature is on or below the dashed line, the frame likely is not thermally broken. In between the two lines, one cannot be sure if the frame is thermally broken. The temperatures must be taken when the window is not seeing direct sunlight. The difference between thermally broken and not thermally broken is small enough that it cannot be detected by hand. An aluminum frame with a thermal break will still be cold to the touch.

Window energy properties may be found on as-built drawings. Or, sometimes, a building owner may have kept the rating label of a window, from when windows were installed (Figure 4.40).

Window ratings may be available from the manufacturer. Window identification can be in several locations, for example, etched in a corner of the glass or printed on the spacer or on a label on the sash or frame (Figure 4.41).[32]

Insulating window shades or blinds offer energy savings through both reduced conduction and infiltration (Figure 4.42).

Retrofit kits are available to allow the low-cost installation of transparent film as a form of interior storm window (Figure 4.43). This film reportedly works well to save energy. As a thin material, it may not last for decades, as it is subject to physical damage, and therefore may require reapplying every few years. It can also interfere with window hardware, and so, for example, may prevent opening a window. It can also appear unprofessional. However, again, this film does work and will save energy.

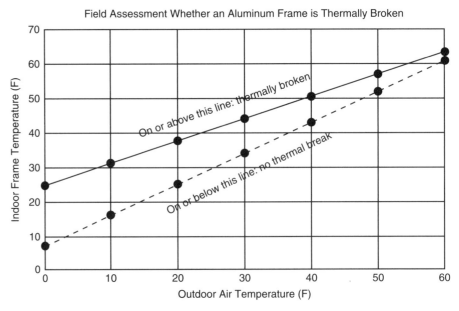

Field Assessment Whether an Aluminum Frame is Thermally Broken

Because operable windows have components that interact extensively with building occupants, there are many behavioral strategies to save energy. For storm windows, building owners and property managers should encourage occupants to close storm windows during winter (Figure 4.44).

Windows themselves should be closed when outdoor conditions are inclement and opened when outdoor conditions allow free cooling. Strategies that have been used to encourage appropriate opening and closing of windows include a visible light inside buildings to indicate when windows should be opened, and a proximity sensor that cuts heating and cooling in a room if a window is open at a time when it should not be.

When examining opportunities for saving energy with windows, we also seek to think outside the box. Many buildings are covered in windows, with far more windows than necessary. Are there any opportunities for replacing windows with insulated wall panels, insulated walls, or other insulated assemblies? To assess these opportunities, we return to the purposes of windows. The purposes of windows are daylighting (natural light), vision glazing (views), and ventilation.

The types of windows required for each purpose are somewhat different. Daylighting is best provided high on the wall, or by small, evenly distributed skylights. The most efficient windows for daylighting are fairly small, optimally comprising less than 10 percent of the wall, to avoid thermal losses being larger than the daylighting gains. This fraction has been dropping as artificial lighting efficiency increases and daylighting savings drop; in many cases, we want less than 5 percent of the wall to be windows for daylighting. Windows for ventilation are used where mechanical ventilation is not available, or where natural ventilation is sought. The ratio of operable window area to floor area, as a means of ventilation, has traditionally been 4 percent, by code, which means less than 4 percent window-to-wall ratio. With small window areas required for daylighting and ventilation, that leaves only the need for vision glazing (views). Vision glazing is extremely dependent on the likes and needs of individual building occupants. A widely quoted standard for vision glazing defines a view as being able to see out within 23 feet of an exterior wall by providing a minimum 20 percent window-to-wall ratio.[33] So, with perhaps 5 percent or less of walls (or roofs) needed for daylighting, typically less than 4 percent for natural ventilation if needed, and 20 percent for vision glazing, we can comfortably look at opportunities for reducing glazing in any building that has more than 30 percent of its surface covered with windows, without sacrificing daylighting, views, or natural ventilation.

This has been affirmed by the most recent version of the International Energy Conservation Code, which limits windows to 30 percent of wall area, and which limits skylights to 3 percent of roof area. Any building with more than a 30 percent window-to-wall ratio is a candidate for replacing windows with a more insulating surface, while remaining within currently recommended standards for daylighting, views, and natural ventilation. In many cases, some windows in buildings with less than a 30 percent window-to-wall ratio might be candidates for replacement with insulated wall panels or other efficient surfaces, especially in unoccupied spaces such as stairwells, corridors, mechanical rooms, utility rooms, basements, and the like.

A similar out-of-the-box strategy for reducing window losses is to replace operable windows with fixed windows, where operable windows are not needed.

Glass block is a window-related architectural specialty, which allows some visible light, but is not transparent (Figure 4.45). It is sometimes referred to as *glass brick*. One large manufacturer reports that its solid glass block (3″ thick) has a U-factor of 0.87, thin hollow glass block (3 $^1/_8$″ thick) has a U-factor of 0.57, standard hollow glass block (3 $^7/_8$″ thick) has a U-factor of 0.51, and energy-efficient hollow glass block has U-factors of 0.34 (unframed) and 0.37 (framed).

Another manufacturer reports U-factors of 0.49 for standard glass block, and 0.26 for energy-efficient glass block. In addition to replacement with high-efficiency glass block, broken blocks should be evaluated for repair. Or, again thinking outside the box, perhaps some portion of an existing glass block can be replaced with insulated panels or wall.

Figure 4.40   Window rating label.

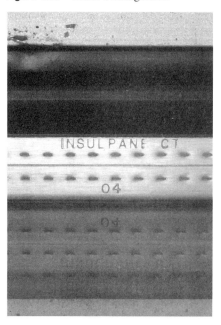

Figure 4.41   Window spacer label.

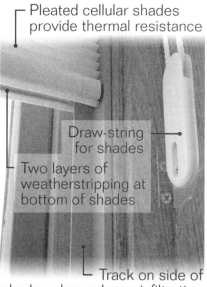

Figure 4.42   Insulated window shades.

Figure 4.43   Retrofit window film.

Figure 4.44   Open storm window in winter.

OTHER ISSUES

Condensation is often found on windows, as shown in Figure 4.46. A frequent conclusion is that if condensation is observed on an insulated window (double-pane or triple-pane), it means that the seals have failed. This is not always the case. For example, condensation on the indoor surface of the glazing system can be the result of frame or edge-of-glass conduction of heat away from the indoor glazing, causing condensation of moist indoor air.

Windows form an important component in the fabric of historic buildings. Many windows have historic characteristics themselves, such as wood or steel frames, or stained-glass glazings. Window replacement is discouraged in historic buildings and historic districts. Many energy improvements are nonetheless possible for existing windows. These include air sealing, storm windows, insulating and sealing window weight cavities (and replacing the counterweights with spring weights), and more.

In evaluating window options, it is important to note that windows installed or painted before 1978 may have lead paint. Lead is a toxic substance that can cause brain damage, among other ill effects. Lead paint was banned in 1978. The chances of a building containing lead paint are 24 percent for buildings built between 1960 and 1978, 69 percent for buildings built between 1940 and 1960, and 87 percent for buildings built before 1940.[34] Consult authoritative resources or qualified professionals before working with windows that may have lead paint.[35]

Windows also play a significant role in allowing infiltration losses, and so improving windows allows reducing infiltration. This subject is covered in more detail in the subsequent discussion on infiltration.

Doors

Exterior doors can be made of metal, wood, fiberglass, glass, other materials, or a combination of materials. New exterior doors must meet energy code requirements. Old exterior doors might not be insulated. It is also not unusual to find uninsulated interior doors used as exterior doors, especially between conditioned and unconditioned spaces, even though they should be insulated exterior doors because they are in alignment with the thermal boundary. Like windows, new doors are rated for energy properties by the NFRC. A steel door can be assessed whether it is insulated by measuring the temperature on the interior surface of the door. See Figure 4.47.

Improvement options include door replacement or the addition of a storm door. Door air-sealing improvements, such as weather-stripping and door sweeps, will be covered in the subsequent discussion on air sealing. Justification for replacement is increased if a door is damaged or bent (Figure 4.48).

Cellar doors, also called *bulkhead doors* or *basement entry doors*, provide direct access from the outdoors to basements (Figure 4.49). Often referred to as *Bilco doors*, a trade name, cellar doors are frequently made of metal, and are usually neither insulated nor weather-stripped.

Cellar doors can be insulated on a retrofit basis, or replaced with doors that are less conductive than metal. Another option is to install an insulated door at the bottom of the stairs.

Overhead doors serve enclosed parking garages, workshops, vehicle service areas, and industrial buildings. In commercial buildings, overhead doors are typically made of steel, and can be uninsulated or insulated. A *pan-style* or *pan* door has a single side of metal. It is said that they are called pan doors because their edges are like a baking pan. A *sandwich* or *thermal* door has two sheets of metal, exterior and interior, with insulation in between. Common insulation is either polystyrene or polyurethane. Polystyrene is rigid and has an R-value of approximately R-4.5 per inch. Polyurethane is injected and has an R-value of approximately R-8 per inch. To visually distinguish between polystyrene and polyurethane insulation, drill a hole at least 1.5″ into the end of the door—yellow insulation is polyurethane. Examples of rated R-values are R-6.6 for a 1-³⁄₈″ polystyrene-insulated door, and, at the high end, R-26 for a 3″ polyurethane-insulated door. Overhead doors fall under the "opaque door" requirements of energy codes and standards. Rated U-factors are similar in definition to those for windows, and account for the entire assembly, including thermal bridging. For example, the door with an R-value of 26 only has a rated U-factor of 0.14. Energy code requirements were introduced in the early 2000s, for example, ASHRAE 90.1–2004 required a maximum U-factor of 1.45 for warmer climates, and 0.50 for colder climates, and IECC adopted these requirements in 2006. IECC 2015 requires a minimum R-value of 4.75 for all climates, with thermal spacers, or maximum U-factors of 0.61 for climate zones 1 through 4, and 0.37 for climates zones 5 through 8. An uninsulated overhead door can be assumed to have a U-factor of 1.2. Retrofit insulation kits are available, but may require adjustments to spring tension or to door opener hardware, to adjust for the changed door weight. Retrofit kits will not eliminate existing thermal bridging. Overhead doors are also rated for infiltration, in CFM/SF.

Infiltration

Figure 4.45 Glass block.

PRINCIPLES

Infiltration is used generally to refer to unwanted outdoor air entering a building or indoor air leaving a building. *Exfiltration* might be more strictly used for indoor air leaving a building, but, practically, infiltration is used for both air entering or leaving a building. Infiltration results in increased heating and cooling energy use, as well as contributing to comfort problems, and a variety of other building performance problems, such as condensation within wall cavities.

Infiltration is driven by air pressure differences between the indoors and outdoors. These air pressure differences are in turn due to a variety of drivers: wind, fans, doors opening and closing, and the stack effect, which is airflow due to the buoyancy of air, resulting from air temperature differences between indoors and outdoors. Whereas infiltration needs air pressure differences to drive the airflow, it

Figure 4.46 Window condensation.

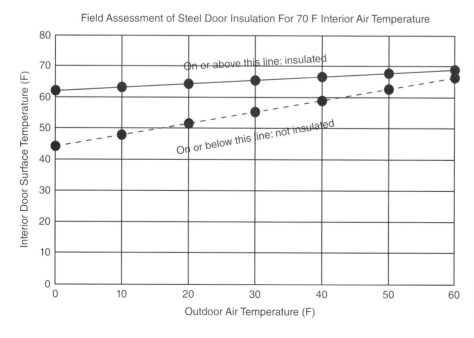

Figure 4.47 The indoor door surface temperature can be used to estimate if a steel door is insulated.

Figure 4.48 Damaged door.

Figure 4.49 Cellar door.

also needs holes in the building envelope through which to flow. We call these holes *infiltration sites*.

Our understanding of infiltration has substantially increased in the past three decades due to the development and multipurpose use of the *blower door test*. A *blower door* is a fan that is placed in an exterior door frame, with a shroud that seals the door opening around the fan. The blower door test involves pressurizing or depressurizing a building, typically to an air pressure of 50 Pascals (positive or negative), relative to the outdoors. To achieve the target pressure, the fan speed is varied automatically. When the target pressure is reached, the airflow through the fan is measured. This airflow must be replaced by infiltration, and so the airflow (CFM) at 50 Pascals (Pa) has become a standard measurement of infiltration, denoted CFM50. Developed largely for single-family homes, the blower door test has come to be used in commercial buildings, even in large commercial buildings. For larger buildings, there is a trend to using a pressure of 75 Pa, for which the infiltration measurement is in units of CFM at 75 Pa, denoted CFM75. The success of the blower door test derives not only from its ability to measure infiltration but furthermore as a tool for diagnostics, as infiltration sites are easier to find and eliminate through the high, detectable airflow induced by the blower door. Savings potential from reducing infiltration is higher in heating-dominated climates.[36]

There is sometimes the misperception that commercial buildings in general are not leaky. However, a survey of 192 large buildings found that "virtually all large buildings, including those built within the last few years, are quite leaky and would not meet current recommendations. ... "[37] Another study found that U.S. commercial buildings are over four times more leaky than commercial buildings in Canada, and over six times more leaky than state-of-the-art buildings.[38]

There is consequently also the misperception that air-leakage reduction in commercial/large buildings is not possible. However, one study showed achieved infiltration reductions of 15 percent average for multifamily buildings, 24 percent for office buildings, and 11 percent for schools.[36] Another study showed that air leakage in high-rise residential buildings can be reduced by 30–40 percent, with heating usage reduced by 7–12 percent.[39]

Infiltration can be used for ventilation, but should not be justified as a reason for not reducing infiltration. Infiltration occurs 24 hours a day, and ventilation needs are rarely 24/7, especially in commercial buildings. For example, a typical office is occupied less than 40 percent of the time. Infiltration is typically more than is needed for peak ventilation. A widely used motto for infiltration control is "seal tight, ventilate right."

STACK EFFECT

The *stack effect* is an important physical phenomenon in which the buoyancy (lower density) of warm indoor air creates an upward movement of air in a building in winter, drawing cold air into the lower levels of the building, and forcing warm air out of the upper levels. It is similar to the effect of the flow of warm combustion gases up a chimney in winter. The stack effect operates in reverse in the summer, in an air-conditioned building, in which airflow moves downward when indoor air temperatures are lower than outdoors.

The resulting air pressure profile in a building in winter is negative on lower floors (relative to outdoors), and positive on upper floors, and vice versa in summer. The level at which the indoor-outdoor air pressure difference changes from negative to positive is called the *neutral pressure plane*. We will primarily direct our attention to the wintertime heating season stack effect (Figure 4.50).

In a hypothetical building with holes evenly distributed around the walls and roof, and with no floors or interior partitions, the pressure rises linearly from the bottom to the top of the building, passing the neutral pressure plane, which is located halfway up the building.

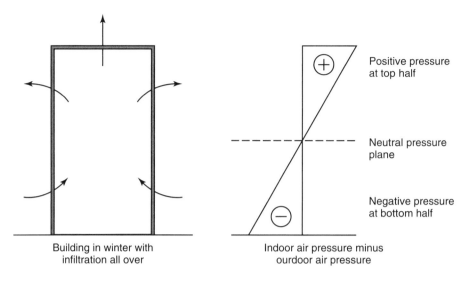

Building in winter with
infiltration all over

Indoor air pressure minus
ourdoor air pressure

Positive pressure
at top half

Neutral pressure
plane

Negative pressure
at bottom half

Figure 4.50 Stack effect.

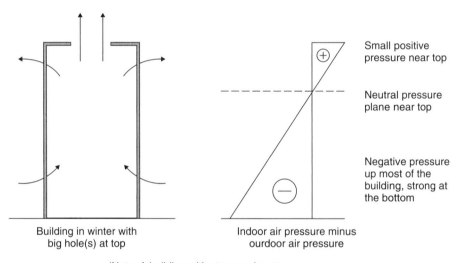

Building in winter with
big hole(s) at top

Indoor air pressure minus
ourdoor air pressure

Small positive
pressure near top

Neutral pressure
plane near top

Negative pressure
up most of the
building, strong at
the bottom

(Note: A building with strong exhaust
ventilation exhibits similar behavior)

Figure 4.51 Stack effect—holes at top.

If a building has large openings at or near the top of the building but does not have large openings near the bottom, the neutral pressure plane rises. This might be characteristic of a building with large vents at the top of the elevators or stairwells, or a building with a vented attic and significant leakage at the attic floor, very common in pitched-roof buildings. The pressure near the top of the building, where there are big holes, becomes smaller, relative to the outdoors (Figure 4.51). However, this low pressure should not fool us. The larger openings will cause significant airflow out of the building, even if the pressure is lower.

Interestingly, this pressure profile is also characteristic of a building with strong exhaust ventilation throughout the building. Exhaust causes the neutral pressure plane to rise. As a result, it makes the negative pressure at the bottom of the building even more negative and increases the airflow into the building to make up for the increased airflow lost from the building. To repeat, increasing the size of a hole at the top of a building, or increasing exhaust airflow, will increase the negative pressure at the bottom of the building, and increase stack effect airflow into, up and through, and out of the building.

If a building has large openings at or near the bottom of the building but does not have large openings near the top, the neutral pressure plane drops below halfway

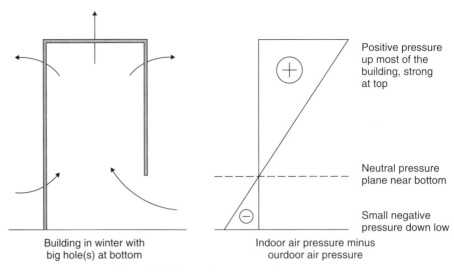

Figure 4.52 Stack effect—holes at bottom.

up the building (Figure 4.52). This might be characteristic of buildings with loading docks, or a basement parking garage, with overhead doors that remain open much of the time. The pressure near the top of the building may still be positive but is less positive, relative to the outdoors.

This pressure profile is also characteristic of a building with strong *makeup air* (or supply air, also called outdoor air) ventilation into building. Makeup causes the neutral pressure plane to become lower, in other words, closer to the ground.[40] Larger openings near the bottom of a building increases the pressure at the top of the building, forcing more air out and, again, stack effect airflow is greater throughout the building: into it, up through it, and out at the top.

In real buildings, each floor serves to disrupt the linearly continuous change in pressure. There are step changes in the pressure at each floor (Figure 4.53).

If a building is poorly compartmentalized, with large openings from floor to floor, these steps are less pronounced. If a building is well-compartmentalized (more airtight from floor to floor), these step changes in pressure from floor to floor become bigger. Note that full-height spaces (stretching from the bottom to the top of a building) will still have a continuous/linear pressure profile, adjacent to the stepped profile in the main occupied spaces, and the two different profiles interact and influence each

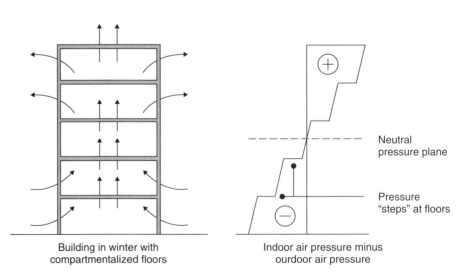

Figure 4.53 Stack effect—how floors affect pressure.

other, based on openings between them. These full-height spaces include stairwells, elevator shafts, and utility chases. Large holes between the full-height spaces and the compartmentalized spaces (such as stairwell doors with large undercuts) will reduce the stepped characteristic of the floor-to-floor pressure profiles.

The interior air pressure characteristics of buildings in response to stack effect can be used in reverse to gain an understanding of a specific building. For example, by temporarily turning off ventilation, we can eliminate ventilation pressure effects. If we then find the neutral pressure plane, if it is above halfway up a building, then holes toward the top of the building are bigger than holes at the bottom of the building. Conversely, if the neutral pressure plane is below the half-height of a building, then holes at the bottom of the building are dominant. If we now turn the ventilation on, if the neutral pressure plane rises, then exhaust ventilation is dominant, but if the neutral pressure plane drops, then makeup ventilation is dominant.

Building pressures can be measured simply by cracking open windows and placing a pressure tap for a manometer out the window, and keeping the reference pressure inside the building. It is important to shield the outdoor probe from wind pressures. The neutral pressure plane is the height at which this pressure difference is zero. A faster approach to finding the neutral pressure plane is to measure the pressure between corridors and stairwells on each floor, or corridors and the elevator shaft, or to use smoke detection to find the neutral pressure plane.[41] However, this approach presumes that the stairwell or elevator shaft runs the full height of the building, and also that it accurately reflects the stack effect pressure profile, where, in fact, they may actually each have slightly different pressure profiles, depending on their own connections to the outdoors, such as stairwell venting.

An understanding of the physical mechanisms of stack effect can lead to a variety of measures to counter it. Air sealing of the exterior envelope will reduce stack effect. Compartmentalizing a building, by weather-stripping stairwell doors and sealing penetrations into chases, will reduce stack effect. Air sealing ductwork in vertical chases will reduce stack effect. We can even go further. For example, removing heat from stairwells, and insulating stairwells, will reduce stack effect. Reduced indoor temperatures in general will reduce stack effect. See Figure 4.54. Insulating hot pipes and ducts in vertical chases and stairwells will also reduce stack effect.

Floor-to-floor pressures can be measured using a daisy-chain approach. For example, we might measure the pressure in a hotel room relative to the corridor, then measure the pressure in the corridor relative to the stairwell, then measure the pressure in the stairwell relative to the corridor above, and finally between the corridor above and a hotel room above. By adding all the pressure differences, we can determine the pressure difference between the two stacked hotel rooms on different floors.

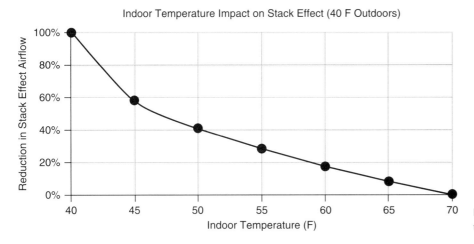

Figure 4.54 Reducing indoor air temperature reduces stack effect.

The stack effect pressure can be estimated by:

$$p = 1905\,h\,[1/(To + 460) - 1/(Ti + 460)]$$

Where:

p = stack effect pressure in Pa
h = height of the neutral pressure plane, feet. A first estimate might be half the height of a building.
To = outdoor air temperature, F
Ti = indoor air temperature, F

Online calculators are available to perform this calculation.[42]

Stack effect pathways are many. Air is drawn into a building through the large openings at the ground floor or basement: Main entrances, other entrances, overhead doors, loading docks. There is a double problem here: The negative stack effect pressure is at its highest, and the openings to the building are at their largest. Air can also enter below the neutral pressure plane, through cracks in windows, through open windows, through balcony doors, through cracks in walls, through pipe and wiring penetrations, and through any other envelope penetrations. Air that enters the building then moves horizontally through the building, down corridors and through rooms, as it seeks vertical pathways through which to rise. Vertical pathways include stairwells, elevator shafts, and utility shafts such as ventilation duct chases and pipe chases. Active chimneys are an important stack effect pathway, because they are warm in winter, exacerbating the stack effect, both during the firing of the combustion equipment served by the chimneys, and between firing cycles. Vertical pathways can also include floor-to-floor penetrations, such as pipe penetrations. Finally, vertical pathways can include wall cavities and the connections between walls and floors, even in buildings with seemingly monolithic floors and walls. As the air rises, its pressure rises, and above the neutral pressure plane, at a positive pressure relative to the outdoors, it then seeks horizontal pathways to start moving outward, to move outdoors, where the pressure is now lower than indoors. It moves horizontally through many of the same kinds of openings: Elevator doors, stairwell doors, pipe penetrations below sinks, electrical receptacles. Finally, the pressurized air seeks ways to leave the building envelope itself, through window cracks, wall cracks, and more. We find the same problem as at the bottom of the building. The pressure relative to outdoors is highest, and many of the building openings are the largest: vents at the top of elevator shafts and stairwells, poorly sealed roof and attic access doors and hatches, poorly sealed attic floors, attic vents, open penthouse windows, and more.

Stack effect is best addressed comprehensively (Figure 4.55). We begin with standard air-sealing approaches: weather-stripping, caulk, and foam at envelope penetrations and at interior pathways. Attention is directed not only to vertical pathways but also, as importantly, to horizontal pathways. Interior doors can help to slow and stop the interior horizontal airflow that is an important part of stack effect airflow, especially basement and attic doors, but also stairwell and corridor doors. Weather-stripping these interior doors helps, but, in addition, self-closing doors are better, for example using spring hinges. Converting combustion systems from natural draft to sealed combustion eliminates the full-height stack of the chimney, as long as the chimney is sealed. Attention must be directed to sealing both the air intake and the chimney, preferably both at the top and bottom. Abandoned chimneys and shafts should also be sealed, both at the top and at the bottom. The reason for sealing both the top and bottom is that a chimney or shaft that is open, for example, at the top only, retains stack effect along its height. It therefore has a strong negative pressure at the bottom (because at the top its pressure is equal to the outdoors), which can contribute to stack effect and infiltration in the rest of the building through air leakage into it.

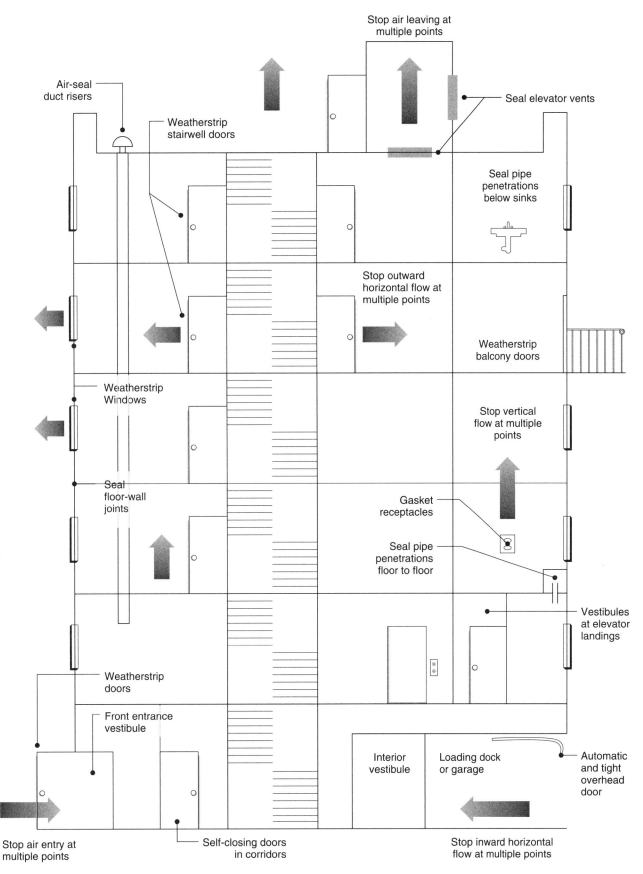

Figure 4.55 Comprehensive strategies for countering stack effect.

Conversely, if it is capped at the top but left open at the bottom, it has an increasingly negative pressure rising up its height.

If a combustion system is not converted to sealed combustion, stack effect losses can still be reduced by installing an automatic vent damper, which prevents stack effect flow between firing cycles of the combustion equipment. Vent damper savings of 12 percent have been reported.[43]

A major question in countering stack effect is the relative merit of interior and exterior air sealing. Some research has concluded that interior air sealing is an important prerequisite to envelope-sealing measures.[44]

FORCED AIR SYSTEMS

Heating and cooling systems that use forced air (ductwork) can impact infiltration by creating positive and negative pressures around a building. For example, a rooftop unit in a small commercial building typically draws in only outdoor air, for ventilation, but does not exhaust air or relieve building pressure. Exhaust is provided separately, usually for bathrooms and kitchens. The rooftop unit fan frequently cycles on and off, in response to calls for heating or cooling. If the exhaust ventilation requirements are small, the building will be at a positive pressure when the rooftop unit fan is on, with air being pushed into the building by the rooftop unit, and out of the building by air leakage. But the building will be at a negative pressure when the rooftop unit fan is off, with air infiltrating the building. A building with strong central exhaust, for example, if there is a commercial kitchen, will be under negative pressure when this fan is on, with continuous significant infiltration, regardless of ventilation air intake into the building.

Duct sizing and location also contributes to pressurization or depressurization. An air handler with undersized return ductwork is more likely to pressurize a space. An air handler with oversized return ductwork is more likely to depressurize a space. Ducts located in exterior walls or roof cavities pressurize or depressurize these spaces, through duct leakage, and therefore contribute to infiltration.

In evaluating the contributions of forced air systems to infiltration, we primarily seek catastrophic situations in which unusual pressures combine with unusual infiltration sites to cause unusual infiltration. Energy improvements include duct sealing, duct reconfiguration, ventilation balancing, and separating ventilation from heating and cooling into dedicated balanced ventilation systems. These will be covered in more detail in a subsequent discussion of heating and cooling distribution systems.

INFILTRATION SITES

Windows and Doors

Windows and doors each present many different infiltration sites, as they themselves are purposeful penetrations of the building envelope. Air leaks in two primary ways through windows and doors:

- Between the moving component and the frame.
- Between the frame and the wall around the window or door.

Each of these two modes requires a different approach to prevent infiltration. Components with relative motion require weather-stripping, which can allow the relative motion and still prevent infiltration. Air leakage sites without any relative motion require a different kind of seal, typically either caulk or foam. Sealing a window or door typically requires treatment of both moving and fixed air leakage sites. To treat one and not the other is a common mistake.

Exterior doors and windows are best examined from indoors, looking for daylight that indicates a potential infiltration site. Door cracks can be seen by viewing the doorframe laterally, from close to the door (Figure 4.56).

Air leakage through windows and doors in poor condition can be one of the largest infiltration sites in a building (Figure 4.57).

Exterior doors present a highly visible infiltration site. Cracks around and below doors show daylight as we leave a building. These sites need and deserve treatment to reduce infiltration. However, a common mistake is to treat these highly visible sites and not seek or treat other infiltration sites, and attribute more infiltration reduction to air sealing these doors than is realistic. The aggregate total infiltration presented by doors is typically not substantial, unless a building has many doors, for example, townhouses or high-rise buildings with balconies. It is important that we not limit infiltration investigation and improvements to doors.

Doors can allow infiltration when they are purposefully open, for example, open doors in buildings with much foot traffic or specialty doors that stay open much of the time, such as overhead doors. For these situations, some form of air lock is needed, such as a vestibule or revolving door. Air curtains or strip curtains have also been used to limit infiltration in doors that need to remain open.

Likewise, windows can allow infiltration when left open, a routine occurrence in overheated buildings or in buildings with heating or cooling balance problems. These problems will be addressed in a discussion on heating and cooling controls.

Pipe and Wire Penetrations

Pipe and wire penetrations are common infiltration sites, either from the outdoors into a building, or at thermal boundaries between conditioned and unconditioned spaces.

Examples include pipe and wire penetrations in exterior walls (Figure 4.58), in the floors of attics, in basement ceilings, and below kitchen and bathroom sinks, where piping penetrates the wall and enters vertical pipe chases (Figure 4.59).

Attics/Roofs

Attic floor sites are complex and have many penetrations, in addition to piping and wiring. Additional attic floor infiltration sites include:

- Attic hatches or walkup doors.
  - Around the frame.
  - Between the door and the frame.
- Where combustion air ducts, or vents, or plumbing vent pipes go up through the attic.
- Uncapped interior walls, especially at the top of plumbing chases.
- Attic ductwork—around duct penetrations to the building below, as well as leaks in the duct itself.
- Around exhaust fans that serve the top floor and are recessed into the top floor ceiling.
- Around light fixtures serving the top floor (recessed into attic).

In some cases, an attic knee wall or other complex wall only has stapled insulation to serve as the thermal and air barriers, without gypsum board or other rigid wall to solidify the barriers. Infiltration occurs freely past the stapled insulation edges.

Many attic sites are complex, with unexpected penetrations. For example, if a wall rises from the building below, up through the thermal boundary at the attic floor, and if there are holes in the wall in the attic, air can rise up the wall cavity and out the holes (Figure 4.60).

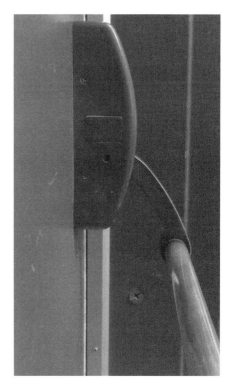

Figure 4.56   Door infiltration.

Figure 4.57   Window in poor condition.

Figure 4.58 Air conditioner piping penetrating an exterior wall.

Figure 4.59 Unsealed pipe penetrations—one escutcheon is not large enough to cover a pipe penetration hole, another escutcheon is loose and not sealed.

Airflow from a heated building up to a vented attic, through attic floor penetrations, not only causes energy losses but also contributes to the formation of ice dams in cold climates, as seen in Figure 4.61.

Typically attributed to inadequate insulation, anecdotal evidence suggests that air leakage is as big, or bigger, a contributor to ice dams. The heat from the building conducts through the roof and melts snow on the roof. The melted snow runs down the pitched roof, and refreezes at the edge of the roof. Ice dams can be used as a form of diagnostic tool—if we observe ice dams, we should look for air leakage. For buildings not inspected in winter, we should poll building management or occupants about whether they have seen ice dams in winter. Reducing air leakage into attics invariably reduces ice dams.

Floors/Basements/Foundations

A basement or crawlspace ceiling typically has similar penetrations as an attic floor: wiring and piping (Figure 4.62), ductwork, access hatch, or stairs.

Basements also have perimeter leakage sites to the outdoors, such as unsealed sill plates, windows and vents, and cellar doors.

Heating and Cooling Equipment

Heating and cooling equipment and distribution systems contain a number of infiltration sites. For example, outdoor air intakes, intended to bring in outside air for ventilation when heating and cooling fans are running, can allow air into a building even when not desired, when the system fan is off (Figure 4.63).

A solution might be a motorized damper in the return duct, interlocked with the main system fan.

Another set of infiltration sites is at or around purposeful openings in a building envelope, such as abandoned chimneys, at connections to rooftop exhaust fans, and at rain hoods for exhaust systems. Active chimneys are an infiltration site whenever the combustion appliances are not firing.

Room air conditioners, either wall-mounted or window-mounted, have a variety of infiltration sites. A recent study found the average room air conditioner contains six square inches of opening.[45] Infiltration happens at the junction between sleeve and wall for through-wall units, where a slide-out chassis meets the sleeve, around wing panels on window-mounted units, and at various components inside the units. Accordingly, air sealing of room air conditioners requires treating each of the infiltration sites, or replacing the units entirely with equipment that does not penetrate the thermal boundary. Room air conditioners pushed into nonmatching sleeves are also not uncommon, increasing the leakage area (Figure 4.64).

Leakage around window-mounted units is also common, as shown in Figure 4.65.

Mechanical dampers can also be a site of significant leakage that merits attention if the dampers are supposed to be closed most of the time. One study found mechanical damper leakage on four pre-1985 buildings to average 64 percent, and somewhat less but still significant at 25 percent leakage for four post-1985 buildings. Another study found similar damper leakage of 27 percent in 14 newer post-2000 buildings.[46] Backdraft dampers in exhaust fans are typically required by code. However, these dampers are not always installed because the fans themselves typically do not include such dampers, so the dampers must be installed separately, and awareness of the code requirement is not high, by designers, installers, or code inspectors. Where backdraft dampers are installed or are shipped with fans, they typically do not seal well (Figure 4.66).

Backdraft dampers also work only to prevent air from flowing back into an exhaust system when the fan is not running. However, they do not prevent air from flowing in the direction of flow, when the fan is not running, and this is usually the direction in which the stack effect promotes airflow. A best practice for ensuring that these locations are not infiltration sites is to install a gasketed damper, powered not by gravity but by an actuator motor, the control of which is interlocked with the fan. When the fan is not running, the damper automatically closes, and the gasket minimizes air leakage, eliminating flow not only in the backdraft direction but in the fan flow direction as well.

Unconditioned Spaces
In addition to attics and basements, other unconditioned spaces, such as attached garages and mechanical rooms, have infiltration sites at both the outer envelope and inner envelope. Mechanical rooms have purposeful openings, such as combustion air openings, that are not always closed when they should be. Inspections should be directed to examining whether combustion air openings have dampers, and, if so, whether the dampers close when combustion equipment is not firing.

Walls
Walls present a challenge because they cannot be inspected as easily as attics, basements, and other unconditioned spaces. Wall infiltration sites include gaps in siding/cladding, unsealed joints between sheathing components, and penetrations in the interior finish, such as electrical receptacles and light switches. Evidence of infiltration can be seen as dust on electrical components or dirty insulation where visible. Removing window and door trim will often reveal unsealed areas in the surrounding walls.

Other
Many existing buildings have openings (vents) at the top of elevator shafts (Figure 4.67) and stairwells.

It was traditionally thought that these openings would relieve smoke, in the case of fire, and this was a code requirement. However, it is now viewed that these openings draw smoke into elevator shafts and stairwells, contribute to stack effect flow of smoke and hot gases from fires, and impede evacuation and firefighting efforts.[47] As of 2015, the venting requirements have been removed from the International Building Code. The openings are also not required for purposes of relieving piston pressure in elevator shafts.[48] In short, vents at the top of elevator shafts and stairwells should be sealed. Vents in elevator shafts can be seen either from the gap between the elevator and the shaft or from the machine room above the elevator shaft.[49]

In seeking infiltration sites, we remain vigilant for what might be called catastrophic infiltration sites, always looking for the unusual (Figure 4.68).

Complex building details are a common site for infiltration. For example, bathtubs on exterior walls are frequently found to have a variety of air movement paths. Air can move up piping chases to the cavity below and around the bathtub (Figure 4.69), and then to the outside wall cavity, and, from there, outdoors.

INFILTRATION MEASUREMENT AND ESTIMATING SAVINGS

In evaluating potential savings from reducing infiltration, two broad approaches are possible:

- To seek and identify infiltration sites and measure hole sizes.
- To pressurize or depressurize a building, or a space in a building, using a blower door, and measure the infiltration rate under a pressure differential between indoors and outdoors, and then make a pressure correction to estimate time-average infiltration.

In practice, a combination of the above approaches works best.

Figure 4.60 Leak path up through wall and out holes in the wall—note how the wall penetrates the thermal boundary, as insulation is at the attic floor.

Figure 4.61 Ice dams.

Figure 4.62   Unsealed pipe penetration.

Advanced approaches, such as tracer gas testing, are also possible, but are generally limited to research-level testing. Tracer gas tests also do not provide much of the useful additional information that blower door tests provide, such as locations of infiltration sites and other diagnostics.

Blower Door Testing

The blower door test, developed and refined since the 1970s, has transformed building science and our overall understanding of infiltration and air movement in buildings. Blower door tests not only quantify air leakage but are also used for diagnostics. The test can be used to identify infiltration sites and to understand how air moves into buildings, through building spaces, and out of buildings, by identifying zone pressures within a building. Furthermore, the test can then be used to guide air sealing, by providing real-time feedback about how well individual air-leakage sites have been sealed, and by assessing when sufficient air sealing is complete. Blower door equipment can further be used for other purposes, such as to pressurize a conditioned space, in conjunction with separate duct pressurization, to measure air leakage from the ductwork to unconditioned spaces or to the outdoors.

A blower door test is done by mounting a blower door (Figure 4.70), a fan within a shroud and frame, in the frame of an open exterior door. The fan motor speed is automatically varied in order to reach a target air pressure in the building, typically depressurizing the building to a target negative pressure, usually 50 or 75 Pa, relative to the outdoors. Interior doors are kept open. Ventilation fans are turned off, and combustion equipment is also turned off. Typically, purposeful vents, such as outdoor air intakes, are taped shut. When the target air pressure is reached, the airflow through the fan is measured, and this airflow represents the building infiltration at the target test pressure. Larger buildings require larger fans or multiple fans. In a small building, the test can be set up and run by a single person in one to two hours. A blower door test was reportedly performed in a 1.1 million square foot building, with 30 blower doors, set up in three hours, and the test complete in nine hours.[50]

Complexities arise when deciding whether to include unconditioned spaces, such as basements. Frequently, the test is repeated, once including the basement, and once with doors to the basement shut. Other challenges arise if outdoor conditions are windy, varying the outdoor air pressure. Providing adequate airflow/pressurization for larger buildings can also be a challenge.

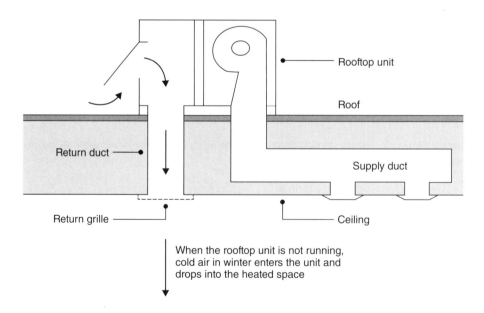

Figure 4.63   Infiltration through heating and cooling equipment.

While the main goal of a blower door test is to measure whole-building infiltration rates, the test can also be used to assess the benefits of air sealing individual components. For example, if a building's infiltration rate at 50 Pa before and after air sealing a sample of windows is measured, the per-window infiltration reduction potential can be estimated: If the building measured 5,000 CFM50 (CFM at 50 Pa) before weather-stripping ten windows, and 4,500 CFM 50 after weather-stripping the windows, we can associate 500 CFM50 with the air sealing, in other words each window has the potential for 50 CFM50 reduction through weather-stripping. This might be applied to a decision whether to weather-strip the remaining windows in a building.

The blower door provides a test result in units of CFM at the test pressure, typically 50 Pa. This result can be converted into a variety of other useful units. By dividing the CFM50 result by the building volume (in cubic feet) and multiplying by 60 (minutes per hour), we obtain the air changes per hour at 50 Pa (ACH50):

$$ACH50 = CFM50 \times 60/VOL$$

ACH50 = air changes per hour at 50 Pa
CFM50 = blower door test result, CFM at 50 Pa
   VOL = building volume, cubic feet per air change

The ACH50 is useful to generically assess, or benchmark, a building's infiltration rate relative to that of other buildings. An ACH50 of 20 is characteristic of old, leaky buildings. An ACH50 of 10 is characteristic of a typical building. An ACH50 of 3 is characteristic of a tight building. And an ACH50 of 0.6 meets the high-performance standards of a Passivhaus building, at the lowest end of infiltration rates.

The ACH50 can be used to estimate the natural, time-average infiltration rate, for purposes of annual energy estimates. A rough conversion used in the early years of blower door testing was: ACHn = ACH50/20, where ACHn is the natural, time-average infiltration rate. More refined calculations are provided by blower door software.

The CFM50 can also be used to estimate the approximate size of holes in the building envelope. The hole size is most closely approximated by the *equivalent leakage area (EqLA)*, adjusting the blower door test results to a reference pressure of 10 Pa. The equivalent leakage area is helpful to allow someone to visualize the approximate size of holes in the building envelope. Very roughly, the equivalent leakage area, in square inches, is equal to the CFM50 divided by 10. The leakage area at a reference pressure of 4 Pa is also sometimes used, referred to as the *effective leakage area* (ELA).

An additional metric divides the measured infiltration by the floor area of the building, for example $CFM50/ft^2$, to give an indication of the infiltration independent of building size.

Yet another metric divides the measured infiltration by the surface area of the building. This is used, for example, by the U.S. Army Corps of Engineers, at a reference test pressure of 75 Pa, to assess if buildings meet their airtightness requirement of $0.25\ CFM75/ft^2$ of building area.

Components such as windows in commercial buildings are routinely field-tested for airtightness during new construction, in order to meet construction specifications. These testing requirements and associated equipment are well suited for energy assessment work. Testing is typically at 75 Pa. Results are usually expressed as CFM75/SF, but can be corrected to CFM50/SF, or CFM50/window. Testing is typically done from indoors but can be done from outdoors if accessibility from indoors is challenging, such as in the case of skylights. Smoke can be used to supplement the tests for diagnostic purposes.

Figure 4.64 Room air conditioner in ill-fitting sleeve

Figure 4.65 Room air conditioner with leakage around window mounting.

Figure 4.66 Leaky backdraft damper in an exhaust fan.

Figure 4.67 Elevator shaft vent.

Figure 4.68 Window air conditioner prevents storm window from closing.

Blower door tests are increasingly being performed on large buildings. Buildings with repeating areas that are limited in size, such as apartment buildings and hotels, can have blower door tests done on individual rooms or apartments. In order to eliminate air leakage to adjacent indoor spaces, *guarded blower door tests* (Figure 4.71) maintain adjacent spaces at the same pressure, so the air leakage measured for the space under test can only be outdoor infiltration.

Guarded blower door tests can also be done by floor (Figure 4.72).

For blower door tests on larger buildings, larger blower door equipment is available, including fans that operate from a flatbed truck. For example, one mobile gas engine blower door is reported to provide 75,000 CFM, big enough to handle a 45,000-square-foot building with 10-foot ceilings, with a leakage of over 7 ACH at 50 Pa.

Visualizing the size of infiltration sites is helpful. Referring back to various infiltration rates, we can translate these into hole sizes for an example 2500 square foot building: A leaky building (20 ACH50) may be visualized as 700 square inches of holes, an average building (10 ACH50) is 350 square inches, a tight building (3 ACH50) is 105 square inches, and a very tight (Passivhaus-compliant) building (0.6 ACH50) is 21 square inches.

Leakage Size Measurement

Measuring the physical size of a leakage site is a valid and useful measurement, for holes such as pipe penetrations, elevator shaft vents, and other visible infiltration sites. Photographs are useful for these measurements. For infiltration sites with long, thin gaps, such as windows, it is possible to measure the width of these small gaps using feeler gauges.

Other Field Techniques

Another tool useful for finding and characterizing infiltration sites is a smoke generator (Figure 4.73). A wet hand can also sometimes detect the low air velocities of infiltration, especially when infiltration is boosted with a blower door or exhaust fan.

Infrared thermography can be useful to find infiltration sites, either with or without blower door testing. Air entering a building in winter will create local cold spots. Likewise, warm air leaving a building in winter will create local warm spots. A blower door test will highlight these local cold and warm spots due to its temporary high airflow.

For larger multiunit buildings, such as apartments or hotels, a blower door method for measuring unit infiltration involves using air-mixing analysis. This relies on an outdoor temperature measurably different than the indoor temperature, preferably at least 20 F colder or warmer than indoors. This method differentiates air exchange with the outdoors from air exchange with the rest of the building. The test is conducted as follows:

- Turn off heat in the unit.
- Depressurize the space using a blower door.
- Measure outdoor air temperature, indoor air temperature elsewhere in building, and the mixed air temperature entering the blower door.
- Use the air-mixing method to calculate effective infiltration from the outdoors (Appendix C).

INFILTRATION AND BUILDING TYPES

Complex low-rise buildings, such as those with pitched-roof attics, are particularly vulnerable to infiltration. Virtually all attic floors have air leakage, and most leakage is significant. Complexity increases the risk of what has been called air bypasses, such as uncapped wall cavities (Figure 4.74), and uncontrolled air leakage due to connections between conditioned spaces and unconditioned spaces, and the outdoors.

Interestingly, we think of low-rise wood-frame buildings as being leaky, and larger buildings as monolithic. However, one study indicates that wood-frame walls are among the least leaky.[37] We also know that the stack effect is more pronounced in taller buildings, and taller buildings reach where wind velocities are higher, so the drivers of infiltration are greater in taller buildings.

INFILTRATION IMPROVEMENTS

Methods and Materials

Historically, we thought of infiltration improvements, in other words, air sealing, as being primarily weather-stripping of windows and doors. However, our new understanding of the many infiltration sites, and interior pathways of unwanted airflow, reveals a broad number of approaches to reducing infiltration.

Many infiltration sites and airflow pathways are large, and need to be addressed with large physical barriers. For example, plywood might be used for holes like uncapped chases. Stone or concrete coping might be used to cap abandoned chimneys. In such cases, the edges still need to be sealed with caulk or foam.

Preventing the horizontal component of stack effect airflow, where air is moving horizontally toward vertical chases that will carry it upward, can be done with doors, vestibules, and other physical barriers. Revolving doors and vestibules are particularly important for large first-floor entrances. Air doors, also called air curtains, are intended to limit infiltration at doors that are opened frequently or that need to remain open, such as at loading docks. Strip curtains or strip doors can serve the same purpose. For pipe and wiring penetrations into chases and wall cavities, pipe escutcheon rings and electrical grommets provide a first air-sealing barrier, supplemented with caulk or foam to complete the seal.

It is particularly important to seal horizontal paths to mechanical rooms and basements, where chimneys serve combustion equipment, to prevent airflow from heated space into the combustion zone, where it is pulled up the chimney by stack effect. This can mean weather-stripping doors, or installing doors where there are none, and sealing other openings as well.

Where weather-stripping and caulk might be appropriate improvements for doors and windows that are already in good condition, replacement doors and windows may well be warranted in cases where the entire door or window assembly is failing or is being replaced anyway.

A wide variety of prefabricated products and assemblies are available to reduce infiltration at specific building components. These include attic hatch covers, room air conditioner covers, diffuser covers for forced air systems not in use (e.g., cooling-only systems in wintertime; see Figure 4.75), covers for recessed light fixtures, gasketed and motor-driven dampers, clothes dryer vent covers, and inflatable chimney balloons.

Sealing ductwork requires its own methods and materials. A variety of duct tapes are generally recognized to be ineffective and not durable, although some specialty tapes can last. Physical fastening of duct connections, with straps and/or screws, is recognized to be a prerequisite to durability for the connections. Beyond that, mastics are recommended for effectiveness and durability. Aerosol-based duct sealing has also been found to be effective and durable, and has the advantage of being able to reach inaccessible locations. Aerosols can be used not only for duct sealing but also for chase sealing.

High-temperature caulks and foams are available to seal around flues and chimneys, for example, where they penetrate the thermal boundary in the attic or at the roof line.

Figure 4.69 Infiltration sites below/around bathtub framing.

Figure 4.70 Blower door.

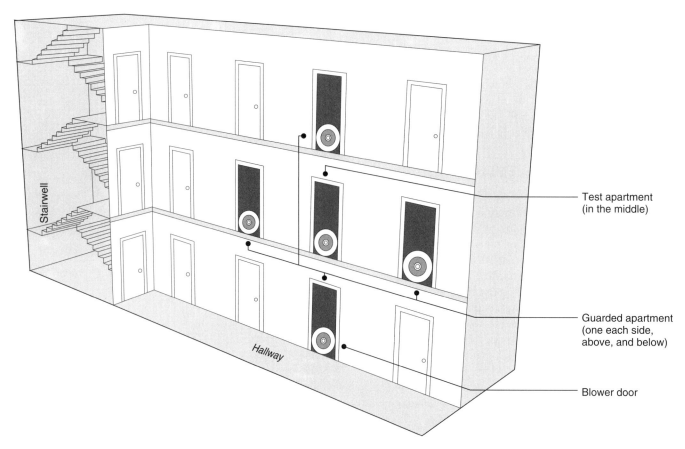

Figure 4.71 Guarded blower door test—by apartment or hotel room.

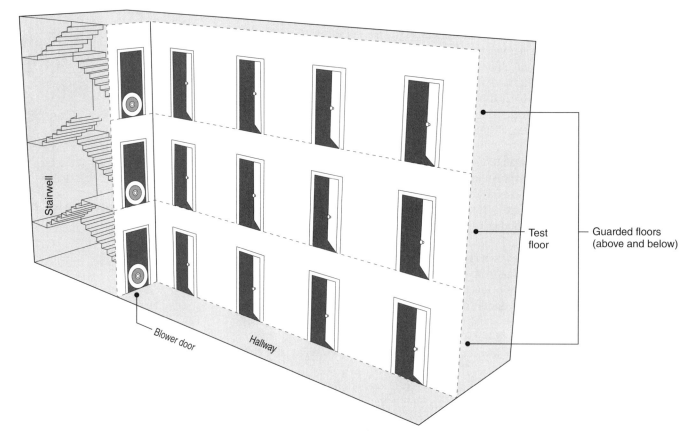

Figure 4.72 Guarded blower door test—by floor.

Increasingly, we look to a variety of insulation to provide both resistance to heat flow and resistance to airflow. We might call this *insulation-based infiltration reduction*. Forms of insulation that resist airflow include rigid insulation (well-taped at the seams), dense-pack cellulose, and spray foam.

And we circle back to the traditional air sealing approaches: caulk and foam for fixed joints, such as window frames, and weather-stripping for joints that see relative motion, such as the joints between window sashes and sills (Figure 4.76). For larger fixed joints and cracks, a backer rod is often used and supplemented with caulk or foam. An emerging approach is to use multiple opportunities to stop infiltration, at any given infiltration site, in order to increase robustness and durability. For example, multiple weather strips can be provided at a single window (Figure 4.77). Weather-stripping can be applied to window and door surfaces, either where two surfaces slide relative to each other or where two surfaces close against each other, or both.

In some cases, infiltration can be reduced without the use of any materials. For example, tightening window and door hardware will frequently allow tighter closure and therefore less infiltration.

In recognizing the many approaches to reducing infiltration and uncontrolled airflow into, continuing on through, and out of building envelopes, we come to see that this work is not simple. We have moved beyond thinking of air sealing as only weather-stripping windows and doors, and beyond seeing infiltration sites as being only those cracks to the outdoors that we can see with our eyes. Infiltration reduction is an advanced discipline that takes professional strategies and skills to quantify, diagnose, specify, and implement. Accordingly, it should be treated as a major focus and major effort of energy conservation work, which frequently requires capital investment. It is not a commodity, or low-cost/no-cost improvement, as it has long been treated. Many of the strategies involve construction trades, such as carpenters for the installation of doors, electricians or heating contractors for the installation of automatic dampers, weatherization specialists for the installation for air-resistant insulation, masons for the installation of coping caps on the tops of abandoned chimneys, and more. When taken seriously, infiltration reduction has great potential to deliver energy savings and comfort to buildings. When treated as a no-cost/low-cost commodity, token infiltration reduction efforts deliver neither measurable savings nor improved occupant comfort, only disappointment.

Modeling Infiltration and Air-Sealing Improvements

Two broad approaches are available for modeling infiltration and air-sealing improvements: whole-building and component-based.

With whole-building models, infiltration is modeled as a single metric. This metric might be volume-based such as ACH50 or ACH75 from a blower door test, or ACHn. For example, a blower door test might have found an ACH50 of 20. Recognizing this as very leaky, and having found a number of infiltration sites, we might assume that this can be reduced to an ACH50 of 10 through a standard package of air-sealing improvements. Alternatives to volume-based metrics are area-based metrics, such as the CFM75 per square foot of surface area used by the Army Corps of Engineers. For example, a blower door test might have found a CFM75 of 25,000; this is divided by the building surface area of 50,000 square feet to obtain an existing leakage rate of 0.5 CFM/SF. We presume that, through air sealing, we can reach the Army Corps goal of 0.25 CFM/SF.

With component-based models, the area of infiltration sites is modeled, before and after air sealing. For example, an elevator vent shaft is measured to be 5 square feet. Using a winter-average outdoor air temperature of 40 F, and a winter-average indoor air temperature of 68 F, and the height of the building, we use a stack effect calculation to calculate a winter-average infiltration airflow. Alternatively, a bin calculation can be done for each five-degree increment of outdoor temperature. As an alternative to area-based component calculations, leakage rates for components

Figure 4.73 Smoke-guided infiltration diagnosis.

Figure 4.74 Uncapped wall cavity.

Figure 4.75 Diffuser cover on a cooling-only forced air system, fastened with a tension fastener in the center of the diffuser, prevents infiltration during heating season.

Figure 4.76 Weather-stripping prevents infiltration at four locations on a window sash—on the front and back edges, and in two locations below the sash.

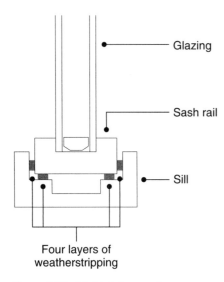

Figure 4.77 Multiple layers of weather-stripping.

are frequently available, such as rated window infiltration. The improvement calculation is still component-based. For example, V-strip weather-stripping has been found to reduce infiltration in double-hung windows by an average of 1.6 CFM50 per linear foot.[51]

Whether whole-building or component-based, infiltration and air-sealing models can further be classified as constant-infiltration or pressure-dependent. Constant-infiltration models are perhaps more typical of component-based calculations, frequently performed with spreadsheets, but are also used by many whole-building models.

Pressure-dependent models account for wind and/or stack effect. Pressure-dependent models can themselves vary in complexity, accounting for building orientation relative to wind (or not), and interior pressure nodes, as we move toward research-level modeling rather than audit-level modeling. Efforts have been made at making these suited to audit-level modeling, for example for spreadsheet analysis.[39]

The component-based approach to modeling and specifying improvements has the benefit of associating infiltration with specific, identified leakage sites. We know ahead of time exactly what we propose to air seal, based on an energy audit. A disadvantage is that the focus of air sealing tends to be visible infiltration sites, ones that are more easily identified during the energy audit.

The whole-building approach is performance-based. Baseline infiltration (infiltration of the existing building, before air sealing) is preferably based on measurement, such as with blower doors. Then, a target is established based on best practice and experience, for the particular type of building. This target encourages air sealing to be done that may require finding less visible infiltration sites, perhaps internal stack-effect pathways, or infiltration sites hidden below attic floor insulation.

References

1. C. J. Schumacher, Thermal Metric Project, slide 17, Westford Symposium, Building Science Corporation August 3, 2011.
2. U.S. Environmental Protection Agency, Agency for Toxic Substances and Disease Registry, Current Best Practices for Vermiculite Attic Insulation, EPA 747-F-03–001, May 2003.
3. http://www.techsupport.weyerhaeuser.com/hc/en-us/articles/201758150-Weyerhaeuser-Forest-Balsam-Wool-Insulation-. Accessed January 22, 2015.
4. Justin Fink, "New Insulation for Old Walls." *Fine Homebuilding, September* 17, 2009, pp. 32–37.
5. 1997 ASHRAE Handbook of Fundamentals, Chapter 24.
6. http://energy.gov/energysaver/articles/insulation-materials. Accessed July 5, 2015.
7. *Thermal Imaging Guidebook for Building and Renewable Energy Applications*, FLIR Systems AB, 2011, p. 16.
8. gSKIN® Application Note: Building Physics, greenTEG AG, Revision 2.06, December 12, 2014.
9. E. R. Vinieratos and J. D. Verschoor, "Influence of Insulation Deficiencies on Heat Loss in Walls and Ceilings," in *Thermal Insulation Performance*, edited by D. L. Elroys and R. P. Tye. ASTM STP 718, 1980, pp. 142–159.
10. Building Science Corporation, "Installation of Cavity Insulation for all Climates," *BSC Information Sheet 501, prepared for the Department of Energy's Building America Program.*
11. Taitem Engineering, "Room Air Conditioner Conduction Losses." *Prepared for: New York State Homes and Community Renewal Weatherization Assistance Program*, 2012.
12. Jonathan Ochshorn, "Determining the Average R-Value of Tapered Insulation." *Part 1. ASHRAE Transactions 117*, 2011.
13. https://courses.cit.cornell.edu/arch262/calculators/tapered-insulation/. Accessed January 18, 2015.

14. U.S. Department of Energy, "Radiant Barrier Fact Sheet." Prepared by the Building Envelope Research Program, Oak Ridge National Laboratory, 2010.

15. Building Science Corporation, "Guide to Insulating Sheathing." Revised January 2007, U.S. Department of Energy Building America.

16. John Straube, Kohta Ueno, and Christopher Schumacher, "Internal Insulation of Masonry Walls: Final Measure Guideline," Building America Report 1105, December 21, 2011.

17. Building Science Corporation, "Basement Insulation Systems." *Building America Report 0202*, 2002.

18. http://www.htflux.com/en/balcony-slab-without-thermal-break-insulation-thermal-bridge/. Accessed January 21, 2015.

19. Ian Shapiro, "Improving Crawlspaces in New York Multifamily Buildings," *Home Energy*, July/August 2006.

20. U.S. Environmental Protection Agency, "Moisture Control Guidance for Building Design, Construction and Maintenance." EPA 402-F-13053, December 2013.

21. Brian Stephens, "Window and Storefront Thermal Performance: What Every Specifier Needs to Know." *The Construction Specifier*, November 25, 2013.

22. http://www.nps.gov/tps/how-to-preserve/briefs/13-steel-windows.htm. Accessed February 3, 2015.

23. http://www.kawneer.com/kawneer/north_america/en/info_page/History_1980.asp. Accessed February 10, 2015.

24. http://www.usgbc.org/articles/green-building-facts. Accessed February 10, 2015.

25. NFRC 100–2004, "Procedure for Determining Fenestration Product U-Factors." 2004 National Fenestration Rating Council, Inc.

26. http://windows.lbl.gov/software/window/window.html. Accessed January 30, 2015.

27. C. C. Sullivan and Barbara Horwitz-Bennett, "Eight Tips for Avoiding Thermal Bridges in Window Applications." *Building Design and Construction*, April 2, 2014.

28. Ian Shapiro, Informal field tests on a variety of storm windows, unpublished, 2015.

29. http://www.lindeus.com/internet.lg.lg.usa/en/images/r_value_analysis_color138_28241.pdf. Accessed February 14, 2015.

30. http://www.archtoolbox.com/materials-systems/thermal-moisture-protection/rvalues.html. Accessed February 14, 2015.

31. THERM 6.3 / WINDOW 6.3 NFRC Simulation Manual, Lawrence Berkeley National Laboratory, July 2013, pp. 2–15.

32. http://www.allaboutdoors.com/article_info.php?articles_id=31. Accessed February 3, 2015.

33. BREEAM. *"BREEAM New Construction: Non-Domestic Buildings."* Technical Manual, SD5073–2.0:2011. Garston: BRE Global Ltd., 2011.

34. U.S. Environmental Protection Agency, "The Lead-Safe Guide to Renovate Right." EPA-740-K-10–001, September 2011.

35. http://www2.epa.gov/lead. Accessed January 24, 2015.

36. Steven J. Emmerich, Tim McDowell, and Wagdy Anis, *"NISTIR 7238, Investigation of the Impact of Commercial Building Envelope Airtightness on HVAC Energy Use."* National Institute of Standards and Technology, June 2005,

37. Gary Proskiw and Bert Phillips, *"Air Leakage Characteristics, Test Methods and Specifications for Large Buildings."* Canadian Mortgage and Housing Corporation, March 2001.

38. Henri C. Fennell and Jonathan Haehnel, "Setting Airtightness Standards." *ASHRAE Journal*, September 2005.

39. "Development of Design Procedures and Air Leakage Control in High-Rise Residential Buildings." Canada Mortgage and Housing Corporation, 1991.

40. John Straube, "Building Science Digests, BSD-014: Air Flow Control in Buildings." Building Science Corporation, May 9, 2008.

41. Mark Limb, "Ventilation and Infiltration Characteristics of Lift Shafts and Stair Wells—A Selected Bibliography." Air Infiltration and Ventilation Center (AIVC), 1998.

42. http://www.wisdomandassociates.com/education/bpi/calculators/Stack%20Effect%20Calculator.htm. Accessed July 24, 2015.

43. Martha J. Hewett, Timothy S. Dunsworth, Michael J. Koehler, and Helen L. Emslander, *"Measured Energy Savings from Vent Dampers, in Low Rise Apartment Buildings."* Minneapolis Energy Office, ACEEE Proceedings, 1986.

44. Peter Armstrong, Jim Dirks and Laurie Klevgard, Pacific Northwest National Laboratory, Yuri Matrosov, CENEf and Institute for Building Physics, Jarkko Olkinuora, IVO

International, Dave Saum, Infiltec, "*Infiltration and Ventilation in Russian Multi-Family Buildings.*" ACEEE Proceedings, 1996.

45. Steven Winter Associates, "There Are Holes in Our Walls: A Report Prepared for the Urban Green Council," April 2011.
46. http://www.rses.org/assets/rses_journal/0314_Energy_Audit.pdf.
47. Stephen Kerber, "*Evaluation of Fires Service Positive Pressure Ventilation Tactics on High-Rise Buildings.*" Gaithersburg, MD: Building and Fire Research Laboratory, National Institute of Standards and Technology.
48. Gregory C. Cahanin, "Change." *Consulting-Specifying Engineer*, May 2005.
49. Steven Winter Associates and the Urban Green Council, "*Spending Through the Roof.*" Report for NYSERDA, 2015.
50. "*Success for the Energy Expert Team in Hessisch Oldendorf*, Germany: The Energiebüro q50 Receives Certification for Its World Record." BlowerDoor GmbH, Press release, June 8, 2013.
51. Victor Shelden and Ian Shapiro, "Weather-Stripping Windows with V-Strip." *Home Energy*, September/October 2012.

Chapter 5

Lighting

Lighting is a major energy load in commercial buildings. Furthermore, reducing lighting energy usage also reduces air conditioning usage. Conversely, reducing lighting energy usage increases heating usage, but the lighting savings typically more than offset the increase in heating usage and cost, due to differences in heating efficiency and fuel cost.

Historically, lighting energy improvements focused on replacing lower-efficiency lamps with higher-efficiency lamps. More recently, improvements have broadened to include what may be termed right-lighting (reduced overlighting) and lighting controls. The three-pronged approach of replacement, right-lighting, and improved control offers far deeper savings than lamp replacement alone, along with frequently providing improved lighting as well.

Lighting energy evaluations are best done on a space-by-space basis. Inventorying existing lighting, identifying deficiencies, evaluating improvements, and listing these improvements to guide the retrofit are all most easily done with a table, such as a spreadsheet, which lists all spaces in a building. The space-by-space approach may seem like it is extra effort, but it typically makes the evaluation and recommendations go faster, with better results and more energy savings. Without space-by-space evaluations, opportunities are missed, and confusion arises when proceeding from evaluation/audit to installation.

Space-by-space evaluations and the three-pronged approach of replacement, right-lighting, and controls are best practices for lighting energy improvements.

Lighting Basics

LIGHTING TYPES

Indoor lighting in commercial buildings has predominantly been *fluorescent*, and exterior lighting has predominantly been *high-intensity discharge* (HID, including primarily *metal halide, high-pressure sodium, and low-pressure sodium)*. There has been a massive new trend toward *solid-state lighting* (SSL, also referred to as *LED lighting*), due to its efficiency, controllability, and durability. *Induction lighting* is another option for high-efficiency and long-duration lighting. Low-efficiency *incandescent* and *halogen* lighting have seen use in commercial buildings, for such applications as display lighting and some general lighting, and are also seeing widespread replacement with high-efficiency options. Older, inefficient *mercury vapor* HID lighting has recently been banned from new use and is expected to be phased out from existing buildings.

LIGHTING POWER DENSITY

Lighting power density *(LPD)* is defined as the electric power consumed by lighting divided by the net floor area of a space in a building, in units of watts per square foot. For example, a 100 square foot office with two 4-lamp fluorescent fixtures, at 170 watts per fixture, has an LPD of $2 \times 170 / 100 = 3.4$ watts per square foot (W/SF).

Codes and standards, such as American Society of Heating, Refrigerating, and Air-Conditioning Engineers (ASHRAE) Standard 90 and the International Energy Conservation Code (IECC), provide maximum allowable LPD for different types of spaces or, alternately, for buildings as a whole. A maximum allowable LPD is also called a *lighting power allowance (LPA)*. For example, IECC 2015 set an LPA of 1.11 W/SF for enclosed offices.

LPD is useful for energy work in a few ways:

- By recording the existing LPD during an energy evaluation, for each space, we have a convenient indication whether a space is a candidate for energy improvements, by comparing the actual existing LPD with standards or best practices. In the example above, the office with two 4-lamp fixtures and an LPD of 3.4 W/SF has a far higher LPD than suggested by the IECC 2015 for enclosed offices (1.11 W/SF), and so the lighting in this office offers significant energy conservation potential.

- By comparing the existing LPD of a space or building with best-practice LPDs for various spaces or whole buildings, we can generate a first estimate of potential lighting energy savings through either/both lamp replacement or right-lighting, without in fact having to go through full design. Best-practice LPDs are usually a good bit lower than LPAs from codes and standards. For example, a best-practice LPD for office lighting is approximately 0.6 W/SF. Compare this to the example of the office with 3.4 W/SF lighting – the potential for energy savings is a substantial 3.4 – 0.6 = 2.8 W/SF.

- Energy projects need to comply with applicable codes and standards.

ILLUMINANCE

Illuminance (sometimes called *illumination*) is a measurement of light reaching a surface. Its units are *foot-candles (FC)*, in English units, and *lux* in the international system of measurement (SI). One lux is 10.8 foot-candles. Illuminance is measured with a light meter. Interestingly, one foot-candle is approximately the illuminance delivered by a candle at a distance of one foot.

Required illuminance depends on the use of a space. For example, a space used for activities such as reading and fine-motor work requires a higher illuminance than do spaces such as corridors or stairwells.

Illuminance measurements are used for energy work to assess if existing light levels are too high, and so provide an indication of potential for energy conservation, as a check on LPD measurements. We can also use illuminance measurements to reassure building owners and managers that lighting modifications will maintain required illuminances for various spaces in a building. Finally, by measuring illuminance, we can identify underlit spaces, and so recommend lighting improvements. Even though this may mean increased energy usage, such recommendations are regarded as important, and add to the credibility of lighting energy work. The majority of spaces are overlit, so finding underlit spaces is less common.

VOLTAGE

Lighting in commercial buildings is used at various voltages. Lighting is always a single-phase load, even if a building has a three-phase service. The most common lighting voltages in U.S. commercial buildings are 120 volts and 277 volts. One hundred twenty-volt lighting is found on "line-to-neutral" for single-phase service in small buildings, or "phase-to-neutral" on the common 120/208-volt 3-phase 4-wire wye service. Two hundred seventy-seven-volt lighting is "phase-to-neutral" for 3-phase 480-volt service in larger buildings. Less common lighting voltages are 240 volts ("line-to-line" on a single-phase service) and 347 volts ("phase-to-neutral" for a 600 volt 3-phase service). The lighting voltage is of interest for specifying replacement lighting.

LUMINOUS EFFICACY

Luminous efficacy is a measure of the energy-efficiency of a particular light source, in lumens per watt (lm/W). Luminous efficacy is useful for comparing types of lighting. For example, one might compare a standard 60-watt incandescent screw-in lamp with a rated luminous flux of 840 lumens to an LED lamp, stated to be a replacement for a 60-watt incandescent lamp, rated at 800 lumens and 8.5 watts. The incandescent lamp has a luminous efficacy of 840 / 60 = 14 lm/W. The LED lamp has a luminous efficacy of 800 / 8.5 = 94 lm/W. If a lamp of 800 lumens is acceptable to replace a lamp of 840 lumens, the savings are 86 percent.

Typical luminous efficacies are provided in Table 5.1. LED lighting is changing rapidly, and the luminous efficacy is expected to rise in the future. Some caution is advised in using luminous efficacy for lighting design decisions for a specific application. Comparing luminous efficacy for two different types of lamps may not accurately reflect energy savings for a specific application, if the efficiency of the light fixture or the type of space is not accounted for. Energy decisions are most accurately based on photometric design, which is addressed below.

COLOR-RENDERING INDEX (CRI)

The *color-rendering index (CRI)* is a measure of a light source's ability to show colors realistically. It traditionally has been regarded as the most important property of lights, although recent discussions have raised some questions about the value of CRI as a property. Common CRI values are shown in Table 5.2. Over 80 is regarded as good, 100 is regarded as ideal. For exterior lighting, traditional lamps (low-pressure sodium, mercury vapor [Figure 5.1], and high-pressure sodium) have been regarded as having a poor CRI, metal halide has had the best CRI, and light-emitting diode (LED) lighting has the promise to improve on metal halide. For indoor lighting, low-efficiency incandescent and halogen have the highest CRI. The common cool-white fluorescent light has a somewhat poor CRI, higher-performance coated fluorescent lights are better, and LED lighting offers the promise of delivering both high CRI and high efficiency.

TABLE 5.1

Luminous Efficacy

Type of Light	Minimum	Maximum	Typical
	Lumens/Watt		
Candle			0.3
Incandescent	10	20	15
Compact fluorescent	40	65	50
Linear fluorescent	50	100	75
Halogen	15	20	17.5
Mercury vapor	30	60	50
Metal halide	50	90	80
High-pressure sodium	85	150	120
Low-pressure sodium	100	200	150
Induction	75	95	80
Solid-state lighting (LED)	60	100	90

TABLE 5.2

Color Rendering Index

Lamp	CRI
Natural sunlight	100
Candle	100
Incandescent	95–100
Halogen	95–100
Tri-phosphor fluorescent	80–95
LED	80–90
Induction	80
Metal halide	65–70
Cool-white fluorescent	62
Mercury vapor (coated)	40–55
High-pressure sodium	20–30
Mercury vapor (clear)	15–25
Low-pressure sodium	–44

Figure 5.1   Mercury vapor light.

TABLE 5.3

Correlated Color Temperature

Source	CCT (Kelvin)
Embers	800
Candle	1,500–1,800
Low-pressure sodium	1,700–1,800
High-pressure sodium	1,900–2,000
Warm CFL	2,700
Halogen	2,800–3,200
Incandescent	2,900–3,100
Warm light	<3,200
Metal halide	3,000–4,000
Cool light	>4,000
Induction	3,000–5,000
Mercury vapor (coated)	4,000
Cool-white fluorescent	4,100–4,200
LED	2,600–6,000
Sunlight	5,500–5,600
Mercury vapor (clear)	6,000
Shade	8,000
Blue sky	10,000

CORRELATED COLOR TEMPERATURE (CCT)

The *correlated color temperature (CCT)* is a property of lamps that relates to its color appearance. The unit of measure of CCT is degrees Kelvin (K). CCTs below 3,200 K are regarded as warm light, and above 4,000 are considered cool light. Common CCT values are shown in Table 5.3.

CODES AND STANDARDS

Codes and standards address lighting in two ways: Minimum illuminance and minimum energy efficiency.

Minimum Illuminance

Recommendations for illuminance are provided by the Illuminating Engineering Society of North America (IESNA).[1] Common recommended illuminance values are provided in Appendix D. It should be noted that recommendations are averages, not minimums. Frequently, a *uniformity ratio*, the ratio of maximum to minimum illuminance, governs the range of acceptable lighting, essentially specifying a minimum illuminance. The *maintained illuminance* is the value below which the average illuminance on a surface should not fall.[2] The implication is that the degradation of lighting over time, for example due to dirt on lamps or fixture lenses, must be accounted for, between maintenance cycles. Best practices for exterior lighting are also provided by IESNA. A summary of exterior lighting requirements is provided in Table 5.4.[3]

Building code requirements for minimum light levels (illuminance) are varied. For example, the International Building Code (IBC) interior environment lighting requirements set an average illuminance of 10 foot-candles at 30″ above the floor, and a minimum of 1 foot-candle at the treads of stairs serving dwelling units. IBC egress requirements are for 1 foot-candle at the walking surface level, except for auditoriums and other performing arts halls during performances, for which the requirement is 0.2 foot-candles. Similarly, the Life Safety Code (NFPA 101) has illuminance requirements that focus on safely evacuating buildings (egress). Its requirement is similar to the IBC, an average of 1 foot-candle at the walking surface of egress, with a minimum of 0.1 foot-candle, and a maximum not more than 40 times as high. Local codes occasionally vary from national standards. For example, New York City requires 2 foot-candles at floor level for egress, instead of 1 foot-candle. It is of note that the IESNA recommendations are not generally code required, other than for select areas such as egress/emergency. However, they are widely accepted as best practice.

Minimum Energy Efficiency

Lighting energy efficiency is covered in the IECC, local variations of this code, and in some cases more stringent variations, such as California's Title 24 requirements. Recent versions of the IECC place requirements on lighting controls, and establish lighting power allowances. Example lighting control requirements include a general requirement for lighting controls for most spaces, occupant sensor controls for many space types (classrooms, conference rooms, copy rooms, lounges/lunch/break rooms, private offices, restrooms, storage/janitor/locker rooms, and warehouses), daylight-responsive controls in spaces with more than 150 watts of lighting that have access to daylight, and controls for exterior lighting.

Despite code and standard requirements for lighting controls, some of these requirements will not save substantial energy. For example, the IECC sets a maximum off-delay for motion sensors of 30 minutes (like ASHRAE 90). However, it has been found that a 30-minute off-delay will not shut lights off at all during the day in occupancies such as multifamily high-rise corridors, where people walk through the corridor more often than 30 minutes.[4] There is no reason for a light to stay on for as long as 30 minutes after a space becomes unoccupied. Similarly, the energy code

TABLE 5.4

Exterior Horizontal Illuminance Requirements

	Minimum Horizontal Illuminance at Surface (foot-candles)	Maximum Max-to-Min Ratio
Parking lots		
Typical conditions	0.2	20:1
Enhanced security (where personal security or vandalism is a likely or severe problem)	0.5	15:1
Parking garages		
Typical conditions	1	10:1
Ramps, day	2	10:1
Ramps, night	1	10:1
Entrance areas, day	50	
Entrance areas, night	1	10:1
Stairways	2	
Service stations		
Approach or driveway, with dark surroundings	1.5	
Approach or driveway, with light surroundings	2	
Pump island area, with dark surroundings	5	
Pump island area, with light surroundings	10	
Building facades or service areas, with dark surroundings	2	
Building facades or service areas, with light surroundings	3	
Landscaping, with dark surroundings	1	
Landscaping, with light surroundings	2	
Walkways/Bikeways		
High pedestrian conflict area, mixed vehicle and pedestrian	2	8:1
High pedestrian conflict area, pedestrian only	1	8:1
Medium pedestrian conflict area, pedestrian areas	0.5	8:1
Low pedestrian conflict area, rural/semi-rural areas	0.2	20:1
Low pedestrian conflict area, low-density residential	0.3	12:1
Low pedestrian conflict area, medium-density residential	0.4	8:1

Horizontal illuminance measured at surface. Consult separate requirements for vertical illuminance, typically approximately twice as high as the required horizontal illuminance, measured 1.8 m above the surface, at point of lowest horizontal illuminance.

requires only that switching be required to allow reducing interior lighting loads by 50 percent, and only requires exterior lighting to be reduced by 30 percent; however, greater reductions are readily possible.

Other lighting energy code requirements are also not stringent, and serve as a reminder that the energy code is a guide for the "worst possible installation allowed by law," and so is not always a best practice guide for energy conservation. As an example, exit signs are limited by code to 5 watts per side, but, accordingly, a two-sided 10-watt exit sign could be met by many fluorescent fixtures, and even some incandescent fixtures, both of which are regarded as antiquated, whereas most LED exit signs today use less than 4 watts.

The main focus of lighting codes and standards requirements are in the area of lighting power allowances. There are typically two methods provided: (1) the building area method and (2) the space-by-space method. The building area method is of less interest to us, as we seek to make energy improvements on a space-by-space basis in buildings. We are not trying to minimally comply with a code requirement; we are trying to go beyond the code to save energy. Of interest to us are the best practices in each type of space.

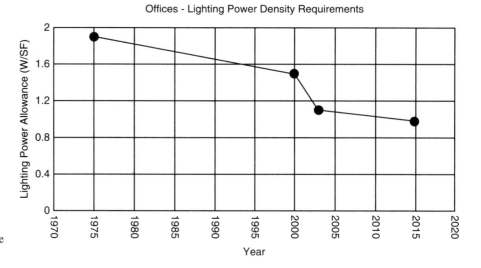

Figure 5.2 Lighting power allowances have dropped by almost half in recent decades.

Lighting power allowances are provided in the IECC, as well as in ASHRAE Standards 90 and 189. As usual, the IECC and ASHRAE Standard 90 have slight differences. Frequently, a new version of the IECC adopts a recent version of ASHRAE Standard 90. And state and local codes usually adopt the IECC at some time after IECC changes are adopted. The 2015 IECC space-by-space lighting power allowances are provided Appendix E.

Over time, the target lighting power allowances have been reduced, as lighting technologies have become more efficient and as lighting design approaches have been refined. For example, the lighting power allowance for offices have been reduced by almost 50 percent over the past 40 years. See Figure 5.2.

As mentioned previously, we can use measurements of lighting power densities on a space-by-space basis in buildings to guide energy improvements. If an existing lighting power density is substantially higher than current standards, this is an indication of good energy conservation potential, and the need to go beyond lamp replacement, and to consider right-lighting, in order to maximize savings. We do not need to stop at current standards. ASHRAE 189, a standard for the design of high-performance buildings, typically has lighting power allowances that are 10 percent less than ASHRAE Standard 90. And best-practice lighting power densities are frequently even significantly lower than the current version of ASHRAE Standard 90 or the IECC, or even ASHRAE 189. Table 5.5 compares some example best-practice LPDs to IECC 2015's targets.

TABLE 5.5

Best-Practice Lighting Power Density[5–9]

Space Type	IECC 2015 LPD (W/SF)	Example Best-Practice LPD (W/SF)
Corridor	0.66	0.40
Lobby	0.90	0.70
Office	1.11	0.40
Parking, interior	0.19	0.10
Stairwell	0.69	0.40

These best-practice LPD's are for fluorescent lighting. LPD's are changing rapidly due to the lower power requirements of solid state (LED) lighting, and so are expected to be 40% or more less than the best-practice LPDs shown in this table.

Field Measurements and Observations

We begin our discussion of lighting fieldwork by identifying common lamps and fixture types.

LAMP TYPES

Fluorescent

Perhaps still the most common type of commercial lamp is the linear fluorescent lamp. Among these, the most common length is 4′. Lamps are also described by their diameter, given in eights of an inch. For example, a T12 lamp is 12/8 of an inch in diameter, or 1.5″. The T stands for *tubular*. A T8 lamp is 1″ in diameter, and a T5 lamp is 5/8″ in diameter.

Older lamps are typically T12. Newer, more efficient lamps are T8, an advantage of which is that the pins have the same size and spacing as T12 lamps, and therefore can be retrofit in existing T12 fixtures. T5 lamps are also efficient, but have more closely spaced pins than T12 lamps, and so require fixture replacement. The difference in energy performance between T8 and T5 lamps is generally considered to be modest.

A fluorescent fixture requires a *ballast* (Figure 5.3). A *ballast* is a control device that regulates the current to the lamp. Older ballasts are magnetic. Newer, more efficient ballasts are electronic. Whether an existing ballast is electronic or magnetic can be identified without physically inspecting the ballast through the use of a *ballast checker*.

Ballasts can also be inspected to obtain make/model by opening light fixtures. They are commonly located below a sheet metal cover, after first removing the main fixture cover (Figure 5.4). The sheet metal cover is squeezed to withdraw a flange from below tabs that hold it in place.

For many years, the replacement of T12 lamps and magnetic ballasts with T8 lamps and electronic ballasts has been a major focus of energy conservation programs. The advent of solid-state (LED) lighting has introduced an even higher-efficiency option.

Variations of the linear lamp are the U-shaped lamp and the circular lamp. A U-shaped fluorescent lamp in a 2′ × 2′ fixture is similar in energy use to a 4′ long linear fluorescent lamp of the same diameter, type, and ballast.

The power consumption of light fixtures is obtained in the following ways, from most accurate to least accurate:

- Measurement. This is not typically done but may be warranted in cases where many lights are being evaluated for replacement in a large facility.
- From the lamp model and the ballast model.
- Rule of thumb.

Other common types of fluorescent lamps are *compact fluorescent lamps* (CFLs). CFLs have built-in ballasts. They can be either screw-in (standard Edison base, to replace incandescent lamps) or pin-type. PL lamps are a type of compact fluorescent lamp, generally pin-type, which are common in commercial buildings. PL stands for Philips Lighting, the company that originally developed the lamp. If two-pin, the ballast is magnetic. If four-pin, the ballast is electronic.

Screw-in CFLs are easier to install, but as a result are also easier to remove, and so are vulnerable to replacement with less efficient lamps. Pin-type lamps are harder to install but are less subject to removal.

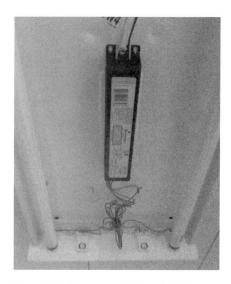

Figure 5.3   Fluorescent fixture ballast.

Figure 5.4   Removing sheet metal cover to access a fluorescent fixture ballast.

TABLE 5.6

Estimated Fluorescent PL Lamp Power Consumption

Approximate Lamp Length (without base) (inches)	Approximate Maximum Overall Length (with base) (inches)	Power Consumption (watts)
2.5	4.0	5
4.0	5.5	7
5.0	6.5	9
6.5	8.0	13
7.5	9.0	18
11.5	13.0	24
15.0	16.5	36

CFL lamps, including PL lamps, generally have their wattages stamped on them. If not, the wattage of PL lamps can be estimated from their length. See Table 5.6. Product catalogs give lamp lengths as their "maximum overall length" (MOL). However, what can be more useful for energy evaluation work is the length of the lamp without the base, to save having to remove the lamp from the socket.

Figure 5.5   Metal halide lamp.

High-Intensity Discharge (HID)

HID lighting is widely used for exterior lighting and also indoors in high-bay and low-bay applications. HID lighting includes metal halide, high- and low-pressure sodium, and mercury vapor. It is identified by an arc-tube inside the main bulb. The most common HID lights are high-pressure sodium (HPS) and metal halide (MH; Figure 5.5). HPS has a thin arc tube, approximately $1/4''$ to $3/8''$ in diameter. Metal halide has a larger-diameter arc tube, typically about $3/4''$ in diameter.

Metal halide is a form of HID lighting with a whitish color that is relatively efficient. It has a reasonably long life, but not as long as mercury vapor or high-pressure sodium: 20,000 hours if vertical, 10,000 hours if horizontal. It is commonly used for high-bay lighting indoors, parking and other exterior applications, and below gas station canopies. Metal halide has a slow *warmup time* (1 to 4 minutes for newer pulse-start lights, 2 to 15 minutes for older probe-start lights), and slow *restrike time* (2 to 8 minutes for pulse start, and 5 to 20 minutes for probe start). *Warmup* is the time for a light to reach full output. *Restrike* is the time a light needs to stay off and cool down before being turned back on.[10] If the power consumption of a metal halide lamps is not marked on the lamp, the power can be estimated from the lamp size. Most 175-watt metal halide lamps are about 5.5″ long, most 250-watt lamps about 8.5″ long, and most 400-watt lamps are about 11.5″ long.

High-pressure sodium is a form of HID lighting that has an orange/yellowish glow. It is more compact than low-pressure sodium, has a more acceptable color, but is not as efficient. High-pressure sodium warmup is about 4 minutes, and its restrike period is 1 to 2 minutes.[11]

Low-pressure sodium is a form of HID lighting that has a very yellow color. It is so monochromatic that low-pressure sodium is strictly limited to outdoor use and makes it difficult to differentiate between other colors, under its light. Lamps are only in the 35- to 180-watt range.[12] Low-pressure sodium is very efficient and has good light uniformity, and maintains its light level well. Its warmup time is 7 to 15 minutes. But it restrikes quickly, even though it may take time to come up to full light output.

Mercury vapor is a form of HID lighting that has greenish-blue color, and skin appears green under the light. For years, the mercury vapor light has been popular for its low cost, well-known as the 175-watt *security light* (Figure 5.6). In recent years, the mercury vapor lamp has been banned in the United States and Europe.

Mercury vapor lamps have a 5- to 7-minute warmup, and 10-minute restrike.[13]

The nominal and actual wattage of HID lamps can be identified by their American National Standards Institute (ANSI) designation, provided in Appendix F.

Differentiating between different HID lamp types is possible by color and warmup/restrike times. Very yellow lights are low-pressure sodium, yellow/orange lights are high-pressure sodium, and the warmup time (slow for low-pressure sodium) and restrike time (fast for low-pressure sodium) might further distinguish it from high-pressure sodium. Metal halide has whiter light. There might be instances where metal halide cannot be distinguished from LED by color; in those cases, turn the lights on and off and back on: LED will turn on and off instantaneously, while metal halide takes time to turn on, and also time to restrike.

Figure 5.6 Mercury vapor light fixture.

Incandescent

Incandescent lamps are perhaps the best-recognized lights, due to their use in homes and some commercial applications. Incandescent lighting dates back to its commercialization by Thomas Edison. It uses a glowing filament to deliver light. The most widely used lamps have an Edison screw base, also developed by Thomas Edison, specifically an E26 base. The E stands for Edison, the 26 is the diameter of the thread, in millimeters. The bulb itself is designated A19, where A stands for *arbitrary*, and 19 is the lamp diameter, in eights of an inch, therefore the common A19 lamp is 19/8, or 2 ³⁄₈″ in diameter.

Other incandescent shapes are also found in buildings, such as the *candelabra* (Figure 5.7) and the *parabolic* downlight, referred to as a *PAR* lamp.

An unfortunate new development is the popular incandescent lamp with a "retro" appearance (Figure 5.8). There are already reports of high-efficiency substitutes to mimic the sought-after look of these anachronistic lamps.

Incandescent lamps are low in efficiency, and are at the early stages of phase-out by legislation. Many have been replaced in recent decades with compact fluorescent lamps, saving approximately 70 percent, for the same light output. A new trend is replacement with LED lamps, delivering savings of approximately 85 percent, for the same light output. LED lamps have other advantages over fluorescent lighting, including not containing mercury and lighting faster at low temperatures, and therefore are expected to dominate incandescent replacements. Incandescent lamps are widely used in recessed downlights, chandeliers, table and standing lamps, porch lights, and decorative lighting (Figure 5.9).

Figure 5.7 Candelabra incandescent lamp.

As LED lamps begin to approach incandescent lamps in shape, the question arises as to how to distinguish them from each other during fieldwork. Incandescent lamps typically have surface temperatures over 250 F (except for smaller sizes, such as 25 watts, which are slightly cooler), whereas LED lamps reportedly have surface temperatures below 120 F. The temperature difference is large enough that an infrared temperature reading can distinguish them. An LED lamp is typically cool enough to hold, whereas an incandescent lamp is too hot to hold.

Halogen

Halogen lamps (Figure 5.10) are a type of incandescent lamp that use a halogen gas, such as iodine or bromine, in the bulb. They are also called *tungsten halogen* or *quartz halogen* lamps. Halogen lamps deliver desirable color properties for some applications and are compact. Halogen lamps also run hot, presenting safety issues in some applications; have short lifetimes; and are inefficient. Halogen lamps present an opportunity to be replaced with more efficient, durable, and safer alternatives.

Figure 5.8 New inefficient incandescent lamp with retro appearance.

Figure 5.9 Decorative incandescent lights.

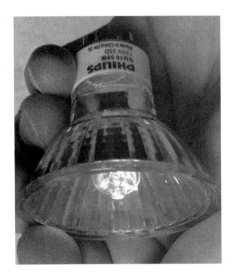

Figure 5.10 Halogen lamp.

Figure 5.11 Induction lamp.

Induction Lighting

Induction lighting (Figure 5.11), also called *electrodeless lighting*, is a high-efficiency and durable form of lighting. With an instant restrike, it is well suited for commercial applications. It is also recognized for its long life, reportedly 100,000 hours. Induction lighting may be identified by its electromagnetic coils (donut-shaped) and tube light. Induction lighting is cool to the touch. Its color is fairly white, although it is also available in a warmer yellow.

Solid-State Lighting (LED)

Solid-state lighting (SSL), more commonly referred to as *LED* for light-emitting diode, is efficient, quick to light even at cold temperatures, and flexible in control. Solid-state lighting is expected to change the world of artificial lighting. LED lights are becoming widely used. LED lamps may be recognized by their small point sources, their generally white light, and their uniformity of light over the lamp. The lenses (covers) of LED fixtures are frequently flat.

FIXTURE TYPES

In addition to identifying lamp types, it is also helpful to identify the type of fixture in which the lamp is found.

Fixture types can be identified by the type of mounting. Indoors, common mounting types are recessed, surface, and pendant. The mounting surface is also of interest: wall mounted or ceiling mounted. Outdoors, common mounting types are wall mounted, pole mounted, and, if under a canopy, the same types as indoors (recessed, surface, pendant).

The most common types of recessed lighting are circular downlights (Figure 5.12) and troffers for linear lights. Circular downlights use a variety of lamps, such as parabolic lamps, PL lamps oriented either downward or sideways, or, increasingly, LED lamps.

A *troffer* is a rectangular fixture, either 2′ × 4′ or 2′ × 2′, that fits into a standard dropped ceiling grid. Often referred to as a *lay-in troffer*, these are the most common type of fixtures in commercial buildings.

Pole-mounted lighting is further classified by the type of head, with cobra (Figure 5.13) and shoebox being common.

The use or application of the lighting is sometimes used as an additional descriptor. Common indoor applications include general lighting, decorative lighting, display lighting, and task lighting. Common exterior applications include parking lighting, entry lighting, pathway lighting, sports or activity lighting, and security lighting.

There are various tricks to opening light fixtures. Dome-shaped fixtures have set screws on the side. The lenses of wraparound fixtures are gently pried apart to pull off the fixture (Figure 5.14). Plastic lenses frequently become brittle with age and crack easily on removal, so care is required. Recessed troffers typically have two latches on their edge that can be pulled downward to open (Figure 5.15).

MEASURING LIGHT LEVELS

Light levels, or illuminance, are measured with a light meter (Figure 5.16). In work areas, illuminance should be measured at the work surface. For example, in offices, illuminance should be measured at desk surfaces. In areas without work surfaces, illuminance should be measured 30″ above the floor. One exception is egress paths and stairwells, where illuminance should be measured at the floor level (or treads in the case of stairs). The other exception is outdoors, where measurements are taken at ground level. A best practice is to measure the peak illuminance, typically below light fixtures, and the minimum illuminance, typically between light fixtures or at the edges or corners of spaces.

Should illuminance be measured in energy audits? Illuminance is a quick added measurement, that gives helpful information. If illuminance is too high, measurement of illuminance will support right-lighting, or reduced overlighting, and can be used to reassure building owners that right-lighting will not compromise safe and code-compliant illuminance. If illuminance is too low, we add to the credibility of the energy audit and do the owner a service in recommending lighting improvements. While already doing a space-by-space lighting inventory, illuminance measurements do not represent a significant added effort. Each measurement only takes a few seconds.

POWER CONSUMPTION

To estimate potential lighting energy savings, we need to know the current usage. Two challenges present themselves for obtaining lighting energy use. Lighting is frequently difficult to reach. For example, high-bay lighting indoors can be over 20′ above floor level, and parking lot lighting can be over 30′ above the ground. And even if light fixtures can be reached, it takes time to reach, open, and inspect them.

Per-lamp energy use can be identified or estimated in a few ways. Facility managers frequently keep spare lamps, which can be examined for make/model and rated power consumption. As-built drawings may have lamp or fixture wattage. If more accurate calculations are required, fixtures and lamps may be opened and inspected, to find make/model/wattage, possibly consulting a manufacturer for rated wattage for specific lamp-ballast combinations. Direct measurement of lamp power use is another option, for the highest accuracy, which may be warranted for large projects. Various electric utilities provide guidance for estimating lighting power consumption, for their energy programs.[14]

A very rough estimate of the wattage of a clean, new exterior lamp on a cobra or shoebox pole may be made by measuring the illuminance at ground level, directly under the light. See Figure 5.17. This plot is for single-luminaire poles, either high-pressure sodium or metal halide. For example, in a parking lot, a 20′ fixture delivers 10 FC at ground level. The closest line *above* 10 FC at 20′ on the graph is 400 watts. Always look for the line at or above the measured point, to account for degradation of light output over time. Again, the graph should only be treated as a rough estimate, and presumes a clean, new fixture, without the influence of other lighting nearby.

DURATION

The duration of existing lighting (hours/day or hours/week) is a critical piece of information needed for evaluating energy savings. The potential for reducing the duration of lighting is also critical. Both the existing duration of lighting, and the potential for reducing the duration, can be improved by measuring the as-found condition of the light, for example on or off, and the as-found occupancy (yes/no) of each space.

In the absence of measurements, the duration of lighting may be presumed according to Table 5.7.[15, 16]

LIGHTING CONTROLS—EXISTING

In inspecting existing lighting controls, the key questions are: How are lights controlled and, for automatic controls, what are the set points?

Controls may be classified as manual or automatic. Within automatic controls, there are three primary strategies:

1. Time control.

2. Photocell control (daylighting).

3. Motion control.

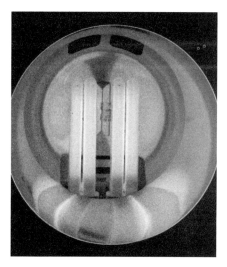

Figure 5.12   Recessed downlight with PL lamps.

Figure 5.13   Cobra head.

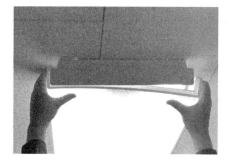

Figure 5.14   Opening a wraparound fixture.

Figure 5.15 Opening a troffer fixture.

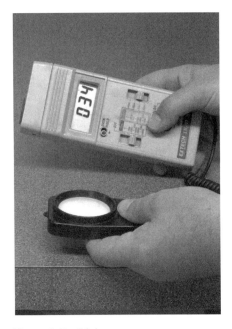

Figure 5.16 Light meter.

For manually switched controls, we are interested to know where the switches are, and which lights are controlled by which switches. Specific attention is directed to whether spaces have more than one light switch to allow multiple levels of general lighting and/or to allow some lights to be turned off. We keep an eye toward possible improvements, with two key questions for any given space:

1. Can the light level be reduced?
2. Can different areas of the same space have their lights turned off?

For time-controlled lights, we want to know the time schedule. For motion-controlled lights, we want to know the duration of the *off-delay* (also called the *lag*, *timeout*, or *time delay*), in other words, how long the lights stay on after the space is no longer occupied. This is best done by test: Enter a space, turn the lights on (whether manually or automatically), leave the space, and see how long the lights stay on. This is a critical measurement. The off-delay is typically far too long and is adjustable: Shorter off-delays deliver significantly more energy savings, with no loss of function or security for lighting in a space.

For photocell-controlled lights, we want to know the light levels at which lights come on and turn off. We specifically want to know if the set point is too high, in other words, if lights are turning on when it is too light outdoors. This is best evaluated late in the afternoon or evening, as light fades, either by measuring the light level in a space or outdoors when the lights come on, or by simple observation: Is it still too light out when the lights come on? How long are the lights on for, unnecessarily, before they are really needed? (See Table 5.8.)

LIGHTING POWER DENSITY

For lighting power density, we need to measure space areas. If there are dropped ceiling tiles, typically $2' \times 4'$ or $2' \times 2'$, a count of the ceiling tiles is an easy way to calculate space areas. Floor tiles may be another option. Pacing is another fast way to estimate space dimensions. Recall that a typical person's footstep is 0.4 times their height. Laser measuring tools are another option. Floor plans can also be used.

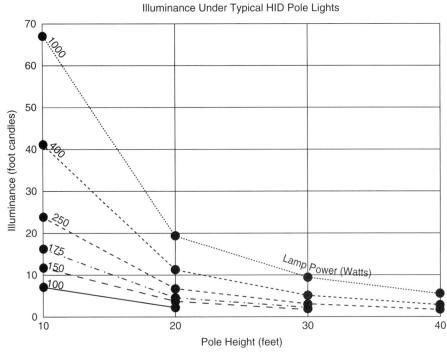

Figure 5.17 Illuminance under HID light poles can be used to estimate lamp wattage.

TABLE 5.7

Duration of Lighting

Space	Duration (Hours/day)	Space	Duration (Hours/day)
Office/Professional	10.3	Strip shopping	11.1
Laboratory	13.9	Enclosed retail	13.7
Warehouse (nonrefrigerated)	9.7	Retail (excluding enclosed)	10.2
Food sales	14.4	Service (excluding food)	9.4
Public order/safety	9.6	Residential	
Health care (outpatient)	9.3	Bedroom	1.4
Warehouse (refrigerated)	10.6	Bathroom	2.0
Religious worship	5.0	Den	2.0
Public assembly	7.3	Hall	2.2
Education	7.6	Garage	2.3
Food service	12.6	Living	2.6
Health care (inpatient)	16.0	Utility	2.6
Skilled nursing	12.0	Yard	3.1
Hotel/Motel/Dorm	10.1	Kitchen/Dine	3.4

TABLE 5.8

Exterior Illuminance

Condition	Illuminance (foot-candles)
Sunlight	10,000
Full daylight	1,000
Overcast day	100
Very dark day	10
Twilight	1
Deep twilight	0.1
Full moon	0.01
Quarter moon	0.001
Starlight	0.0001
Overcast night	0.00001

HEALTH AND SAFETY

While doing lighting fieldwork, it is important to be vigilant for unsafe lighting conditions. These include undersized wires, poor wiring connections, and incorrect lamps for specific sockets and fixtures. Evidence includes hot wires, charred sockets (Figure 5.18) and fixtures, and premature lamp failure. Unsafe conditions should be brought to the attention of building owners and property managers.

OTHER

Lighting power use can also be measured directly. Frequently, lighting is on dedicated circuits, so a measurement can be taken at the breaker panel.

An example space-by-space lighting inventory data sheet is shown in Table 5.9.

The data sheet can be extended with added columns for proposed lighting. Frequently, it is helpful to begin to describe the proposed lighting even while still in the building, especially for nonstandard spaces. For energy auditors, it is difficult to remember every space in a building after the field work is complete—the more that can be done while still in the building, the better. The space-by-space data sheet can be used for comparisons to best-practice LPD for different space types, and can then be used for energy savings calculations.

Advantages to the space-by-space approach include:

■ Customized by space.

■ Combines field data collection with calculations.

■ Methodical.

■ Can serve as the basis for installed work.

There is some merit to doing exterior lighting fieldwork in the afternoon and through dusk. At dusk, we can see the time of day and light level at which outdoor lights turn on, to assess if lighting is coming on too early, and to allow estimating savings from correcting this. After dusk, we can examine the lighting if the type of light is not known: yellow is likely low-pressure sodium, yellow/orange is likely high-pressure sodium, white is likely metal halide, blue/green is likely mercury vapor.

Figure 5.18 Charred light socket is evidence of an unsafe condition.

After dusk, we can also measure light levels, to assess if areas are overlit or underlit. After dusk, we can identify which fixtures have failed lamps. The light level below fixtures can also be used to estimate wattage for pole-mounted fixtures that are too high to inspect. (See Figure 5.17.) At dusk, we can check photocell settings, and possibly adjust them as well. After dusk we can inventory the purpose of exterior lights, whether for access, security, parking, recreation, decorative, or other purposes, to allow us to compare the need for the lighting to its existing control strategy, and to develop associated possible control energy improvements.

Lighting Improvements

We reiterate the fundamentals of lighting improvements:

- Replacement
- Right-lighting, in other words, reduced overlighting
- Controls, primarily to reduce the duration of lighting, but in some cases to adjust light levels. Controls comprise one or more of the following: increased manual control, timer control, photocell control (daylighting), and motion control.

REPLACEMENT

Lighting replacement might include only lamp replacement, or replacement of an entire fixture. Low-efficiency lighting includes incandescent, halogen, mercury vapor, and T12 fluorescent with magnetic ballasts. High-efficiency lighting includes solid-state lighting (LED) and induction lighting. Lighting that falls in a middle category of efficiency, until recently regarded as high efficiency but increasingly viewed as replaceable, include T8 fluorescent, compact fluorescent, metal halide, high-pressure sodium, and low-pressure sodium. The ability of LED lighting to turn on and off quickly, and to operate well at low temperatures, offer additional advantages relative to HID lighting.

It is important to recognize that some fixtures are intrinsically higher in efficiency than other fixture types. For example, linear fixtures are significantly more efficient

TABLE 5.9

Space-by-Space Lighting Inventory Data Sheet

Space	Type	Area (SF)	Fixture Description	Fixture Quantity	Watts per Fixture	Lighting Power Density (W/SF)	Hours Per Day	Existing Control*	Quantity of controls	Notes
Example: Room 101	Office	165	T12/magnetic - 4' 2-lamp	2	86	1.04	9	M	1	No weekend hours.

* M - Manual   P - photocell   O - occupancy   T - timer

than recessed downlights (Figure 5.19). So we try to think outside the box: instead of evaluating simple lamp replacements for fixtures like recessed downlights, we ask, "Could we replace the downlight fixtures with linear fixtures?"

Task lighting involves placing lighting close to tasks, such as the use of desk lamps in offices, or under-counter lamps. While we might think that task lighting makes sense, research has found that task lighting does not deliver energy savings.[17] There are in fact risks, as some people will use both task lighting and overhead ambient lighting, increasing energy usage. In some instances, task lighting has been observed to be installed but not used.

RIGHT-LIGHTING

Right-lighting, or reduced overlighting, is a cornerstone of lighting energy conservation. We know, from our space-by-space lighting power density survey, which spaces are using more energy than necessary. We know from codes and standards and best practice what lighting power densities are possible. This comparison alone can form the basis for an energy survey.

Right-lighting takes more effort than simple lamp replacement, or one-for-one fixture replacement, but the benefits are far broader and last far longer. The energy savings significantly amplify the savings of high-efficiency replacement lighting. For example, an energy audit of a large outpatient psychiatric facility, with five buildings built between the 1930s and the 1990s, found the average lighting power density in offices to be 1.2 W/SF, the average in restrooms to be 2.6 W/SF, and the average in stairwells to be 1.1 W/SF. By changing lighting layouts and replacing with high-efficiency lighting, savings over 60 percent were found possible, far higher than by replacing lights alone. The energy savings of right-lighting last longer than simple replacements, because the revised layout is permanent: it outlasts any single lamp. Fewer lamps or fixtures means savings in the cost of future replacement lamps, as well as reduced labor to replace lamps, savings that will last decades into the future.

Right-lighting is done in several ways:

- Delamping, for example, removing two lamps from a four-lamp fixture
- Changing existing fixture layouts
- Changing both fixture type and layout

We can go a step beyond an estimate based on lighting power density, and design the new lighting. This is best done with photometric software. Many programs are available, either online or as downloads, to do this. A selection of lighting software is provided in Appendix G. Choosing new fixtures, or a new layout using existing fixtures, can be done in minutes for a simple space, on the basis of space height and size, and target foot-candles. Lighting vendors can also provide photometric design.

When designing lighting, we consider designing to the low end of IESNA-recommended illuminance, unless individuals using a space are visually impaired of have other special visual needs. The prevalence of computer screens and other self-lit equipment in our buildings have brought a new need to avoid glare and over-lighting. We also note that the difference in illuminance between, for example, 30 and 70 FC (the recommended range for offices) is small, whereas the difference in energy use is large, typically requiring more than twice the energy to deliver 70 FC than 30 FC. See Figure 5.20 for example illuminances.

Lighting design goes beyond simply choosing efficient fixtures for a space. Other approaches, such as changing the height of lighting, can be used to reduce energy use. When performing lighting design calculations, surface *reflectance* (Figure 5.21) can also be used to substantially reduce lighting energy use. *Reflectance* is light that is reflected from a surface, such as a floor or ceiling, expressed as a percentage of the light that hits the surface. In other words, if one measures the illuminance leaving a

Figure 5.19   Recessed downlighting is less efficient than linear fixtures.

Figure 5.20   Example illuminances, from top to bottom: 70, 50, 30, and 10 foot-candles.

surface with a light meter, and then one turns the light meter around and measures the incoming illuminance hitting the same surface, the ratio of the two measurements is the reflectance. For example, if 60 foot-candles is measured reflected off a wall, and 100 foot-candles is measured reaching the wall, then the wall's reflectance is 60 percent. Lighting design computer programs use default reflectance of 80 percent for ceilings, 50 percent for walls, and 20 percent for floors, and most lighting is designed using these default values. However, higher reflectance is readily available. For example, ceiling tiles are available with 90 percent reflectance, walls can be painted in a variety of colors with reflectance over 60 percent and reaching 90 percent, and many floor options (tile, lighter woods, polished concrete) offer reflectance over 50 percent. Appendix H provides lighting reflectance for a variety of surfaces. In situations where changing the reflectance of a space is an option, such as repainting, replacing ceiling tiles, changing the flooring, or even where furniture is being replaced, increasing the reflectance can allow substantial reductions in artificial lighting. For example, surface reflectance of 90 percent for ceilings, 70 percent for walls, and 40 percent for floors saves a substantial 28 percent in lighting energy use. Note also that the higher surface reflectance means proportionally fewer light fixtures, which means lower installation cost.

Once familiar with lighting design software, it is possible to develop *lighting patterns*, which are frequently-used strategies for lighting in spaces that are common to one's building stock. Example lighting patterns are provided in Table 5.10. Lighting patterns are also available from a variety of sources.[18]

Maintenance improvements can form a part of right-lighting efforts. Cleaning lighting reflectors, lamps, and lenses increases light output and might allow delamping or a reduction in the number of light fixtures, if periodic cleaning is planned in order to maintain light levels. *Reflectors* comprise the area inside fixtures behind the lamps. *Lenses* are the covers of light fixtures. Cleaning should not be expected to substantially increase light output, but increases of a few single percent are reportedly possible. One example reported shows an increase in luminaire efficiency from 58.4 percent to 66.7 percent, through cleaning.[19] Retrofit *specular reflectors* are another approach to increase light output, to allow removing lamps or fixtures. Specular reflection means highly polished reflection, as contrasted with diffuse reflection that is delivered by dull surfaces such as reflectors that are painted white. Research has shown that specular reflectors do increase light output, and so can allow delamping or light fixture removal.[19, 20] One study found an increased illuminance, ranging from 9 percent to 35 percent. However, specular reflectors generally direct light in a more downward direction, and so increase the variation of light within a space. The study concluded that specular reflectors may be better suited to spaces where the lighting is placed directly above occupants, such as above work spaces.

CONTROLS

Along with lighting replacement and right-lighting, improving lighting controls forms the third major strategy for effective lighting energy conservation.

Control improvements can either be improved manual controls, or automatic controls, or combined manual-plus-automatic controls. Automatic controls primarily include motion controls, timer controls, and photocell control.

Manual Control

Recall the two questions we asked when we inspected existing lighting:

- Can the overall lighting level be turned down?
- Can lights in different areas of a single space be turned off?

If automatic controls are not provided for a space, the key is to provide more than one manual light switch. A single manual light switch serving a space with more

TABLE 5.10

Example Lighting Patterns[5–9]

Space Type	Example Lighting Pattern
Corridor	Single-lamp 4′ T8, or equivalent LED, spaced at 20′
Office, 10′ ceiling, 80/50/20 reflectance	40-watt LED 2 × 4 fixture, one per 100 SF (10 × 10), 30 foot-candles
Parking garage, enclosed, 60′ wide	Four-lamp 4′ T8, or equivalent LED, in two rows, each 15′ from each side and 15′ from the center of the driving lane, spaced 40′ along the length of the driving lane
Stairwell (typical scissors configuration)	Single-lamp 4′ T8, or equivalent LED, above each landing

Figure 5.21   Use of surface reflectance to reduce artificial lighting lamps, fixtures, and energy use. Note the reflective floor and ceiling. This design uses two-lamp T8 fluorescent troffer fixtures, delivering high illuminance at only 0.6 watts per square foot, 60 percent less than the typical lighting power allowance for a sales area.

than one light fixture is an immediate indicator of a need for control improvement (Figure 5.22).

We want at least two light switches per space, even for small spaces, to at least either allow the light level to be turned down or to allow lights in different areas of the same space to be turned off. However, ideally, we want both, which requires four switches. Three-lamp fixtures with inboard-outboard control offer great flexibility for manual control. See Figure 5.23.

Consider a two-person office with two existing light fixtures. By replacing each light fixture with a three-lamp fixture with inboard-outboard control, and providing separate wall control for each of the two light fixtures, we have provided not only four levels of lighting per light fixture (off, one lamp on, two lamps on, or three lamps on), we have provided the ability to turn off light to each area of the office, essentially providing eight levels of control, with four light switches. We now have the flexibility to turn off the light in either of two areas of a space, and then to further adjust the light level of whichever area that remains lit. See Figure 5.24. The goal is to go beyond 25 percent energy savings, and to approach transformational 60 percent, or 70 percent energy savings, or more.

The strategy of allowing flexible control of both the light level and the spatial delivery of a light to a space is extended from small spaces to larger spaces. More switches are better. In larger spaces, we also seek to disaggregate the different functions of a room and allow lighting control accordingly. In a conference room, for example, we seek to allow separate control of lights over the conference table, over perimeter seating areas, and over areas with projection screens.

Motion Control

Motion control, also called *occupant* or *occupancy* control, turns lights on and off in response to human motion. Motion control can also be used in combination with manual control: *vacancy sensors* (also called *manual-on control*) require an occupant to manually turn a light on, and then automatically turn the light off after motion is no longer sensed. See Figure 5.25.

Regular motion sensors that require motion to turn lights both on and off are more appropriate for public areas where occupants may not know where the light switches are or do not have access to them. Vacancy sensors are appropriate for spaces where occupants know where the light switches are but may not need or want to have the light on. For example, a light above a photocopy machine in an office corridor is appropriately controlled by a vacancy sensor. If controlled by a regular motion sensor, the light would come on each time someone walks down the corridor, whether or not the copy machine is being used. Vacancy sensors save more energy than regular

Figure 5.22   A single manual light switch serving more than one light fixture should be regarded as a red flag.

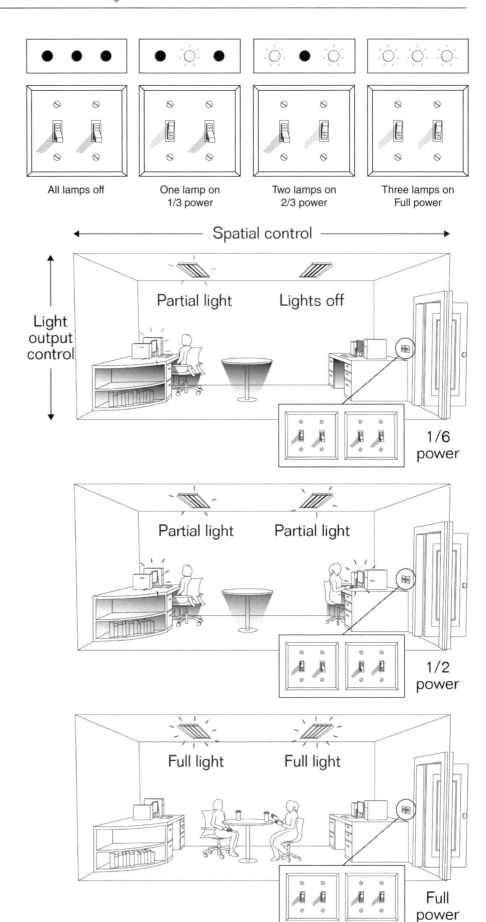

Figure 5.23   Inboard-outboard lighting.

All lamps off

One lamp on
1/3 power

Two lamps on
2/3 power

Three lamps on
Full power

Spatial control

Light output control

Partial light   Lights off

1/6 power

Partial light   Partial light

1/2 power

Full light   Full light

Full power

Figure 5.24   Spatial control plus output control.

motion sensors, since they require the occupant to turn them on, and so should be used wherever appropriate.

The two main types of motion control are *passive infrared (PIR)* and *ultrasonic*. Passive infrared sensors work by detecting heat from occupants in a space and therefore are suitable for where the sensors have a direct line of sight to occupants. Ultrasonic sensors work by detecting responses to ultrasonic signals sent out by the sensors, so they do not require a direct line of sight to the occupant. However, ultrasonic sensors can be mistakenly activated by motion in adjacent spaces. *Dual-technology motion controls* incorporate both passive infrared and ultrasonic sensors.

Motion control sensors can be integrated with light fixtures, integrated with light switches, or stand-alone. When integrated with light fixtures, motion sensors allow controls such as bilevel lighting, operating at a low light output during unoccupied mode and at a high level during occupied mode. When integrated with light switches, they allow convenient retrofit of existing switches, lowering installation labor costs. When stand-alone, they lower material costs. Stand-alone controls are typically ceiling mounted, to provide the best coverage and reduce the risk of line-of-sight issues. When wall mounted, such as when integrated with wall light switches, motion sensors present higher risks of either nuisance trips off (if the sensor does not see motion of occupants) or nuisance trips on (if sensors see motion of occupants in adjacent areas).

We previously mentioned the importance of the *off-delay* (also called the *lag*, *timeout*, or *time delay*), in other words, how long the lights stay on after the space is no longer occupied. Research has shown that shortening the off-delay substantially increases energy savings. With the advent of LED lighting, and the possibility of even shorter off-delays without risking shortening lamp life, even higher savings are possible, see Figure 5.26.

In some applications, savings can be more than doubled if off-delays are shortened from 20 minutes to 1 minute or less. See Table 5.11.

Motion sensors can be integrated with multilevel manual control, either with multiple motion sensors, or using motion sensors that themselves allow multiple levels of control. See Figure 5.27.

Timer Control

Timer control refers to any timed control of lights. Timer control can use stand-alone timers or can be part of a central computerized control system. Timer control can be used to turn lights on and off, or to turn lights off after being turned on manually, or to turn lights off after being turned on by a separate automatic control, such as a photocell control that turns on lights at dusk before being turned off by a timer

Figure 5.25 Vacancy sensor.

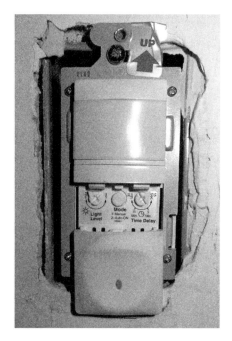

Figure 5.26 Control settings inside a wall-mounted vacancy sensor.

TABLE 5.11

Motion Sensor Energy Savings with Shorter Off-Delay [4, 21]

	Off-Delay		
	1 Minute	5 Minutes	20 Minutes
Break room		29%	17%
Classroom		58%	52%
Conference room		50%	39%
Private office		38%	28%
Restroom		60%	47%
Corridor	72%	59%	34%
Stairwell	77%	74%	65%

Figure 5.27   Motion sensor with two levels of control.

Figure 5.28   Crank timer.

at the end of the evening. Another example of a manual-on/timer-off control is a typical crank timer (Figure 5.28), used in some bathrooms to control both the light and exhaust fan.

Historically, timer control has been used to control exterior lighting, and lighting in large buildings such as offices. Increasingly, motion sensors are viewed as providing more energy-efficient control than timers.

Photocontrol

Photocontrol for indoor lighting is synonymous with *daylighting*. We seek to reduce or eliminate indoor artificial lighting when either *sidelighting* (light from windows on walls) or *toplighting* (light from skylights) can provide some or all of the required light indoors. Photocontrol for exterior lighting is addressed separately below in the discussion of exterior lighting control.

Energy savings from daylighting is decreasing as lighting efficiency (replacement and non-daylighting controls) increases. Therefore, interactive effects between daylighting savings and lighting efficiency need to be accounted for. For example, if existing inefficient T12 fluorescent fixtures near windows are being replaced with LED lighting, the savings from daylighting should be calculated relative to the new LED lighting, in order to avoid double-counting savings. Even more importantly, if other controls are being considered that reduce the duration of lighting, or that reduce the level of lighting, in response to motion, it may not make sense to consider daylighting at all. For example, if a stairwell light next to a window can be replaced with bilevel lighting, such that its power draw is low for over 99 percent of the time,[4] it does not make sense to also install photocell daylighting control to dim the light from its high power setting for what essentially will be single minutes per day.

When we examine daylighting in the context of possible envelope improvements, if there are ever opportunities to replace portions of windows with components that are more thermally efficient, such as insulated wall panels, the benefits of daylighting need to be examined relative to the benefits of envelope improvements. For most scenarios, the benefits of envelope thermal improvements outweigh the benefits of daylighting. In cases where glazing can be reduced, daylighting should not be blindly considered over envelope thermal improvements without analyzing both.

Commissioning of photocontrols can be challenging. Anecdotal reports of malfunctioning daylighting controls are widespread, with either undesirable variations of light levels or nuisance switching of lights. Commissioning is preferably done at dawn or dusk, and during both full-sun and cloud-cover conditions.

Photosensor control can either use outdoor illuminance as an input, or indoor illuminance. If outdoor illuminance is used, this is called *open loop control*. With open loop control, the sensor is not measuring the illuminance in the space and therefore cannot tell if, for example, window shades are open or closed. However, open loop control is simpler. If indoor illuminance is used, this is called *closed loop control*. For closed loop control, placement of the control sensor is important, and more sensors are required, generally one per space. For open areas, multiple sensors are used. One manufacturer recommends one sensor per 30 feet of window wall. It is important to note that the sensor is measuring indoor illuminance at the location it is installed, generally at the ceiling, and this is typically not identical to the location of interest, typically a work surface. The difference in illuminance between the point of measurement and the target area of interest should be accounted for in commissioning. Indoor sensors also need to avoid placement directly in the line of sight of indoor lighting. Hybrid sensors are available that measure both indoor illuminance and the light entering windows.

Photocontrol can either be *switched daylighting* (on/off), turning lights off when adequate daylight is available, or the control can be *continuous daylighting* (*dimmed*)

TABLE 5.12

Relative Light Output and Dimming Ballast Power[22]

Relative light output	0%	20%	40%	60%	80%	100%
Fluorescent ballast power	23%	38%	55%	69%	86%	100%

on a variable basis, in response to available daylight. Switched daylighting cannot provide as fine control as continuous daylighting. Specifically, with switched daylighting, just before the lights are turned off and just after the lights are turned back on, a space is typically overlit, with a combination of too much daylighting and artificial lighting. Energy savings are ostensibly not as high as with dimmed control, although the continued power draw of dimming ballasts can offset energy savings of continuous daylighting systems, and should be accounted for. Switched daylighting can be done with multilevel switching, such as bilevel control, to reduce overlighting and increase energy savings, without using continuous control. Continuous control requires that the lighting be dimmable. For fluorescent lighting, this has meant dimming ballasts. As ballasts and light output are dimmed, energy use is reduced, but not by as much as the light is dimmed, for fluorescent ballasts. See Table 5.12. It is also important to note that at zero light output, the ballast still consumes almost a quarter of its fully power. So if power to the ballast is not switched off when no light is needed, it will continue to draw power throughout the period of zero light output, which will offset savings from daylighting. There have been reports of dimming control relays making objectionable noise, when the lights are turned off at the end of dimming.

Dimming LED lights saves more energy, as power use more closely tracks lighting output. However, many dimming issues appear to not have been fully resolved, so commissioning of LED dimming is still essential.

Energy savings from daylighting have been found to be less than predicted, especially for side-lit daylighting applications.[23] Caution is urged in modeling and implementing daylighting, to account for a variety of common daylighting problems.

Exterior Lighting Control

Historically, the main approach to exterior lighting control has been to use photocell control, to ensure that lights turn on at dusk and turn off at dawn. This is good, but falls far short of achieving the potential savings by best practices in exterior lighting control. Few lights in fact need to stay on all night. Lights do not need to stay on all night for any of the following exterior lighting applications: parking, evening recreational activities, evening entry/exit/access, evening decorative applications, signage, and more. For all these applications, all-night lighting means that the lights are operating for many hours more than necessary.

Priority should be given to motion-sensing control for exterior lighting. Motion-sensing control means that lights are on only when they need to be, reducing operation from the many hours of all-night operation to perhaps a few minutes per night. Even for security applications, motion sensing is likely more effective than all-night lighting in deterring intruders when lights go on due to motion. A strong case has been made that all-night lighting reduces security.[24]

Even manual control, or timer control, can be better than all-night photocell control.

Photocell control does have its place and should be used as a secondary control for all exterior lighting. The function of the photocell control is to keep lights off during the day. Whichever control is selected as a primary control (motion, manual, or timer), it is always best to use this primary control *in conjunction* with the secondary photocell control. The function of the primary control is to turn lights off at night, to complement the secondary photocell control that keeps lights off during the day.

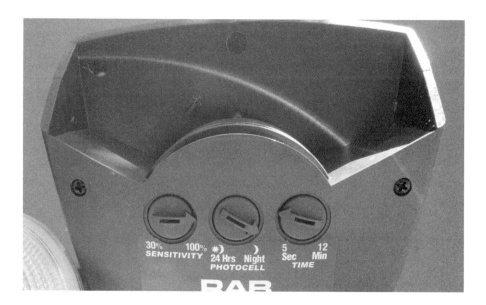

Figure 5.29 Combined motion and photocell control.

When used in conjunction with each other, the two controls can provide lighting when it is needed, prevent lighting when it is not needed, and reduce exterior lighting from hours per night to minutes per night.

In summary, there is a hierarchy for energy-efficient exterior lighting control. The best is motion-plus-photocell. Second best, depending on the application, is either timer-plus-photocell or manual-plus-photocell control. Photocell-only control and manual-only control are the least efficient strategies.

Built-in controls are available that provide combined control, such as motion-plus-photocell control (Figure 5.29). Controls are also available that integrate all three control functions: photocell, timer, and motion. During the day, the control keeps the light off with the photocell. At dusk, the light comes on for a period of time ("evening hours"), which the user can set. After the timer period is over, the motion control takes over, during night-time hours, until the photocell takes over again in the morning, to keep the light off during the day.

For motion sensors outdoors, like indoor lighting controls, short off-delays increase energy savings. This should be specified in design/purchasing, and tested after installation.

Photocell control for exterior lighting has traditionally been on/off. However, the advent of solid-state (LED) lighting has made dimming control more feasible. Dimming LED lighting reduces power usage in fairly close proportion to reduced light output, although, as mentioned, LED dimming is still an emerging technology, with some unresolved issues. Dimming controls are also available for motion controls for exterior lighting, to allow a slow ramp-down in light output after the end of motion, and to allow a rapid (but not instantaneous) ramp-up, and therefore prevent what might be perceived as the more startling on-and-off cycling of exterior lighting.[25]

Correct specification and testing is also needed for exterior photocontrol. Historically, testing of photocells has simply involved covering the sensor during daylight, to see if the light comes on. This type of testing does not assess if the illuminance level at which the light comes on is correct. For example, photocontrols are available with factory-shipped default set points of 200 foot-candles or higher. If left at this level, lights will turn off too late in the morning, will come on too early in the evening, will come on in shaded areas such as porch lights, and will come on with passing clouds. Illuminance cut-in and cut-out set points (when the lights turn on and off) should be specified and tested. Testing is best done at dawn or dusk and does not necessarily require a light meter, although having a light meter is preferable. At the desired level of outdoor light (for example, 1 foot-candle), adjust the photocell

TABLE 5.13

Lighting Duration Datasheet

	Example Classroom		Space 2		Space 3		Space 4	
	Occupants	Lights	Occupants	Lights	Occupants	Lights	Occupants	Lights
12–1 A.M.	0	Off						
1–2 A.M.	0	Off						
2–3 A.M.	0	Off						
3–4 A.M.	0	Off						
4–5 A.M.	0	Off						
5–6 A.M.	0	Off						
6–7 A.M.	0	Off						
7–8 A.M.	1	On						
8–9 A.M.	1	On						
9–10 A.M.	20	On						
10–11 A.M.	20	On						
11–noon	20	On						
Noon–1 P.M.	5	On						
1–2 P.M.	20	On						
2–3 P.M.	0	On						
3–4 P.M.	5	On						
4–5 P.M.	1	On						
5–6 P.M.	0	On						
6–7 P.M.	0	On						
7–8 P.M.	0	On						
8–9 P.M.	1	On						
9–10 P.M.	0	Off						
10–11 P.M.	0	Off						
11–12 midnight	0	Off						

sensitivity so that the light just turns on. This ensures that the light turns on at the right light level, and turns off slightly above this level, so the light will always at a minimum be at the 1 FC.

Central Controls

Control settings for existing central controls may be obtained from the user interface, or from control sequences provided in user manuals for the installation, or from control system vendors or service contractors. If not available, they are readily obtained by observing the operation of lights. Of key interest is whether lights are on at night and other periods of low or no occupancy. This is another area in which collaboration between an energy manager and an energy auditor can be productive. The energy manager can collect information on the lighting control sequence and occupancy, and the energy auditor can use this information to estimate savings for improved controls. Once the baseline/existing control sequence is known, the space-by-space survey can be used to guide proposed modifications to the central control.

Modeling Energy Savings for Lighting Controls

Lighting energy use is simply:

$$\text{Energy use (kWh)} = \text{watts} \times \text{hours} / 1{,}000$$

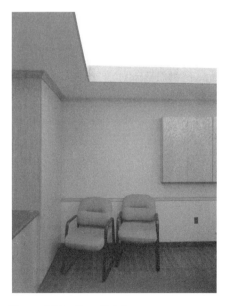

Figure 5.30   Cove lighting.

Figure 5.31 LED track lighting.

Figure 5.32 Decorative lighting.

Figure 5.33 Decorative uplighting.

Corrections for interaction with heating and cooling are sometimes performed or are performed automatically by whole-building models.

Lighting control savings are most readily modeled by the space-by-space approach that we use for lighting data collection, analysis, and specification. A single spreadsheet allows the data to be entered during fieldwork, modeled for energy savings, estimated for installed cost, and given to installers for purposes of implementation. Working space-by-space further allows each space to have the most appropriate control evaluated and recommended.

Most lighting can be modeled in such a spreadsheet on the basis of existing (pre-retrofit) power and lighting duration (hours per day or hours per week), and proposed (post-retrofit) power and lighting duration. Attention should not only be directed to the proposed lighting and duration but also to the existing lighting and duration. The accuracy of the savings projection depends strongly on getting the existing power consumption and duration right.

In the absence of such measurements, duration can be estimated using Table 5.7. Duration of lighting for different spaces is best obtained on a building-specific basis. For a large office building, for example, it is worth doing a count of occupants at their desks, for each hour of the day, and to also inventory if lights are on or off in these offices, at the time of the occupancy count, in order to estimate savings. A sample data collection form for such an inventory is provided in Table 5.13.

For proposed lighting that will be controlled at multiple discrete levels, such as automatic bilevel lighting or multilevel manual control, calculations need to account for the estimated hours at each lighting level. As always, we seek to avoid overestimating savings, as much as we seek to avoid underestimating savings.

For daylighting, more complex modeling is required, accounting for the full range of contributing factors: size and location of side-lighting or top-lighting, target indoor light levels, indoor light fixture locations, light fixtures and lamps, control strategy (such as dimming or multi-level control), dimming ballast power consumption, and any manual or automatic over-rides such as toggle switches or timer control.

OTHER

It should be noted that if fluorescent lamps are removed from a fixture, without disconnecting the ballast or otherwise removing power to the fixture, the ballast will continue to consume electricity.[26] For example, an energy-efficient magnetic ballast is reported to use 4 watts when the lamps are removed but the fixture remains energized.[27] Other types of ballasts, such as metal halide, reportedly also draw power if the lamp is removed but the fixture is still energized.

We have discussed how daylighting is synonymous with photocontrol of indoor lighting. Our prior discussion focused on adding photocontrol to existing lighting in cases where existing windows or skylights can displace the use of some indoor artificial lighting. There might be instances in which such windows or skylights do not exist, but could be added to a building, for purposes of delivering daylighting. Such cases should be carefully evaluated, assessing the potential lighting energy savings, and comparing it to the added thermal load. The best possible applications are large single-story buildings with high levels of indoor lighting that see much daytime use (long daytime duration of lighting), where judicious top-lighting (skylights) can be added to meet most of the lighting load, without adding significant thermal load. For example, big-box retail stores meet most of these criteria. In some cases, light tubes can be installed to deliver daylight to top-story spaces below attics, or to floors below top-story spaces. The high installed cost of windows and skylights needs to be carefully examined, relative to potential savings. These costs are reduced if, for example, skylights are installed at a time when a flat roof on a single-story building is being replaced anyway.

A variety of health and safety issues arise, when addressing lighting. For indoor lighting, these include, most importantly, glare and insufficient lighting. For outdoor lighting, these include light pollution (lighting of the night sky), light trespass (lighting of neighboring property), security, and insufficient lighting. It is also important to be aware that lighting needs are different for the elderly, and others whose vision might be compromised.

Specific Lighting Applications

HIGH- AND LOW-BAY LIGHTING

High-bay and low-bay lighting refers to lighting used in spaces such as gyms, warehouses, parking garages, and other taller spaces. Low-bay is generally for installation less than 20'' from the floor (to the bottom of the fixture), and high-bay lighting is for over 20'. Historically, low-bay and high-bay lighting has been HID, with larger fixtures (22–28'' diameter) for low-bay lighting, and smaller fixtures (15–18'' diameter) for high-bay lighting. In recent years, T5 fluorescent has been a popular high-efficiency replacement for HID indoor high-bay and low-bay lighting, both for its efficiency but also for its quick warmup and restrike times, when contrasted with slow HID lighting. Increasingly, LED lighting is also being used as a replacement for high-bay and low-bay lighting.

COVE LIGHTING

Cove lighting (Figure 5.30) is almost always single-lamp fluorescent, laid end-to-end in the cove.

If access is difficult, reasonable assumptions are that if the ballast is magnetic, power consumption is 11 watts per linear foot of cove, and if the ballast is electronic, power consumption is 8 watts per linear foot of cove. If the cove lighting has two lamps side by side, double these estimates.

TRACK LIGHTING

Halogen is the most common type of track lighting usually with an "MR" designation, which stands for *multifaceted reflector*. Lamp designation is given in eighths of an inch across the lamp diameter. For example, MR 16 is 16/8'', or 2'', in diameter. The most common track lighting is 50-watt MR16 halogen, although lamps range from 20 to 75 watts. Fifty-watt halogen MR16 lamps are in the 12- to 15-lumen/watt efficacy range. If the wattage is not available from boxes of replacements, the wattage may be printed on the lamp itself. To remove a track lamp, turn off the light and let it cool. If a spring clip holds the lamp at the front, use needle-nose pliers to remove the clip. If there is a ribbed edge around the perimeter, turn counterclockwise to unscrew, sometimes just a half-turn. If recessed, remove the entire fixture, unplug, and remove the lamp from the back of the fixture. In the absence of any solid information on the lamp wattage, a reasonable presumption is 50 watts, the most common wattage.

The ANSI standard designations for MR16 lamps are:

- ESX: 20-watt, 10-degree beam (20MR16/10°).
- BAB: 20-watt, 35-degree beam (20MR16/40°).
- EXT: 50-watt, 15-degree beam (50MR16/15°).
- EXZ: 50-watt, 25-degree beam (50MR16/25°).
- EXN: 50-watt, 40-degree beam (50MR16/40°).
- FNV: 50-watt, 60-degree beam (50MR16/60°).
- FPA: 65-watt, 15-degree beam (65MR16/15°).
- FPC: 65-watt, 25-degree beam (65MR16/25°).

Figure 5.34 Decorative lighting primarily serves to light objects, rather than to light spaces.

Figure 5.35 Decorative lighting frequently includes many lamps for a single application.

Figure 5.36 Avoid reluctance to examine decorative lighting because of its perceived importance.

Figure 5.37 Standards are changing regarding how the importance of various decorative lighting is viewed.

Figure 5.38 Decorative lighting energy improvements are the same as for any lighting: high-efficiency, right-lighting, controls.

Figure 5.39 Exit lighting.

- FPB: 65-watt, 40-degree beam (65MR16/40°).
- EYF: 75-watt, 15-degree beam (75MR16/15°).
- EYJ: 75-watt, 25-degree beam (75MR16/25°).
- EYC: 75-watt, 40-degree beam (75MR16/40°).

LED replacements are available, at about 60 to 70 lumens/watt. See Figure 5.31.

DECORATIVE LIGHTING

Decorative lighting (Figure 5.32) covers a wide variety of lighting applications, in which the purpose is to highlight objects, rather than to light working spaces, recreation, way finding, and other human needs. Decorative lighting includes chandeliers, building exterior façade uplighting, retail store displays, and decorative views of building from outdoors at night. It can also include signage, religious lighting, nonessential supplemental lighting, artwork accenting, and art that incorporates lighting itself.

Decorative lighting can be indoors or outdoors. Decorative lighting is frequently inefficient incandescent or halogen lighting. Decorative lighting frequently includes many lamps for a single application.

In doing energy work, we must avoid any reluctance to examine decorative lighting because of its perceived specialty or importance. All lighting is candidate for reducing energy use. In fact, decorative lighting deserves added attention because, in many cases, it is nonessential. Our collective standards for the importance of decorative lighting are changing rapidly. The uplighting (Figure 5.33) of building facades is increasingly viewed as light pollution, instead of as desirable. The interior lighting of buildings at night, for decorative purposes only, is increasingly viewed as wasteful, rather than beautiful. The scalloped illumination of interior walls, for show only, is increasingly viewed as unnecessary. Decorative lighting is undergoing a revolution in perception, as appreciation for the beauty of the dark night sky grows, in parallel with concerns about wasteful electricity use. Decorative lighting is a next frontier in lighting energy reduction. See Figures 5.34 through 5.37.

Improvements for decorative lighting are identical with other lighting: replacement with high-efficiency lights, right-lighting, and controls (Figure 5.38). We might add another option to right-lighting, especially for decorative lighting, and that is eliminating a specific decorative light. For each decorative light, the first question we ask is, "Is this light really needed?"

EXIT LIGHTING

Exit lights (Figure 5.39) are on 24 hours a day. Existing exit lights are most commonly incandescent, with a smaller number of fluorescent lights. LED lighting has become popular and is more efficient than either incandescent or fluorescent, with lamps that last longer. Photoluminescent exit lighting is also available that does not use any power.

Most energy audits overestimate the existing energy use of exit lights. Commonly used reference sources overestimate the existing incandescent exit light energy use incorrectly as being in the 40-watt range.[28–30]

Two test-based sources indicate that incandescent exit lights are more accurately in the 12- to 16-watt range.[31, 32] We recommend assuming 14 watts for incandescent exit signs and 18 watts for fluorescent exit signs.

LED exit signs use varying amounts of power. Research has shown that LED fixtures use 4 watts on average, but can use as much as 15 watts. Recent versions of the energy code only require that they use less than 5 watts per face (10 watts total). We recommend specifying LED exit signs that use less than 4 watts. Caution should be exercised when using retrofit LED exit kits. Measurements have shown that these retrofit kits use widely varying power, as high as 23 watts, and averaging 6 watts.[31]

Existing incandescent or fluorescent exit lights are typically brighter where their lamps are located, and typically have two lamps, although one or both lamps are often blown. The lamps can be identified by opening the fixtures (the covers typically open easily or the lenses slide off), or by looking at the fixture from the bottom, where the lamps are often visible. See Figure 5.40.

To save the time and effort to stand on a ladder and open fixtures to identify the lamps and wattage, an indirect measurement was developed, by measuring the maximum surface temperature of the exit light fixture with an infrared thermometer. By pointing the thermometer at the brightest spot of the fixture, the power usage can be estimated as $0.4 \times DT + 2$, where DT is the temperature difference between the fixture and the surrounding air.[32] For example, if the maximum fixture surface temperature is 100 F, and the air temperature is 70 F, the power draw of the fixture is $0.4 \times (100 - 70) + 2 = 14$ watts.

THEATER LIGHTING

Theaters and other performing arts venues use much light for stage lighting (Figure 5.41). Typical lights are halogen or incandescent, and are in the 750- to 1,000-watt range. LED replacements are available, that reportedly provide not only significantly lower energy use but better control. Existing lamp wattage is printed not only on the lamps but on the fixtures. Spare fixtures are typically stored backstage, allowing convenient inspection.

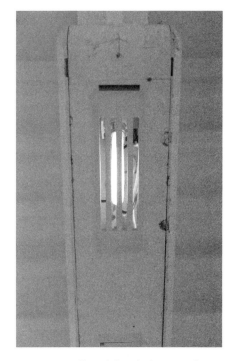

Figure 5.40   Examining the bottom of an exit light to identify the lamp type.

References

Figure 5.41   Theater lighting.

1. Illuminating Engineering Society (IES), New York, NY, *The Lighting Handbook*, 10th ed., 2011.
2. Waldemar Karwowski, *International Encyclopedia of Ergonomics and Human Factors*, 2nd ed., 2012, CRC Press, Boca Raton,FL.
3. http://www.acuitybrandslighting.com/sustainability/External_Light_Levels.htm#Car_Dealerships0. Accessed August 18, 2015.
4. C. Rubin, T. Ruscitti, and I. Shapiro, "Boosting Multifamily Energy Savings through Lighting Control Settings." *Home Energy*, August/September 2013.
5. Taitem Engineering, "Savings Opportunity—Corridor Lighting." *NYSERDA Multifamily Performance Program Technical Topic—Existing Buildings*, December 2008.
6. Visual simulation, Lithonia lighting A-2GTL4LP835 fixture, http://www.visual-3d.com/tools/interior/. Accessed July 18, 2015.
7. Taitem Engineering, NYSERDA Multifamily Performance Program Technical Topic—Existing Buildings, *Lighting Power Density*, October 2009.
8. Taitem Engineering, "Parking Garage Lighting." *NYSERDA Multifamily Performance Program Technical Topic—Existing Buildings*, September 2008.
9. Taitem Engineering, "Savings Opportunity—Stairwell Lighting." *NYSERDA Multifamily Performance Program Technical Topic—New Buildings*, March 2009.
10. http://www.lrc.rpi.edu/programs/nlpip/lightinganswers/mwmhl/restrikeTimes.asp. Accessed August 2, 2015.
11. http://www.lightingassociates.org/i/u/2127806/f/tech_sheets/High_Pressure_Sodium_Lamps.pdf. Accessed August 2, 2015.
12. http://smud.apogee.net/comsuite/content/ces/?utilid=smud&id=1176. Accessed August 2, 2015.
13. http://www.wisdompage.com/SEUhtmDOCS/3SE12.htm. Accessed August 2, 2015.
14. http://www.xcelenergy.com/staticfiles/xe/Marketing/Lighting-Wattage-Guide.pdf. Accessed August 2, 2015.
15. Navigant Consulting, Inc., "U.S. Lighting Market Characterization, Volume I: National Lighting Inventory and Energy Consumption Estimate." *Final Report, prepared for U.S. Department of Energy, Office of Energy Efficiency and Renewable Energy, Building Technologies Program*, September 2002, p. 41.

16. Heschong Mahone Group, "Lighting Efficiency Technology Report: Volume I. *California Baseline. Prepared for the California Energy Commission*, September 1999.

17. G. Newsham, C. Arsenault, J. Veitch, A. M. Tosco, and C. Duval, "Task Lighting Effects on Office Worker Satisfaction and Performance, and Energy Efficiency." NRCC-48152, National Research Council Canada, Institute for Research in Construction, 2005.

18. http://algonline.org/. Accessed July 18, 2015; http://www.lrc.rpi.edu/patternbook/. Accessed July 18, 2015.

19. Lighting Research Center, Rensselaer Polytechnic Institute, *"Specular Reflectors." Specifier Reports* 1, no. 3 (1992). National Lighting Product Information Program.

20. Daiva E. *Edgar, "Performance of Specular Reflectors Used for Lighting Enhancement."* U.S. Army Corps of Engineers, Construction Engineering Research Laboratories, 1994.

21. Dorene Maniccia, Allan Tweed, Bill Von Neida, and Andrew Bierman. "The Effects of Changing Occupancy Sensor Timeout Setting on Energy Savings, Lamp Cycling, and Maintenance Costs." *Lighting Research Center, Rensellaer Polytechnic Institute, and the U.S. EPA ENERGY STAR Buildings Program*, IES Paper #42, 2000.

22. Lighting Research Center, Rensselaer Polytechnic Institute, "Dimming Electronic Ballasts." *Specifier Reports* 7, no. 3 (1999), National Lighting Product Information Program.

23. J. Mardaljevic, L. Heschong, and E.S. Lee. *"Daylight Metrics and Energy Savings."* Berkeley, CA: Lawrence Berkeley National Laboratory, Lighting Research + Technology, 2009.

24. http://www.darksky.org/light-pollution-topics/lighting-crime-safety. Accessed July 7, 2015.

25. Lithonia Lighting, *"Motion Sensing Guide for Outdoor LED Luminaires."* Acuity Brands Lighting, Inc., 2012.

26. "Delamping and Reduced Wattage Lamps." Technology Information Sheet, Saskatchewan Energy Management Task Forces, TIS-L0002, 02/96.

27. http://www.naturallighting.com/cart/store.php?sc_page=50. Accessed April 1, 2015.

28. "Exit Signs: Energy Saving—Fact Sheet." North Carolina Energy Office, March 2010.

29. "Save Energy, Money, and Prevent Pollution with LED Exit Signs." Energy Star.

30. "Life Cycle Cost Estimate for 1 Exit Sign." Environmental Protection Agency, Department of Energy.

31. *Specifier Reports: Exit Signs*. National Lighting Product Information Program, March 1998.

32. Taitem Engineering, "Exit Lighting." Unreleased study, 2015.

Chapter 6
Heating

Heating systems may be classified by their fuel. The most common fuel is natural gas. Other common fossil fuels include oil (Figure 6.1) and propane. Electricity is used to power heat pumps, and for electric resistance heat. Emerging biomass fuels include wood pellets and wood chips.

Heating systems are also classified by their main *plant*, including boilers, furnaces, or heat pumps. And heating systems can be classified by their *distribution system*, including forced air, hot water, steam, and refrigerant.

The most common types of heating systems in commercial buildings are *hydronic* (boiler) systems, generally in larger buildings, and *furnaces* or heat pumps, generally in smaller buildings.

Hydronic systems heat water, and then the water, in turn, heats air in buildings. Furnaces directly heat air, as do heat pumps and electric resistance heat.

In evaluating heating systems, we keep in mind the full spectrum of energy improvements:

- Replace a heating system with a *high-efficiency* system.
- Change the heating system to a *different system* with a higher-efficiency fuel or approach.
- Change *components* of a heating system to improve efficiency.
- Modify or change the system *controls*.
- Improve the efficiency of the heating *distribution system*.

Figure 6.1 Heating fuel oil tank.

Hot Water Boilers

Hot water boilers are typically found in mechanical rooms, basements, or occasionally in rooftop penthouse rooms. Natural gas is the most common fuel for boilers, followed by fuel oil, on occasion propane in rural areas, and more rarely other sources, such as electricity. Three types of distribution systems are common: Perimeter radiators, hot water heat exchangers (frequently referred to as *coils*) in central air handlers, and coils in *fan coils* (as it sounds, a coil with a fan) that are distributed throughout a building.

Hot water boilers range from vast machines, over 10 feet tall, to small wall-mounted devices that can be located in closets.

Occasionally, boilers are used to heat domestic hot water, also sometimes called service hot water, in other words, consumable hot water for sinks, showers, kitchens, and the like. The boilers can either be dedicated to domestic hot water, or the domestic hot water can be integrated into a boiler system that also provides space heating. Domestic hot water is covered separately in this book.

HOT WATER BOILER RATINGS

Nameplate capacity ratings on boilers, provided in units of Btu/hour, typically include:

Input. This is the fuel consumption rate, sometimes referred to as the burner capacity.

Heating capacity (**also referred to as the** *gross capacity*). This is the rated output of the boiler as seen by the water entering and leaving the boiler.

Net rating (**also referred to as an** *IBR or I=B=R rating*). The net rating is the capacity of the boiler and distribution system as ostensibly seen by the delivered heat to the space, in other words, accounting for presumed piping losses between the boiler and the radiators or other delivery means. The net rating is obtained by dividing the gross capacity by 1.15. This presumption is sometimes referred to as representing 15 percent piping losses, but the presumed losses (gross capacity minus net capacity, divided by the gross capacity) are actually .15/1.15, or 13 percent (as a percentage of the gross capacity). As a function of the *input*, the presumed losses are 13 percent times the thermal efficiency. For example, for an 82 percent thermal efficiency, the presumed losses are $0.13 \times 0.82 = 10.7$ percent, when presumed as a percentage of the input. Again, the losses are presumed. They are a safety factor, to be used to compare to the radiation capacity, if the actual losses are not known. IBR or I=B=R refers to the old Institute of Boiler and Radiator Manufacturers.

From the nameplate, the rated *thermal efficiency* can be obtained, as the heating capacity divided by the input.

Hot water boilers (which do not actually boil water) are derivative of steam boilers (which do boil water), which will be discussed below. Some currently-manufactured boilers are still convertible from steam to hot water.

Residential boilers, below 300,000 Btu/hr input, are rated for efficiency with an *Annual Fuel Utilization Efficiency (AFUE)*. The current federal requirement is 82 percent AFUE for gas-fired boilers, which increased from 80 percent in 2012. These boilers are often referred to as residential boilers, even though they see use in many small commercial buildings, and in some larger commercial buildings when installed as a bank of multiple boilers.

The conditions for the AFUE test are 120 to 124 F entering water temperature, for noncondensing boilers, and 118 to 122 F for condensing boilers, with 140 F supply water temperature. Separately, the boilers are rated for thermal efficiency, with lower entering water temperatures, 35 to 80 F for noncondensing boilers, and 80 F for condensing boilers.

The relationship between AFUE, thermal efficiency, and combustion efficiency is shown schematically in Figure 6.2.[1]

In 1992, the federal requirement for boilers under 300,000 Btu/hr input was set at 80 percent AFUE.[2] Prior to this, the requirement was 70 percent. In 2001, well over 90 percent of newly installed boilers were in the 80 to 84 percent AFUE range, with an average of 82 percent, and only 6 percent were over 85 percent AFUE (met ENERGY STAR requirements). By 2007, over 34 percent were over 85 percent AFUE.[3]

The initial ENERGY STAR requirement for boilers under 300,000 Btu/hr was 85 percent when the program was introduced in June 1996. These requirements did not change until October 1, 2014, when the ENERGY STAR requirements increased to 90 percent AFUE for gas-fired boilers, and 87 percent for oil-fired boilers. As with other products, finding the ENERGY STAR label on a boiler in a building should not be assumed to represent current standards for high-efficiency. An old 85 percent ENERGY STAR labeled boiler found in the field may well be replaced with a higher-performance boiler and deliver measurable energy savings.

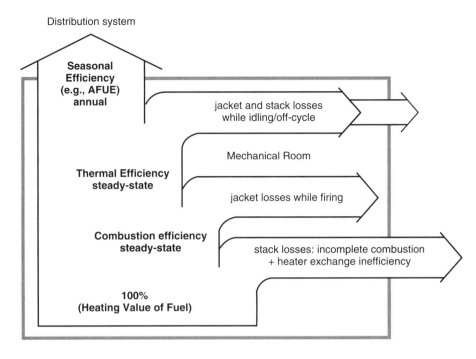

Figure 6.2 Relationship between boiler efficiencies.

TABLE 6.1

Estimates of Boiler AFUE (Smaller than 300,000 Btu/hr)

Vintage	Pre-1980	1980	1990	2000	2010
Efficiency (AFUE)	70%	75%	80%	82%	84%

In the absence of rated efficiency information, Table 6.1 provides typical efficiencies for boilers under 300,000 Btu/hr input.

The age of a boiler is frequently available from its serial number. For example, for Dunkirk boilers, the third and fourth characters are the year of manufacture. For Hydrotherm, digits 3 through 6 are the year of manufacture. For Marley, the two digits at the end of the serial number are the year of manufacture.[4] For Utica Boilers, from 2003 to the present, the year of manufacture is the first and third characters, for example UXC is 2003; from 1979 to 2002, the second character is the year of manufacture, starting with I in 1979, and prior to 1979 the third and fourth characters were the year of manufacture.[5]

A common mistake is to underestimate AFUE for older boilers. One web site provides an AFUE estimate of 60 to 70 percent for boilers that are 10 to 30 years old.[6] However, the average boiler AFUE 10 years ago was 82 percent, the minimum 20 years ago was 80 percent (by law), and the minimum 30 years ago was reportedly 70 percent, so a better estimate for boiler efficiencies 10 to 30 years old is probably 75 to 85 percent, far higher than 60 to 70 percent. The Department of Energy suggests that natural draft heating systems have AFUE's in the 56 percent to 70 percent[7] range—this is likely also too low.

Boilers between 300,000 and 2,500,000 Btu/hr input, sometimes referred to as small commercial boilers, are rated for efficiency with a thermal efficiency, with a current federal requirement of 80 percent for gas-fired hot water boilers, which went into effect March 2, 2012.[8]

Looking at historical values of efficiency, American Society of Heating, Refrigerating, and Air-Conditioning Engineers (ASHRAE) 90–75 set a minimum thermal efficiency of 75 percent for gas boilers, in 1975, and ASHRAE 90.1–1989 set a minimum combustion efficiency of 80 percent (presumed equivalent to a thermal efficiency of 77.5 percent), in 1989.[9] Ten years later ASHRAE Standard 90.1–1999 changed the requirement to 75 percent thermal efficiency. The federal Energy Policy Act (EPCA) set a requirement of 80 percent combustion efficiency in 1999.[10] The average thermal efficiency of small commercial boilers in 2005 was 80.9 percent, and the average combustion efficiency was 83.5 percent.[10] Accordingly, we estimate historic average boiler thermal efficiencies in Table 6.2.

However, the energy auditor does not need to rely on estimates for thermal efficiency, as most boiler nameplates show the input and output required to calculate the rated thermal efficiency.

Large boilers (over 2.5 million Btu/hr input) are energy-rated using *combustion efficiency*, with a current federal requirement of 82 percent. The thermal efficiency and combustion efficiency tests are run with 180 F supply water temperature, and with the water flow rate adjusted so that the entering water temperature is 80 F, for condensing boilers, and in the range of 35 to 80 F for noncondensing boilers. The thermal efficiency is based on the rate at which heat is delivered to the water, divided by the fuel input rate. The combustion efficiency is based on flue gas analysis, and so does not account for losses between combustion and heat delivered to the water.[11]

The relationship between the thermal efficiency and combustion efficiency has been estimated by the Department of Energy, as shown in Table 6.3.

Is there any simple relationship between AFUE and thermal efficiency? Knowing this would be helpful for situations where we want to compare a residential boiler rated in AFUE with a small commercial boiler rated in thermal efficiency, for example if an existing building has small commercial boilers, and are being considered for replacement with a bank of residential boilers. Recall that AFUE is based on entering water temperature of approximately 120 F, and thermal efficiency is based on entering water temperatures of 80 F or below. Examining 2015 ratings for over 1,000 boilers in the AHRI directory, the typical AFUE is almost identical to the thermal efficiency, for efficiencies below 90 percent. At efficiencies over 90 percent, there is greater variation between the two rated efficiencies, with the typical thermal efficiency dropping to as much as almost 4 percent below the AFUE. See Figure 6.3.

TABLE 6.2

Estimated Boiler Thermal Efficiencies (300,000 to 2,500,000 Btu/hr)

Vintage	Pre-1970	1980	1990	2000	2010
Thermal Efficiency	73%	75%	77%	79%	81%

TABLE 6.3

Estimated Difference between Combustion Efficiency (Ec) and Thermal Efficiency (Et).

Equipment Class	Difference between Et and Ec (%)
Small gas-fired hot water	1.5
Small gas-fired steam all except natural draft	2.0
Small gas-fired steam natural draft	3.5
Small oil-fired hot water	2.5
Small oil-fired steam	2.5
Large gas-fired steam all except natural draft	1.5
Large gas-fired steam natural draft	2.0
Large oil-fired hot water	2.5

Combustion efficiency is higher than thermal efficiency.[12]

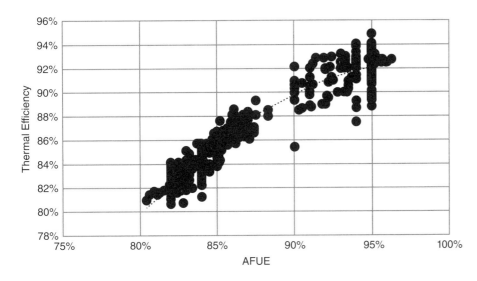

Figure 6.3   Boiler efficiency ratings.

HOT WATER BOILER ENERGY IMPROVEMENTS

High-Efficiency Boilers

The primary improvement for hot water boiler systems is to replace low-efficiency boilers with high-efficiency boilers, typically condensing boilers.

Condensing boilers (Figure 6.4) extract more heat from combustion gases than noncondensing boilers, by using added heat exchange surface, and by using lower entering water temperatures, so that moisture in the flue gas is cooled below its dew point, condenses, and releases its heat to the boiler water. Efficiency depends strongly on entering water temperature. Condensing boilers will not deliver full savings unless the entering water temperature is low.

Condensing boilers can vent into low temperature piping, such as plastic piping (Figure 6.5). This venting can be routed through side walls of a building, or through the roof.

The potential for energy savings with high-efficiency boilers is strong. For example, a 2014 study in Massachusetts found that 84 percent of the existing boilers in commercial and industrial buildings are still low-efficiency noncondensing boilers.[13]

In order to establish the baseline existing boiler efficiency, it is helpful to know not only the existing rated boiler efficiency, but also whether the existing control is varying the water temperature. Variation of the water temperature is also referred to as *outdoor reset* because the input to the control is the outdoor air temperature. With outdoor reset control, if the outdoor air temperature is cold, the supply water temperature is raised, typically to about 180 F. If the outdoor air temperature is mild, the supply water temperature is reduced. A typical system in which the boiler water temperature is *not* varied has a water temperature control, or *aquastat*, that is accessible, with a visible temperature set point, usually set at 180 F (Figure 6.6). As another way to distinguish systems that have outdoor reset control from ones that do not, if the water temperature *is* being varied with outdoor reset, and if the outdoor air temperature is anything other than its midwinter, coldest design day temperature, the supply water temperature will be less than the typical 180 F. This temperature is usually visible on a thermometer near the boiler.

Aquastats should not be confused with high-temperature limit safety controls, which often look similar, but are typically set at 220 F or higher, and can be visually distinguished by a reset button. See Figure 6.7.

Efficiency may also be measured in the field, most readily with a combustion flue gas test. This provides the combustion efficiency. It is important to also measure the return water temperature, so corrections can be made for return water temperature (see below).

Figure 6.4   Condensing wall-hung boiler.

Figure 6.5   Plastic vent piping (upper pipes are combustion exhaust, lower pipes are combustion air intake).

Figure 6.6 Boiler-mounted water temperature control, or aquastat, with a fixed supply water set point of 180 F.

Figure 6.7 Boiler-mounted water temperature high-limit safety switch.

Flue Gas Economizers

If a boiler is not being replaced, another option is to recover heat from the flue gases. This is done with a *flue gas economizer*, a heat exchanger installed in the flue gas stream.

Vent Dampers

Vent dampers (Figure 6.8) close when a boiler or furnace is not firing. Typically viewed as a heating improvement, vent dampers are primarily an infiltration reduction improvement.

One detailed research study of vent dampers found savings ranging from 4 percent to 23 percent in six tests over two years, averaging 12 percent, but with some concern about the statistical significance of the tests.[14]

Distribution

An improvement in hydronic systems is to increase radiation in a building, and further reduce boiler water temperatures below what might already be done with outdoor reset. Increasing radiation and reducing water temperature allows further reduction in piping losses, and increase in boiler thermodynamic efficiency. This improvement allows further gains for a building that might already have a condensing boiler, or that seeks to maximize gains from a new condensing boiler. Additional savings of 3 percent-5 percent are possible, for example, if a building has a condensing boiler with a design water supply temperature of 180 F (and outdoor reset down from 180 F), and it increases the radiation to allow a design water supply temperature of 125 F along with outdoor reset. Even further temperature reductions are possible if envelope improvements are also made (insulation, air sealing, and window and door improvements). Even further reductions in supply water temperature, and increases in efficiency, are possible if *radiant floor heating* is used, which frequently uses a low supply water temperatures in the 90 F range.

Increasing radiation can be done in many ways: length can be added to perimeter radiation, or, if length cannot be increased, higher-capacity radiation may be substituted for existing radiation, or fan coils can be replaced with higher capacity units. For example, three-tier commercial fin-tube radiation can deliver the same heat at 140 F as single-tier radiation at 180 F. See Table 6.4.

TABLE 6.4

Typical Fin Tube Radiation Capacity

Supply Water Temperature (F)	Residential 3/4″ tube	Commercial 1.25 copper tube, 4.5″ square aluminum fin, 40 fins/ft		
		Enclosure Height		
	9″	9-3/8″	15-3/8″	21-3/8″
	Heating Capacity (Btu/hour)			
215	1,081	1,660	2,680	3,200
200	930	1,428	2,305	2,752
180	746	1,145	1,849	2,208
160	573	880	1,420	1,696
140	433	664	1,072	1,280
120	281	432	697	832
100	162	249	402	480

Similar heat transfer increase can be made to allow temperature reductions to other hydronic delivery systems, including heat exchangers (also referred to as coils) in air handlers, cabinet heaters, and unit heaters. For example, cabinet radiators frequently have space available to install additional fin tube radiation. See Figure 6.9

A related improvement is to lower the water flow rate through a boiler. A lower flow rate will result in a lower return water temperature, delivering a higher combustion efficiency, along with possible pump power savings.[15]

Significant envelope improvements, such as insulation, air-sealing, or replacement windows, can also allow lowering heating system water temperatures.

Operation and Maintenance

The maintenance of high-efficiency boilers has been found to be important. If combustion air filters on condensing boilers are not periodically cleaned, incomplete combustion produces unburned gas, and much of the hard-earned energy savings are lost. Anecdotal reports indicate that this is a common occurrence.

Figure 6.8   Vent damper.

Steam Boilers

DISTINGUISHING STEAM FROM WATER SYSTEMS

Steam boilers can be visually distinguished from hot water boilers in a few ways. If all the radiators have air vents, it is most likely a steam system. If the pressure gage on the boiler reads below 5 psi, it is most likely a steam system. When the boiler is firing, if the temperature gage on the boiler reads over 200 F, it is more likely a steam system. A steam boiler typically has a water level indicator. A steam boiler will not have an expansion tank, which is common on hot water systems. See Figure 6.10.

STEAM SYSTEM INEFFICIENCIES

Steam boilers are regarded as an old technology, but they are still surprisingly prevalent in old buildings, and occasionally even see application in new buildings. Steam systems have been found to be unusually inefficient, resulting in high energy and water usage. There are many losses:

- Boiler and piping losses due to the intrinsically high temperatures
- Low combustion efficiency due to the high return water temperature
- Steam losses throughout the system, at:
  - Pipe joints.
  - Failed or malfunctioning air vents.
  - Condensate tanks.
  - Failed heat exchangers in the boilers.
- Failed steam traps
- Overheated buildings, or areas of buildings, due to system imbalance or other inadequate temperature control

Steam losses are particularly problematic because they are difficult to detect. A steam leak dissipates into the air in a building, or up the chimney if the leak is in a boiler room. This is unlike a water leak in a hot water system, which is typically detected immediately because the leak is visible. Steam losses represent not only an energy loss but also a water loss.

A study of heating use in multifamily buildings in New York State found that the overall energy use was 24 percent higher in steam-heated buildings than in buildings not heated with steam.[16] The study also found that water use in steam-heated buildings is far higher than water consumption in buildings that are not heated with steam.

Figure 6.9   Cabinet fin tube radiator (cover removed).

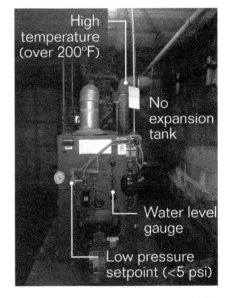
Figure 6.10   How to identify a steam boiler.

A separate study confirmed that steam-heated buildings have significantly higher heating energy use than buildings not heated with steam.[17] Multiple studies have found that converting steam systems to hydronic (hot water) reduces heating energy use fairly consistently by over 40 percent, and by as much as 50 percent.[16, 18–20]

HOW STEAM SYSTEMS WORK

Steam systems use a central boiler to boil water into steam. The steam expands rapidly through the system, displacing air in the supply pipes. The air is allowed to leave the system through air vents, typically located at radiators. Supplementary *master* vents are required, located at the ends of piping mains, in order to allow steam to rapidly reach all risers.

Steam systems are classified as either two-pipe or single-pipe. This can be identified at the radiation: A single pipe connected to a radiator means that it is a single-pipe system (Figure 6.11), and conversely two pipes connected to a radiator means that it is a two-pipe system (Figure 6.12).

In the radiation, the steam gives up its heat to the air, and condenses back to water phase. In single-pipe systems, the condensed water flows by gravity back down the same pipe through which the steam enters the radiator. In two-pipe systems, the water

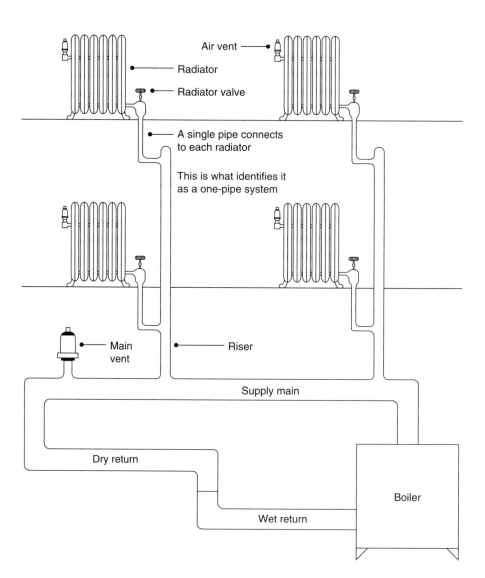

Figure 6.11 One-pipe steam system.

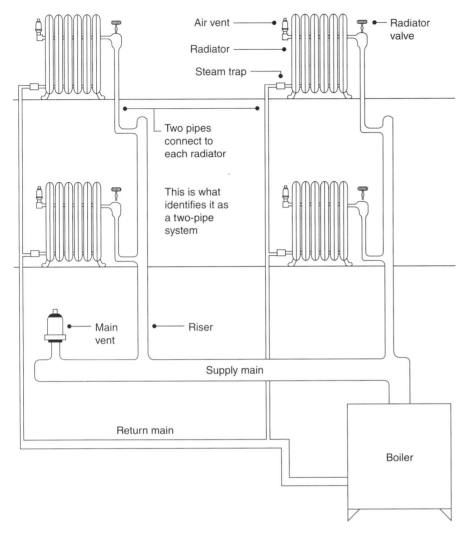

Air vent

Radiator valve

Radiator

Steam trap

Two pipes connect to each radiator

This is what identifies it as a two-pipe system

Main vent

Riser

Supply main

Return main

Boiler

Figure 6.12   Two-pipe steam system.

flows back to the boiler through a steam trap at each radiator. The steam trap serves to prevent steam from flowing back to the boiler, only allowing water to flow.

Steam is also used in heat exchangers other than radiators, such as in coils in central air handlers, and in fan coils.

A particular type of steam system that operates at low pressures and low temperatures is called *vacuum steam*. Vacuum steam systems do not leak because they operate at a vacuum. Their lower temperatures also result in lower pipe losses. For these reasons, vacuum steam systems are more efficient than pressurized steam systems. Vacuum steam systems are a small minority of existing steam systems.

Steam systems sometimes heat hot water in steam-to-water heat exchangers, and the hot water is then used in a hydronic distribution system. In other words, the steam stops in the boiler room and is not distributed throughout the building, but, rather, hot water is distributed. These heat exchangers are referred to as *generators*. See Figure 6.13.

Steam systems sometimes heat domestic hot water (DHW) as well. The year-round requirement for domestic hot water means that a central large steam boiler stays hot year-round, wasting even more energy.

Figure 6.13   Hot water generator (steam-to-water heat exchanger).

STEAM SYSTEM ENERGY IMPROVEMENTS

Steam systems are readily converted to hot water. Successful system conversions have consistently reported retaining the existing radiators.[16, 18, 19] For cast iron radiators,

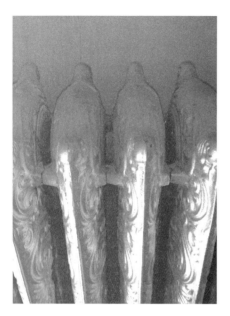

Figure 6.14 Cast iron steam radiator without pipe connection across the top—cannot be converted to hot water.

Figure 6.15 Cast iron steam radiator, with pipe connection across the top—can be converted to hot water.

a conversion requires that the radiators are connected across both the top and bottom of the radiator sections, in order to allow effective parallel flow through the radiator. See Figures 6.14 and 6.15.

Converted systems need to be pressure tested, in order to ensure that any leaks are repaired prior to use with water.

Another option is to convert steam systems to heat pump systems. For example, ductless heat pump fan coils are generally smaller than radiators of the same capacity. Advantages of this approach include the elimination of boiler room losses, freeing up the boiler room space, providing cooling if desired, and converting to an energy source that can be supplied through renewable energy such as photovoltaics.[21]

Individual components within a steam system can also be improved. Steam traps can be evaluated for failures and replaced. Some work has been done to replace the function of traditional steam traps with orifice plates. Uninsulated piping and fittings can be insulated. Failed air vents can be replaced. Leaks can be identified and repaired. In order to eliminate the different and substantial losses, a wide set of improvements are needed (insulation, trap replacement, air vent replacement, leak repair, etc.), that may approach or exceed the cost of simply replacing the steam system with a hydronic system.

Estimates of the individual losses of steam heating systems include[21]:

Pipe losses: 13 percent.

Steam traps: 15 percent.

Building overheating: 8 percent.

Combustion efficiency (relative to hot water systems): 2 percent.

Steam leaks (other than steam trap losses): 12 percent.

Furnaces

FURNACE TYPES

Furnaces, like boilers, generate heat through combustion. However, whereas boilers heat water that in turn heats air in buildings, furnaces instead heat air directly.

Furnaces are typically used with ductwork, referred to as a forced-air system. However, stand-alone furnaces are also available, for example, as suspended heaters or as floor-mounted or wall-mounted room heaters (Figure 6.16).

Furnaces are typically vented to the outdoors. However, although not common, furnaces are available that vent into the indoor air. These are called *unvented* or *direct-fired* furnaces. This is undesirable for indoor air quality and humidity. Unvented furnaces range from room-sized heaters to large rooftop units.

Furnaces typically come with a fan that moves air that is heated. If the fan is upstream of the heat source, it is referred to as *blow-through;* if the fan is downstream, it is called *draw-through.* If the furnace is packaged separate from the system fan, it is called a *duct furnace.* Old furnaces are available without a fan, called *gravity furnaces* or *octopus furnaces,* so named because of the multiple branch ducts leading away from a central furnace in a basement.

Furnaces can be classified by the orientation of airflow. Most common is *upflow,* but *downflow* and *horizontal flow* are also available.

Residential furnaces are common in small commercial buildings. The size range is typically 40,000 – 120,000 Btu/hr input, although formally they are classified as below 225,000 Btu/hr input. Sometimes, two furnaces are operated together in tandem, in an approach called *twinning.* Commercial furnaces are defined as furnaces over 225,000 Btu/hr input.

Furnaces are extremely common in packaged rooftop heating and cooling units. These will be addressed separately in the chapter on cooling and integrated heating/cooling systems.

Low-efficiency furnaces have high-temperature flue gases, and so must vent into a brick or lined chimney, or into metal venting. High-efficiency furnaces deliver low-temperature flue gases, and so can vent into plastic piping. They can be identified by the plastic vent piping, typically white, which is often adjacent to a separate plastic pipe that is used to bring in combustion air. Medium-efficiency and high-efficiency furnaces use a small fan called an *inducer* for the combustion air stream. See Figure 6.17. Low-efficiency furnaces, without an inducer, are referred to as *natural-draft furnaces*. See Figure 6.17.

Furnaces draw combustion air either from the space surrounding the furnace or from outdoors. Combustion air from outdoors can either be supplied near the furnace or can be routed directly into the furnace, in which case it is referred to as *sealed combustion*. Sealed combustion is typical of high-efficiency furnaces.

Furnaces can be located in mechanical rooms, basements, attics, closets, on rooftops, on walls, or directly within the spaces they serve.

Furnace systems sometimes draw in air for general building ventilation, and sometimes do not. Ventilation air is outdoor air brought into a building to improve indoor air quality, as distinguished from combustion air, which is air brought to a furnace for purposes of combustion with the furnace fuel. Ventilation air, when present, is drawn in at the return air side of a furnace, where the air pressure is low. If a furnace system *only* draws in ventilation air, and does not recirculate any indoor air, it is referred to as a *makeup air system* or *100 percent outdoor air* system (Figure 6.18).

FURNACE EFFICIENCIES

Commercial warm air furnaces in the United States have been regulated by the U.S. Department of Energy (DOE) since 1994. The DOE's regulations cover furnaces with an input of 225,000 Btu/hr or more. Since 1994, the requirements have not changed, and remain 80 percent minimum thermal efficiency for gas-fired commercial warm air furnaces, and 81 percent for oil-fired furnaces. The thermal efficiency is the rate of heat delivered to the air stream divided by the rate of fuel input. The test is a steady-state test and does not account for cycling losses. The thermal efficiency also does not account for electrical energy use, for example, for the blower motor, inducer motor, or controls. The efficiency requirements for commercial warm air furnaces are expected to increase in 2019.[22] All commercial warm air furnaces currently certified fall in the range of 80 percent to 82.2 percent thermal efficiency.

As with residential boilers, the efficiency of residential furnaces is provided as an annual fuel utilization efficiency (AFUE), which accounts for cycling losses. The AFUE does not account for electrical energy use, for example, for the blower motor, inducer motor, or controls. The AFUE test is based on entering air temperature between 65 and 100 F (between 65 and 85 F for condensing furnaces), and an air temperature rise within 15 F of the maximum air temperature rise.

Figure 6.16 Wall-mounted vented gas room heater. It is identifiable as gas fired by the gas pipe on the outside wall, and as vented by the vent on the outside wall.

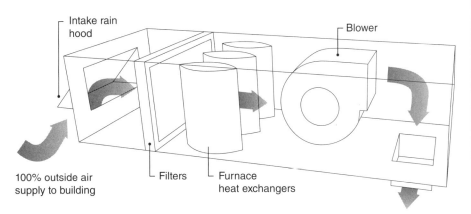

Figure 6.18 Gas-fired makeup air unit.

Intake rain hood

Blower

100% outside air supply to building

Filters

Furnace heat exchangers

Inducer, which is an indication that the furnace is either medium-efficiency or high-efficiency

White PVC vent pipe confirms that the furnace is high-efficiency

Figure 6.17 Identifying the type of furnace.

The federal requirement for minimum AFUE went into effect in 1992, at 78 percent. It was raised in 2015 to 80 percent.

The effect of varying entering air temperatures on the efficiency of furnaces would be good to know, in order to make corrections for furnaces located in cold or warm spaces, for example, furnaces used to maintain industrial spaces at lower or higher temperatures than 70 F, or for furnaces operating during unoccupied-mode setback, or for furnaces that are subject to warm supply air bypass to the return airstream. Unfortunately, these data are not published and have not been found in research literature. We know that boiler efficiency varies strongly with varying entering water temperatures. If the variation in gas furnace efficiency as a function of entering air temperature is similar to the variation in boiler efficiency with entering water temperature, we might expect a 1 percent increase in efficiency for each 10 F decrease in entering air temperature, and vice versa.

The actual annual efficiency of furnaces has been found to not degrade over time, and so the rated AFUE of an existing furnace is a good basis for baseline energy calculations.[23]

A study of residential-size furnaces found that seven of nine had airflow rates that were lower than recommended by the manufacturer (air temperature rise higher than recommended maximum). The resulting 6.4 percent average efficiency penalty was attributed primarily to this lack of airflow. With an average airflow shortfall of 33 percent in the study, we might estimate a 2 percent efficiency penalty for every 10 percent airflow shortfall. This is significant enough to warrant evaluation of airflow adequacy in the field. Because it is hard to measure airflow, an indirect measurement is recommended, using the *air temperature rise* (supply air temperature minus return air temperature). Measure the air temperature rise, and estimate the airflow as:

Airflow (CFM) = (Rated output in Btu/hr) / (Temperature rise × 1.08)

Compare this to the rated airflow, nominally the rated output divided by the maximum temperature rise (listed on the nameplate) minus 15 F (the target temperature rise for the AFUE test).

Airflow might be inadequate for several reasons, including a dirty air filter, a high-efficiency filter, fan speed set too low, or undersized ductwork. If the ductwork is undersized, the ductwork does not necessarily need to be replaced; the furnace may be replaced with a smaller furnace (if oversized, which is common, or if separate envelope load reduction is being done), or the furnace may be replaced with one that has a variable speed motor.

It should be noted that high-efficiency air filters are not a likely cause of low airflow. High efficiency air filters have been found to have little impact on energy use.[24]

A rule-of-thumb relationship between AFUE and thermal efficiency can be helpful, for example if seeking to compare an existing large commercial furnace (rated in thermal efficiency) for energy savings by replacing it with multiple smaller high-efficiency residential furnaces (rated in AFUE). Examining a sample of furnaces for which both AFUE and thermal efficiency are available, the average difference is 1.4 percent. In other words, to convert a thermal efficiency to an AFUE, subtract 1.4 percent from the thermal efficiency.

As with boilers, assumed efficiencies for furnaces tend to be lower than they are in reality, which means that savings are generally over-predicted. One source reports that furnaces 20 to 40 years old have AFUE's that are 65 to 70 percent, another source reports 68 percent in the mid-1980s.[25, 26] However, the federal minimum for AFUE has been 78 percent for over 20 years, several of the larger states had AFUE requirements over 70 percent before 1992, and industry statistics indicate an average AFUE of 74 percent in 1985.[27] And, again, AFUE has been found to not deteriorate over time. Industry guides are also available that list furnace makes, models, and rated efficiencies for old furnaces.[28]

Furnace efficiency may also be measured in the field. A combustion efficiency measurement is more readily taken than a thermal efficiency measurement. A combustion efficiency measurement requires a combustion analyzer, and the measurement is taken in the flue, before the draft diverter. A thermal efficiency measurement requires measuring airflow, temperature rise, and fuel input rate. Of these measurements, measuring airflow accurately is difficult in the field, and virtually rules out an accurate efficiency measurement.

In the absence of rated efficiency data or field measurements, existing furnace efficiencies (AFUE) may be estimated based on the furnace type: natural draft, standing pilot—70 percent; natural draft, ignition other than standing pilot—75 percent; induced draft—80 percent; condensing—90 percent.

ENERGY IMPROVEMENTS

A primary improvement for furnaces is to replace a low-efficiency furnace with a high-efficiency one. Residential gas furnaces originally qualified for ENERGY STAR in 1995. The initial requirements were for a minimum AFUE of 85 percent.[29] In 2006, the ENERGY STAR requirement increased to 90 percent for gas furnaces. As of 2013, ENERGY STAR requirements are 95 percent AFUE for gas furnaces in the northern United States (including and north of Oregon, Idaho, Utah, Colorado, Kansas, Missouri, Illinois, Indiana, Ohio, West Virginia, Pennsylvania, and New Jersey), and 90 percent for the southern United States. Additional requirements are 2 percent maximum furnace fan power (as a ratio of fan electrical consumption to total furnace energy consumption), and 2 percent maximum air leakage of the furnace cabinet. There is no ENERGY STAR label/requirement for commercial furnaces (over 225,000 Btu/hr input).

Models are available that are more efficient than the ENERGY STAR requirements, not only more efficient in combustion, but also lower in fan power, due to high-efficiency variable-speed motors. Variable-speed furnace motors go by the generic term *brushless permanent magnet motors* (BPMs) or brushless DC electric motors, or by the trade name *electronically commutated motors* (ECMs). Variable-speed motors have been available for residential furnaces for over 25 years and are seeing increased use. Variable speed motors have started seeing use in commercial furnaces, such as packaged rooftop systems. This is a positive trend.

It is important to note that while converting from a non-condensing furnace with a conventional permanent split capacitor (PSC) motor to a condensing furnace with a BPM motor will save electricity, research has shown that converting to a condensing furnace with a PSC motor can risk *increasing* electricity use, even while gas use is decreased.[30] This is likely because condensing furnaces use higher airflow, and have greater flow resistance due to additional heat exchanger surface. For purposes of energy estimating, it has been found that non-condensing furnaces with PSC motors use 6 kWh/year per MMBtu of delivered heat, condensing furnaces with PSC motors use 9 kWh/year per MMBtu, and condensing furnaces with BPM motors use 5 kWh/year per MMBtu.

When specifying high-efficiency furnaces, other important aspects include:

- Specify the required efficiency, rather than generally specifying "condensing" or "minimum 90 percent efficiency," which does not capture the full possible savings available. Many furnaces are available that are over 95 percent efficient. An 80 percent efficient furnace replaced with a 90 percent efficient furnace delivers savings of 11 percent. An 80 percent efficient furnace replaced with a 95 percent efficient furnace delivers savings of 16 percent.

- Specify variable speed motors. As previously indicated, a condensing furnace that uses a traditional PSC motor may well result in *increased* electricity use.

Figure 6.19 Gas-fired suspended unit heater.

Figure 6.20 Poorly sealed filter housing.

■ Require that the vent or chimney previously used for the removed furnace be well sealed, including caulking of the perimeter and capping and sealing the top, in addition to the bottom, to minimize infiltration and conduction losses. Require that existing combustion air openings be sealed, too.

■ Encourage building owners to use setback temperature control. Lower indoor air temperatures not only reduce envelope losses in heating, but also increase the combustion efficiency of furnaces.

ENERGY STAR provides a spreadsheet for calculating energy savings of high-efficiency furnaces.

Zoning of furnace systems is possible on a retrofit basis, using zone dampers. These systems can deliver improved comfort and some energy savings due to zone control. However, savings are offset by losses due to a common approach to use air bypass to relieve air pressure as dampers close. For energy efficiency, zoning is not advisable for constant volume systems using air bypass and should be limited to systems with modulating capacity and variable-speed blowers, in other words, systems without air bypass.

Thinking out of the box, one need not simply replace a low-efficiency furnace with a higher-efficiency furnace of the same type. For example, a low-efficiency commercial furnace might be replaced with two high-efficiency condensing residential furnaces with variable speed motors. Recall that the highest-available thermal efficiency for commercial warm air furnaces is currently only 82.2 percent, whereas residential furnace efficiencies are available well over 90 percent. Likewise, a low-efficiency suspended unit heater (Figure 6.19) might be replaced with a condensing furnace.

Improvements other than high-efficiency furnaces include:

■ Increasing airflow, as described earlier.

■ Switching fuels. For example, gas furnaces have lower carbon emissions than oil furnaces. Converting to heat pumps is another option, and one that supports a net-zero goal, if a renewable form of electricity such as photovoltaics is being considered.

■ Control-related improvements. These will be addressed in a subsequent discussion on controls.

■ Where possible, one might consider relocating furnaces from unconditioned spaces to within the thermal boundary, to reduce or eliminate losses to unheated spaces, which can be as high as 10 to 20 percent or more. A variety of improvements relating to distribution and ducting will be addressed in a separate discussion, below. However, a solution that goes beyond trying to reduce distribution losses is, instead, to eliminate these losses entirely, by relocating furnaces within the thermal boundary.

■ Tuning combustion. If a furnace is not being replaced, combustion analysis should be performed, and tuning undertaken, to maximize efficiency.

A variety of health and safety deficiencies are often identified during energy work, and should be remedied. Conditions include inadequate air for combustion, backdrafting due to low air pressure in the combustion zone (common if return air is open to the space, from return ductwork, a poorly sealed filter housing (Figure 6.20), or poorly sealed furnace fan section), blocked venting, and poor combustion.

Electric Resistance Heat

Electric resistance heat is common and available in a wide variety of types: electric baseboard radiators, unit heaters (Figure 6.21), cabinet heaters, electric coils in air handlers or fan coils, electric radiant panels, stand-alone plug-in heaters, and more.

Electric resistance heat is generally not desirable, because it is a highly inefficient form of heating, in terms of the source of electricity typically used, from the electric grid. It is also high in cost. With most electricity still coming from power plants, and most power plants still fired by fossil fuels, electricity used for electric resistance heat typically represents approximately three units of energy consumed for each unit of energy used in a building for heat.

The most common type of electric resistance heat is electric baseboard (Figure 6.22). Similar in appearance to hot water baseboard radiation, electric baseboard heaters can be visually distinguished in several ways: The electric element in the middle of the fin is smaller in diameter than a typical hot water pipe, the electric element is typically silver in color rather than the copper that is typical of hot water pipe, no pipes enter or leave the radiator, and the most common thermostats for electric baseboard radiators are either mounted on the unit or are rectangular wall-mounted line-voltage thermostats. Electric resistance heaters typically deliver 250 watts of heat per linear foot.

Of particular and unusual note are the large number of stand-alone plug-in heaters. The annual sales volume of such heaters in the United States is believed to be over 10 million units. We find such heaters everywhere. These heaters are usually intended to solve comfort control problems: the cold office, the cold kitchenette, the cold entry. Where we rarely find these heaters is in spaces that have their own temperature control, or any spaces with fan coils, or dedicated separate heating systems. Significant use of supplemental electric resistance heat can be seen on utility bills for buildings that are not otherwise electrically heated, as an increase in wintertime electricity usage. The problem is exacerbated because the temperature control provided by electric resistance heaters is itself not refined. The thermostats are not calibrated, and so the typical electric resistance heater is overheating its space. There is also no setback control. We need to actively seek out and eliminate supplemental electric resistance heat. The problem can sometimes be solved by rebalancing hydronic or forced air systems.

Electric resistance heat is best replaced with higher-efficiency forms of heat. Heat pumps make sense in many situations, to take advantage of the existing electricity supplied to the electric resistance heat. In some cases, fossil fuels such as natural gas may lower energy consumption, energy costs, and carbon emissions. For plug-in heaters that have been added to solve comfort problems with a poorly controlled central heating system, a solution is to add active temperature control to the spaces in question, to shift the energy use to a high-efficiency heating system, preferably with programmable control.

Specialty Heating Systems

Air doors (also called *air curtains;* Figure 6.23) provide heat to a sheet of blown air, supplied overhead, at entry doors. The intent of air doors is to reduce infiltration, and the heat is intended to provide comfort. Most air doors use electric resistance heat, whereas some use gas heat.

Infrared heaters deliver heat through the heat transfer mechanism of radiation, in other words, by glowing. They are used to deliver local comfort in industrial spaces and vehicle service spaces, and other colder spaces, such as parking garages. They are occasionally even used to deliver heat outdoors, for example, outside building entrances (Figure 6.24) or at outdoor seating in restaurants in colder weather.

Infrared heaters are available either as gas fired or electric. Gas-fired infrared heaters are typically tube-shaped, routed below the ceiling around the perimeter of industrial buildings and workshops. Electric radiant heaters are more typically rectangular, and serve smaller areas (Figure 6.25).

Indoor use of infrared heaters can conserve energy, as they allow a space to operate at lower temperatures, reducing heating loads. The outdoor use of heat is

Figure 6.21   Electric unit heater.

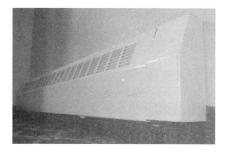

Figure 6.22   Electric baseboard resistance heater.
Note the absence of pipes below the heater, an indication that it is electric heat and not hot water.

Figure 6.23   Air door/curtain.

Figure 6.24 Infrared heater used outdoors.

Figure 6.25 Electric infrared heater.

not viewed as being conserving of energy, and should be evaluated for elimination whereever possible.

Destratification

Ceiling fans have been used for *destratification* to counter the vertical variation of temperatures in a space, in an attempt to move warm air from ceiling level down to floor level. By reducing the air temperature below the ceiling, and supplying heat from below the ceiling to the occupied floor level, destratification ostensibly reduces heat loss and raises the temperature of occupied areas in order to reduce the need for heating.

One older study found that ceiling fans were not effective for destratification.[31] A second study produced somewhat inconclusive results, with savings at one building but no savings at another building.[32] A third study by a manufacturer of ceiling fans found savings of 26.4 percent (19,542 dekatherm) in reduced gas use, in a 29.5' high warehouse, offset by an increased 1,068 kWh in the destratification fan motor use, over a three-month period.[33] The same author offers a rule of thumb of 3 percent winter heating savings for each meter (3.3'), just under 1 percent savings per foot of ceiling height, if destratification is provided.[34]

References

1. Derived from "Boiler Efficiency Definitions." NYSERDA Multifamily Performance Program Technical Topic, April 17, 2008, p. 1.
2. NYS Energy Conservation Construction Code, 1991, p. 29.
3. Science Applications International Corporation, "Development of Issues Papers for GHG Reduction Project Types: Boiler Efficiency Projects." *Prepared for California Climate Action Registry*, January 7, 2009, p. 26.
4. From http://www.nachi.org/forum/f20/weil-mcclain-boiler-ages-829/. Accessed December 20, 2014.
5. From http://inspectbeforebuying.com/hrc/serialnumbers.asp. Accessed December 20, 2014.
6. http://naturalgasefficiency.org/residential/heat-Gas_Boiler.htm. Accessed March 18, 2015.
7. http://energy.gov/energysaver/articles/furnaces-and-boilers. Accessed March 18, 2015.
8. http://www1.eere.energy.gov/buildings/appliance_standards/product.aspx/productid/74. Accessed March 20, 2015.
9. M.A. Halverson and W. Wang, "Simulation Analyses in Support of DOE's Fossil Fuel Rule for Single Component Equipment and Lighting Replacements." Prepared for the U.S. Department of Energy under Contract DE-AC05–76RL01830, PNNL-22887, October 2013, p. 6.
10. U.S. Department of Energy, Office of Energy Efficiency and Renewable Energy Building Technologies Program. Appliances and Commercial Equipment Standards. Technical Support Document (TSD): "Energy Efficiency Program for Commercial and Industrial Equipment. *Efficiency Standards for Commercial Heating, Air Conditioning, and Water Heating Equipment*. Including: Packaged Terminal Air Conditioners and Packaged Terminal Heat Pumps Small Commercial Packaged Boilers Three Phase Air Conditioners and Heat Pumps." March 2, 2006, p. 14.
11. BTS-2000, Testing Standard, Method to Determine Efficiency of Commercial Space Heating Boilers, Hydronics Institute Division of AHRI, Air Conditioning, Heating, and Refrigeration Institute, 2007.
12. http://www1.eere.energy.gov/buildings/appliance_standards/commercial/pdfs/ch_5_ashrae_nopr_tsd.pdf, pp. 5–19 Accessed March 20, 2015.

13. Massachusetts PAs and EEAC Consultants, *"Massachusetts Commercial and Industrial Evaluation. Massachusetts Boiler Market Characterization Study."* DNV GL, July 17, 2014.

14. Martha J. Hewett, Timothy S. Dunsworth, Michael J. Koehler, and Helen L. Emslander, *"Measured Energy Savings from Vent Dampers, in Low Rise Apartment Buildings."* Minneapolis Energy Office, ACEEE Proceedings, 1986, p. 41.

15. http://energy.gov/sites/prod/files/2013/12/f6/condensing_boilers.pdf. Accessed March 20, 2015.

16. I. Shapiro, "Water and Energy Use in Steam-Heated Buildings." *ASHRAE Journal*, May 2010.

17. E. Guerra and R. Leigh, *"Tales from the AMP Database: Energy Consumption in a Selection of New York State Multifamily Buildings."* Presented at the NYSERDA Multifamily Building Conference, New York City, 2006.

18. Mary Sue Lobenstein and Martha J. Hewett, "Converting Steam Heated Buildings to Hot Water Heat: Practices." *Proceedings of the American Council for an Energy Efficient Economy*, 1986 Summer Study, Vol. 1, pp. 183–196, Washington, DC.

19. Personal correspondence with Michael Wellen, Community School of Music and Art, Ithaca, NY, March 7, 2013. Savings of a steam-to-hydronic conversion were 40 percent and 44 percent, relative to two different baseline years.

20. Mary Sue Lobenstein and Timothy S. Dunsworth, "Converting Two Pipe Steam Heated Buildings to Hot Water Heat: Measured Savings and Field Experience." Minneapolis Energy Office, Report MEO/TR89–4-MF, September 1989.

21. I. Shapiro, "Blowing Off Steam." Final Report for the Syracuse Center of Excellence, 2015.

22. http://www.appliance-standards.org/blog/doe-proposes-first-update-commercial-furnace-standards-20-years. Accessed April 5, 2015.

23. L. Brand, S. Yee, and J. Baker, "Improving Gas Furnace Performance: A Field and Laboratory Study at End of Life." *Prepared for the National Renewable Energy Laboratory, on behalf of the U.S. Department of Energy's Building America Program, Office of Energy Efficiency and Renewable Energy, Golden, CO, NREL* Contract No. DE-AC36–08GO28308, Prepared by Partnership for Advanced Residential Retrofit, Gas Technology Institute, Des Plaines IL, February 2015.

24. Iain S. Walker, Darryl J. Dickerhoff, David Faulkner, and William J. N. Turner, "System Effects of High Efficiency Filters in Homes." *Environmental Energy Technologies Division, Ernest Orlando Lawrence Berkeley National Laboratory*, March 2013. Submitted for presentation at ASHRAE Annual Conference, June 2013.

25. http://naturalgasefficiency.org/residential/heat-Gas_Furnace.htm. Accessed April 4, 2015.

26. https://www.energystar.gov/products/certified-products/detail/furnaces. Savings Calculator, accessed April 4, 2015.

27. http://buildingsdatabook.eren.doe.gov/TableView.aspx?table=5.3.2. Accessed August 4, 2015.

28. http://www.prestonguide.com/. Accessed April 4, 2015.

29. http://life.gaiam.com/article/7-energy-saving-appliances. Accessed April 4, 2015.

30. "Gas Furnace Electricity Usage." NYSERDA Multifamily Performance Program, April 15, 2013.

31. J. Ashley, *"Hangar Destratification Investigation."* Naval Civil Engineering Laboratory, April 1984.

32. Joel C. Hughes, "Technology Evaluation of Thermal Destratifiers and Other Ventilation Technologies." Naval Facilities Engineering Service Center, undated.

33. Richard Aynsley, "Saving Heating Costs in Warehouses." *ASHRAE Journal*, December 2005.

34. Richard Aynsley, *"Circulating Fans for Summer and Winter Comfort and Indoor Energy Efficiency."* *Environment Design Guide*, Australian Institute of Architects, November 2007.

Chapter 7

Cooling and Integrated Heating/Cooling Systems

Air conditioning is primarily the process of cooling air. *Dehumidification* is a secondary but important process that occurs when air is cooled below its *dew point temperature*, by definition the temperature at which moisture condenses out of an airstream.

Cooling equipment is rated in Btu/hr, representing the sum of the capacity to reduce temperature (*sensible cooling*) and the capacity to remove moisture (*latent cooling*). Another metric for capacity is *tons of cooling*, where one ton of cooling is equal to 12,000 Btu/hr.

We introduce various cooling systems in this chapter but also discuss integrated heating/cooling systems, because whereas heating systems often stand alone, cooling systems typically provide options for heating with one of the heating systems already described.

As with heating, in evaluating buildings, we keep in mind the full spectrum of possible cooling energy improvements:

- Replace a cooling system with a *high-efficiency* system.
- Change the cooling system to a *different system* with a higher-efficiency approach.
- Change *components* of a cooling system to improve efficiency.
- Modify or change the system *controls*.
- Improve the efficiency of the cooling *distribution system*.

Cooling System Types

Whereas heating systems seem to be classified fairly simply into boilers, furnaces, and a few specialty systems, the proliferation of cooling systems is often more confusing to new energy auditors and energy managers. Some of the challenge derives from confusing nomenclature, where terms like *split systems* might be classified as a type of "packaged unit," alongside "single-package" systems, which are also a type of packaged unit, but which are different than split systems. We will attempt to bring some clarity to these definitions.

A helpful parallel is that *chillers* are like boilers: Chillers first cool water, and then the water cools air, much like boilers first heat water, and then the water heats air. The function of chillers may perhaps be remembered by referring to them as *liquid chillers*. Similarly, a broad class of air conditioners known as *direct expansion* (or DX) systems are like furnaces: DX systems cool air directly, without using water as an intermediate medium, just as furnaces heat air directly without using water. The function of DX systems may perhaps be remembered by recalling that *direct expansion* systems *directly* cool air.

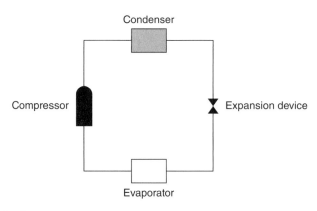

Figure 7.1 Cooling system.

A typical cooling system has four main components: a *compressor*, two heat exchangers (*condenser* and *evaporator*), and a small *expansion device*. See Figure 7.1.

We refer to these as refrigeration components because they are the same main components as are used in refrigeration. The compressor compresses refrigerant gas from a low pressure to a high pressure, at which it is at a high temperature. In the condenser, because it is at a high temperature, the refrigerant is able to give up its heat, typically to the outdoor air, and the refrigerant condenses to a liquid, still at a high pressure. The high-pressure liquid flows through the expansion device, which is simply a flow restriction (either a small copper tube called a *capillary tube*, or a modulating mechanical valve called a *thermostatic expansion valve*, or TXV, or a modulating *electronic expansion valve*, or EXV), and becomes a very low-pressure liquid. At a low pressure, it is also cold. In the evaporator, the cold liquid refrigerant absorbs heat, from the indoors, which is the main sought cooling effect provided by the air conditioner. Evaporated into gas phase, the refrigerant returns to the compressor where the cycle continues.

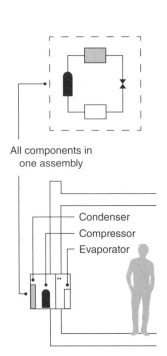

All components in
one assembly

Condenser
Compressor
Evaporator

Figure 7.2 Room air conditioner.

Room Air Conditioners

Room air conditioners comprise *window air conditioners* and *through-wall air conditioners*. Room air conditioners have all four refrigeration components in one unit, as shown in Figure 7.2.

Room air conditioners are a type of DX system. We note that while DX systems are sometimes used to only refer to split systems, they more broadly can be used to describe any cooling system that directly cools air, without first cooling water. At the small end, direct expansion systems include room air conditioners, such as window-mounted and through-wall units. These units are not ducted.

Window air conditioners range from 0.5 tons to over 2 tons, but 0.5 to 1 ton are most common (Figure 7.3). Controls are mounted on the unit, and electricity is provided with a plug-in cord. Small through-wall units are similar to window air conditioners. These small units sometimes provide heating as well, through electric resistance heaters, or as heat pumps.

Room air conditioner cooling efficiencies are rated in *energy efficiency ratio (EER)*. The EER is the capacity of the air conditioner, in Btu/hr, divided by its power input, in watts, at a test rating condition of 80 F air temperature indoors and 95 F outdoors. The indoor temperature test condition is typical of all air conditioner tests, but note how it is slightly warm. An air conditioner used to maintain a significantly lower temperature will use more energy, and therefore can provide more energy savings if changes are made to the equipment efficiency. Minimum energy efficiency requirements were established for room air conditioners by some individual states

Figure 7.3 Room air conditioners are used in a wide variety of commercial building applications.

TABLE 7.1

Weighted-Average Room Air Conditioner Efficiencies

Year	1976	1980	1984	1988	1992	1996	2000	2004
Efficiency (EER)	6.3	7.0	7.6	8.2	8.9	9.1	9.3	9.7

in the 1970s and 1980s. In 1990, federal efficiency requirements went into effect, varying by capacity. Federal and ENERGY STAR efficiency requirements have both increased over the years. Federal and ENERGY STAR efficiency requirements since 1990 are provided in Appendix I.

Table 7.1 shows estimated U.S. nationwide weighted-average efficiencies (EER) for room air conditioners from the 1970s to the early 1990s[1]:

We note that because the federal requirements have always varied by capacity, these averages should be primarily associated with a typical capacity, approximately 10,000 Btu/hr.

A form of through-wall room air conditioner is the *packaged terminal air conditioner*, or PTAC (pronounced "p-tack") (see Figures 7.4 and 7.5). Common in hotels and motels, the PTAC also sees some use in other types of commercial buildings. The PTAC assembly can be slid out a wall-mounted *sleeve* for convenient service and replacement. PTACs date back to the 1930s, with *Popular Mechanics* reporting a "room-size air conditioner fits under window sill" in its June 1935 issue.

PTAC sleeve sizes have been standardized in the industry, with the most common size being 42″ wide and 16″ high, representing over 90 percent of shipped PTACs.[2] PTAC capacities typically range from 7,000 to 15,000 Btu/hr (roughly 0.5 ton to 1.25 tons). Like room air conditioners, PTAC energy efficiency are rated in EER, but PTACs have their own efficiency requirements. There are no ENERGY STAR requirements for PTACs.

Room air conditioners and PTACs frequently also provide heat. The heat can be electric resistance heaters, heat pumps, hot water coils, or steam coils. Heat pump versions of PTACs are called *packaged terminal heat pumps* (PTHPs). Current PTHP models typically only operate as heat pumps down to the 20 to 25 F range of outdoor air temperatures. Below this temperature range, the compressor is locked out, and the units only heat with electric resistance heat. Older PTHPs switch from heat pump to electric resistance at even higher outdoor temperatures. PTHPs also frequently use the backup electric resistance heat under various other conditions, such as coming out of setback, or when the compressor has failed.

Minimum efficiency requirements for PTACs are shown in Table 7.2. Even though there are no ENERGY STAR requirements for PTACs or PTHPs, a number of higher-efficiency products are available. For example, American Society of Heating, Refrigerating, and Air-Conditioning Engineers (ASHRAE) Standard 189.1–2009 set minimum efficiency requirements of 11.9 EER for units below 7,000 Btu/hr, 11.3 EER for units 7,000–9,999 Btu/hr, 10.7 EER for units 10,000 to 12,999 Btu/hr, and 9.5 EER for units 13,000 Btu/hr and larger.[3]

As with room air conditioners, the challenge of fitting equipment in a small chassis limits the efficiencies of larger units.

A few tips may assist in identifying the type of heating system in a room air conditioner or PTAC:

- Call for heat with the thermostat. If there is a compressor hum, it is a heat pump.
- Examine the nameplate (typically located inside the supply air grille). Heating ratings in Btu/hr indicate a heat pump. Heating ratings in kW indicate electric resistance heat.

Figure 7.4 Packaged terminal air conditioner (PTAC).

Figure 7.5 PTAC architectural grille viewed from outdoors.

TABLE 7.2

PTAC Efficiency Requirements

Type	Federal Requirements	Capacity (Btu/hr)			Notes
		<7000	7000–15000	>15000	
PTAC, Standard Size (EER)	1/1/1994	8.88	10 − (.16*cap)	7.6	4, 5
PTAC, Non-Standard Size (EER)	"	8.88	10 − (.16*cap)	7.6	4, 5
PTHP, Standard Size, Cooling (EER)	"	8.88	10 − (.16*cap)	7.6	4, 5
PTHP, Standard Size, Heating (COP)	"	2.72	10 − (.16*cap)	2.52	4, 5
PTHP, Non-Standard Size, Cooling (EER)	"	8.88	1.3 + (.16*EER)	7.6	4, 5
PTHP, Non-Standard Size, Heating (COP)	"	2.72	1.3 + (.16*EER)	2.52	4, 5
PTAC, Standard Size (EER)	9/30/2012	11.7	13.8 − (.3 * cap)	9.3	1, 2, 3
PTAC, Non-Standard Size (EER)	"	9.4	10.9 − (.213 * cap)	7.7	1, 2, 3
PTHP, Standard Size, Cooling (EER)	"	11.9	14 − (.3 * cap)	9.5	1, 2, 3
PTHP, Standard Size, Heating (COP)	"	3.3	3.7 − (.052 * cap)	2.9	1, 2, 3
PTHP, Non-Standard Size, Cooling (EER)	"	9.3	10.8 − (.213 * cap)	7.6	1, 2, 3
PTHP, Non-Standard Size, Heating (COP)	"	2.7	2.9 − (.026 * cap)	2.5	1, 2, 3

Notes:
1. "cap" means rated cooling capacity, in kbtu/hr
2. EPA, ENERGY STAR Market & Industry Scoping Report, Packaged Terminal Air Conditioners and Heat Pumps, December 2011.
3. Standard size is greater than 16″ high or greater than 42″ wide, and a cross sectional area greater than or equal to 670 square inches.
4. https://www.law.cornell.edu/cfr/text/10/431.97, accessed 4/8/15
5. http://www.gpo.gov/fdsys/pkg/CFR-2010-title10-vol3/xml/CFR-2010-title10-vol3-sec431-97.xml, accessed 4/8/15

- Sometimes electric resistance coils are visible through the supply grille.
- An electric resistance heater is often identified by a faint odor on an initial call for heat, as dust on the heater coils is burnt off.
- PTACs sometimes have hot water or steam heating coils. These are indicated by piping connections to the unit, or valves inside the unit.

Room air conditioners and PTACs are factory-sealed refrigerant assemblies, and so are less prone to the refrigerant-related issues (leaks, overcharge) that more often occur in split systems, which are field-assembled. They also are not ducted, and so do not have duct-related problems. They nonetheless might have efficiency degradation due to bent heat exchanger fins, dirty filters or heat exchangers, the occasional refrigerant leak, indoor or outdoor air obstructions, or air leakage between the indoor and outdoor sections. A preliminary assessment of performance degradation may be made by measuring the supply air temperature. In cooling, the supply air temperature should be at least 20 F colder than the return air (or room) temperature in typical climates (at least 15 F colder in humid climates, at least 25 F colder in dry climates). If the supply air temperature is higher than these values, this is an indication of degraded capacity and likely degraded efficiency. An efficiency measurement can be done in the field, for example by using the rated airflow, and measuring supply and return conditions, and measuring the input power.

Specific problems may also be examined and diagnosed. For example, air leakage from the outdoor section to the indoor section of a room air conditioner or PTAC may be identified and roughly quantified by running the fan, only, without cooling or heating, and using the air mixing method (Appendix C). We would like the outdoor air flow rate to be zero. If not, outdoor air is being drawn in through a ventilation damper that is stuck open, or through air leakage through the interior partition of the unit. Air leakage or air bypass may also be identified by removing the front cover, covering the indoor coil with plastic wrap, and turning on the fan. Any airflow supplied by the unit under these conditions is either air bypass (air that is not moving through the indoor coil, as it should) or air leakage—preferably, there should be no airflow in this test.

For energy estimating, the existing efficiency may be obtained from the FTC (EnergyGuide) label, if still on/near the unit, from measurement as described above, or assumed from the historic federal efficiency requirements above. The California Energy Commission's appliance database has data on approximately 16,000 archived room air conditioners and over 20,000 PTACs/PTHPs, going back to the 1970's. The efficiency of an existing unit can also be obtained by the rated capacity and power (watts), which are frequently provided by manufacturers, and sometimes provided on the nameplate. Again, EER = (capacity in Btu/hr) /(power in watts). If the power (watts) are not provided, a rough estimate of efficiency may be obtained from the capacity and current (amps) on the nameplate: EER = (1.1 × capacity in Btu/hr) / (volts × amps). This estimate is very approximate, due to inconsistencies in the current rating, and should be treated as only accurate within 10 percent. Be careful not to use the "locked rotor amps," but rather only the rated load amps, or full load amps. For recently installed units, or for the efficiency of new replacement units, the Association for Home Appliance Manufacturers (AHAM) maintains a directory of room air conditioner energy ratings, and the Air-Conditioning, Heating and Refrigeration Institute (AHRI) maintains a directory of PTACs.[4, 5]

As another energy improvement option, we might consider replacing room air conditioners with ductless split or central air conditioners. The efficiencies of these products are generally higher than room air conditioners, and envelope losses are eliminated. Ductless split air conditioners are a better option because zone control is retained, and there are no duct losses. If we are considering replacing room air conditioners with central air conditioners, we need to compare the EER rating for room air conditioners to the seasonal energy efficiency ratio (SEER) rating for central air conditioners. We can get an idea of the relationship between EER and SEER by graphing a few randomly selected ratings from the AHRI directory.[6] See Figure 7.6. All points are from different manufacturers. The average difference between EER and SEER is 2.1 (SEER is higher), but a better fit is: SEER = 1.23 × EER − 0.62.

Room air conditioners and PTACs are subject to infiltration because they penetrate the building envelope. They are also a site for conduction losses. Both of these issues have been examined previously in the envelope chapter. When considering replacement of room air conditioners and PTACs with central systems, one can take credit for this reduction in infiltration and conduction losses.

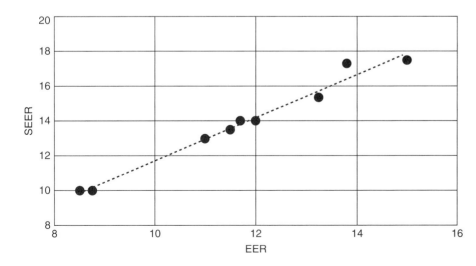

Figure 7.6   SEER is slightly higher than EER.

Figure 7.7 Fin damage.

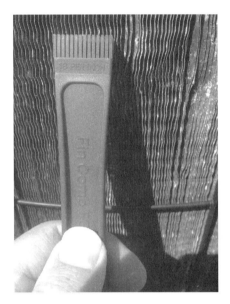

Figure 7.8 Fin comb.

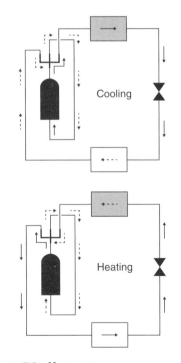

Figure 7.9 Heat pump.

Similarly, in heating, we might consider replacing PTHPs with central heat pumps. To allow comparison of PTHPs rated with a coefficient of performance (COP) with central heat pumps rated with a heating seasonal performance factor (HSPF), we may use a relationship developed by the U.S. Department of Energy (DOE)[7]:
HSPF = 1.5843 × COP + 2.4909.

Room air conditioners frequently have fin damage, especially if the condenser or evaporator is inadequately protected. Based on studies of reduced airflow in air conditioners, we can assume that damaged evaporators lose 1.5 percent efficiency for each 10 percent loss in airflow due to fin damage (Figure 7.7), and damaged condensers lose 4 percent efficiency for each 10 percent loss in airflow due to fin damage.[8] Fin combing can be used to regain lost efficiency (Figure 7.8).

Room air conditioners and PTACs lend themselves well to behavioral/educational measures to control energy use. These include:

- Keep ventilation closed if not needed.
- Set the unit for automatic fan operation, which runs the fan only when the compressor is running, rather than continuously.
- Turn off when not in use.
- Set thermostat at appropriate settings.
- Keep the air filter clean.

Heat Pumps

We take a detour from cooling systems to discuss heat pumps. Heat pumps are related to cooling systems, and use the same principle. Just as an air conditioner moves heat from inside a building to the outdoors, so a heat pump moves heat from the outdoors to the indoors. It seems somewhat counterintuitive, but again, the same principle is at work. An air conditioner cools the air inside a building by removing heat, and the heat has to be put somewhere, so it is put outdoors. In winter, a heat pump cools the air outside a building, and in so doing has to put the heat somewhere, so it puts the heat where we want it, inside the building.

A *heat pump* (Figure 7.9) works by using the same four components as an air conditioner: A compressor, two heat exchangers (coils), and an expansion device. A heat pump in summer works as an air conditioner, as described above: A compressor compresses refrigerant into a condenser (outdoors), the condensed refrigerant liquid expands to a low pressure, and evaporates in an evaporator (indoors), where it removes the heat from indoor air. In winter, the heat pump cycle is reversed, with the compressor compressing refrigerant gas into the *indoor* heat exchanger, where it condenses and supplies heat to the indoor air. The condensed refrigerant is expanded to a low pressure, moves to the outdoor heat exchanger, where it evaporates, *absorbing heat from the outdoors*, even though the outdoors is cold.

A *reversing valve*, also called a *four-way valve*, is used to reverse the flow between summer and winter. This is the reason that heat pumps are sometimes referred to as *reverse cycle air conditioners* (Figure 7.10).

Heat pump capacity and efficiency, in heating, reduces as the outdoor air temperature reduces. The high efficiency of heat pumps, in heating, at milder outdoor temperatures has made them popular in the U.S. South and West. However, they have seen increasing use in the North as well. Originally developed with backup heat, most commonly electric resistance heat, but on occasion with fossil fuel heating in *dual fuel heat pumps*, the development of low-temperature heat pumps, with variable speed compressors, now allows heat pumps to operate down to below –10 F outdoor temperature, without backup heat.

Because the performance of heat pumps depends so strongly on outdoor air temperature, there is a greater sensitivity to outdoor air temperature when performing energy calculations. Of particular note, in addition, are conditions under which backup heat sources start working. For example, most PTHPs do not operate in heating mode below the range of 20 to 25 F, and so rely on electric resistance heat in the depth of winter. Other conditions under which backup heat like electric resistance are used are in recovery from setback mode, on startup, during defrost, in the event of a compressor failure, and if the indoor temperature is at risk of freezing. These conditions can add up to a significant contribution of the inefficient backup electric resistance, unless precautions are taken to prevent it. The best precaution is to avoid any use of backup electric resistance heat, wherever possible.

There are several different kinds of heat pumps. *Air-source* heat pumps draw heat from the outside air. *Water-source* heat pumps draw heat from water, either water that is circulated through the ground, or rivers or lakes, or from an interior water loop that, itself, is cooled by a cooling tower. Water-source heat pumps that draw heat from the ground, rivers, or lakes are sometimes called *geothermal, geoexchange*, or *ground-source* heat pumps. Mostly, heat pumps are used to heat and air condition indoor air. Sometimes, they are used to heat indoor water only, for example, used in a radiant floor system. Heat pumps that draw heat from air, and in turn heat air, are called *air-to-air heat pumps*. Heat pumps that draw heat from water, and in turn heat air, are called *water-to-air heat pumps*. Heat pumps that draw heat from air, and in turn heat water, are called *air-to-water heat pumps*. And heat pumps that draw heat from water, and in turn heat water, are called *water-to-water heat pumps*.

Figure 7.10 Heat pump compressor and reversing valve.

Split Systems

A *split system* (Figure 7.11) has two units, one outdoors and one (or more) indoors, connected by refrigerant piping. The outdoor unit combines the compressor and condenser in a *condensing unit*.

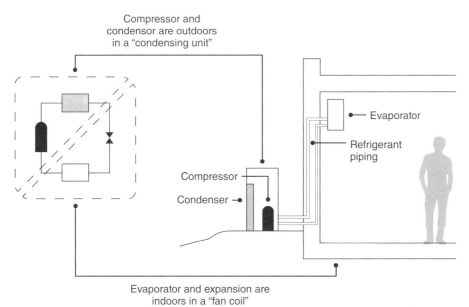

Figure 7.11 Split system.

Figure 7.12 Residential split system outdoor units (condensing units) used in commercial buildings.

Figure 7.13 Large commercial split system outdoor unit (condensing unit).

Figure 7.14 Wall-mounted ductless indoor unit.

The indoor heat exchanger, often referred to as an *indoor coil*, can be located in an *air handler*, as part of an assembly with a fan that also has heating coils for heating. Air handlers typically serve large zones of a building, comprising multiple rooms and even multiple floors. Or the indoor coil can be part of a *fan coil*, typically not ducted, typically serving just one room. A fan coil is composed of a fan and a coil. A fan coil is typically not ducted, whereas an air handler is ducted. There are exceptions to these rules; for example, small ducted air handlers are sometimes referred to as fan coils.

A split system is typically visually distinguished by its outdoor unit. It has no ductwork connecting the outdoor unit to the building. It typically has two copper refrigerant pipes connected to it, which then are routed inside the building. It has a large fin-tube heat exchanger, typically aluminum fin on copper tube, which covers much of the surface of the unit. Some older units have aluminum fins on aluminum tubes. A split system outdoor unit typically has one or more compressors inside the unit, which are frequently visible by peering into the unit. It has one or more condenser fans.

Split systems range from units as small as 1.5 tons up to as large as 160 tons. Residential systems are considered to be 1.5 to 5 tons, and mostly use single-phase electric service. Residential systems are frequently used in clusters to serve commercial buildings (Figure 7.12). Sometimes, a single residential unit will serve a dedicated need in a commercial building, such as a computer room.

Commercial systems are considered to be over 5 tons, and mostly use three-phase electric service (Figure 7.13).

The traditional American split system has been a ducted system. More recently, *ductless systems*, developed overseas, have been gaining popularity. Well suited for retrofits, where space for ductwork is not available, ductless systems have other advantages such as excellent zone control. *Mini-split* ductless systems pair up one condensing unit with one fan coil. *Multi-split* ductless systems pair up one condensing unit with multiple indoor fan coils. Ductless split system condensing units (outdoor units) are typically horizontal discharge (the fan blows air horizontally), whereas ducted split system condensing units are typically vertical discharge. These fan coils are available as floor-mounted, wall-mounted (Figure 7.14), ceiling recessed (sometimes referred to as a cassette; Figure 7.15), ceiling surface-mounted (Figure 7.16), and ducted.

Split systems can also be integrated with heating, in a variety of ways. The most common is through the use of a furnace, in which case the furnace typically provides the fan for the forced air system, and the air conditioner indoor unit is simply a coil. The shape of this coil (often an inverted V) is the basis for the frequent reference to the air conditioner "A-coil" in such systems. The other common integration of heating is as a heat pump. Like air conditioners, heat pumps can be ducted or ductless. Ductless heat pumps are predominantly manufactured overseas, use variable speed compressors, and are higher in efficiency than ducted heat pumps, which are mostly manufactured in the United States, primarily with constant-speed compressors.

Split system air conditioners over 65,000 Btu/hr have historically been rated for efficiency with the EER (Btu/hr per watt) and, more recently, with the integrated EER (IEER). Residential-size split systems, commonly used in small commercial buildings, are rated in SEER. And so we have the anomaly of the smallest systems (room air conditioners and PTACs, mostly in the 0.5 to 1 ton capacity range) being rated in EER, medium-size systems (1.5 to 5 tons) being rated in SEER, and larger systems being rated in EER again (and also IEER).

In the 1980s and early 1990s, a few states instituted minimum efficiency standards. In 1992, uniform federal standards went into effect, requiring 10 SEER for air conditioners under 65,000 Btu/hr, 8.9 EER for units 65 to 135,000 Btu/hr, and 8.5 EER for units over 135,000 Btu/hr. In the mid-2000s, the requirements were increased to 13 SEER for units under 65,000 Btu/hr, 10.3 EER for units 65 to 135,000 Btu/hr (10.1 EER for units with heating other than electric resistance), and progressively lower efficiencies in the less common ranges of 135,000 to 240,000, 240,000 to 760,000, and over 760,000 Btu/hr. Effective 2016 the requirements are 14 SEER for units below 65,000 Btu/hr, 11 EER and 12.8 IEER for units 65,000 to 135,000 Btu/hr (subtract 0.2 EER for units with heating other than electric resistance), and, again, lower efficiency for larger units.

A subset of small split systems is the high-velocity system. Designed around small ductwork, these systems use larger fan motors to overcome the resistance of the small ductwork. They are low in efficiency and the DOE has accommodated this deficiency by lowering their required rating, with SEER requirements far lower than regular air conditioners.

AHRI maintains a directory of current energy ratings for split systems, at www .ahrinet.org.

ENERGY STAR requirements only cover products smaller than 65,000 Btu/hr. They were initially introduced in 1995. In the early years of the program, the SEER requirement was 12, and the HSPF requirement was 7. In 2006, the requirements were set at 14 SEER and 11.5 EER, and, for heat pumps, 8.2 HSPF. In 2009, the requirements increased to 14.5 SEER and 12 EER, but remained at 8.2 HSPF. In 2015, the efficiencies increased to 15 SEER, 12.5 EER, and 8.5 HSPF.

As we examine the performance of DX air conditioners in cooling, it is helpful to be aware that:

Figure 7.15   Ceiling-recessed ductless indoor unit (cassette).

- As indoor air temperature goes down, or as outdoor air temperature goes up, efficiency goes down.
- In cooling, capacity is more sensitive to indoor air temperature, whereas power and efficiency are more sensitive to outdoor air temperature. For heat pumps in heating, the opposite is true.

Air conditioners are rated at 80 F indoor air temperature and 95 F outdoor air temperature, for capacity and efficiency. The actual operation of most air conditioners is typically lower than 80 F indoors, frequently closer to 75 F, and the outdoor temperature varies widely, from perhaps 60 F to 100 F or higher. In a commercial building, with internal gains such as lighting and people, the outdoor air temperature might well average closer to 80 F over operation in cooling, rather than 95 F. The lower typical indoor air temperature means a slightly lower efficiency than the rated efficiency, perhaps 5 percent lower, but the significantly lower annual-average outdoor air temperature means a measurably higher annual-average efficiency, perhaps 25 percent higher. And so a typical air conditioner in a commercial building operates perhaps 20 percent more efficiently, on average, over a year, than its rated efficiency.

Examples of efficiency as a function of indoor and outdoor air temperature, for an 11.5 EER split system air conditioner, are shown in Figures 7.17 and 7.18.[9]

The energy use of split system air conditioners can be reduced by replacing them with high-efficiency systems. High-efficiency systems not only have high-efficiency compressors and heat exchangers, but also sometimes have variable speed compressors and fan motors. Some of the highest-efficiency systems are in the residential range of 1.5 to 5 tons, so, where possible, one may consider downsizing into this range to take advantage of these efficiencies. Ductless systems bring multiple additional energy benefits, including the elimination of duct losses and better zone control, so a conversion from a ducted system to ductless is another option.

Split systems are highly vulnerable to energy inefficiency due to having too little refrigerant charge, or too much refrigerant charge. These problems arise because split systems are field assembled, and are frequently field charged, so mistakes can be made in charging them with the incorrect quantity of refrigerant. Field assembly also makes them more vulnerable to refrigerant leaks, at piping connections made in the field. Studies have shown over 50 percent loss in efficiency if a system is

Figure 7.16   Ceiling surface-mounted ductless fan coil.

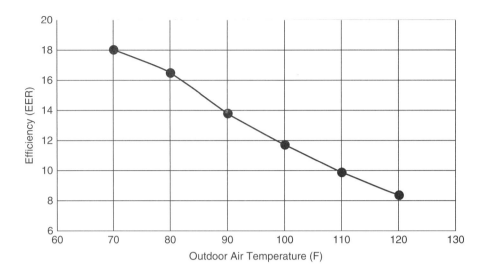

Figure 7.17 Air conditioners operate at lower efficiency at higher outdoor air temperature.

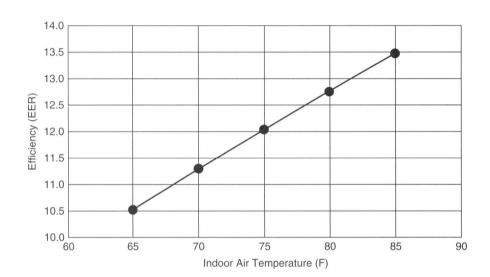

Figure 7.18 Air conditioners operate at lower efficiency at lower indoor air temperature.

60 percent undercharged, and a 10 percent loss in efficiency of a system is 30 percent overcharged.[10] Problems with refrigerant charge are very frequent, reportedly in over 50 percent of installed systems. The problems are particularly severe for systems with a fixed expansion device (mostly older residential-size systems) but also occur for systems with variable expansion devices (TXV or EXV), for example, if systems are more than 30 percent undercharged and the expansion device reaches its wide-open position. Deficiencies in charging may be indicated by a high-supply air temperature in cooling, or a low-supply air temperature for heat pumps in heating, because refrigerant charge problems also reduce capacity. Another possible symptom of refrigerant charge problems is if a building cannot be adequately cooled on a midsummer day, or heated on a midwinter day (for heat pumps). For a definitive diagnosis, refrigerant pressures and temperatures need to be measured by a trained and certified technician, as does adding or removing charge to correct charging deficiencies.

Inadequate airflow, either for the evaporator or the condenser, is another problem. Causes of inadequate airflow include dirty filters, dirt on the coils, bent coil fins, vegetation blocking the condenser, condensers located too close to buildings, failed fan motors, and fans that have come loose from their motor shafts. These deficient conditions can be identified through inspection. Like all other DX equipment (and unlike chilled water systems), inadequate airflow reduces the thermodynamic efficiency of the refrigeration cycle. In estimating savings, we can assume that inadequate evaporator airflow causes a 1.5 percent efficiency reduction for each 10 percent loss in airflow, and condensers lose 4 percent efficiency for each 10 percent loss in airflow.[8] Most of these conditions can be remedied. For example, conditions like bent fins can be fixed through fin combing.

Air conditioning and heat pump outdoor units need to have sufficient clearance from adjacent structures and vegetation to prevent obstruction of airflow (Figures 7.19 and 7.20). This clearance can be fairly forgiving, but only to a point. If only one side of a unit is obstructed, the clearance can be as small as 4″ to 6″, depending on the manufacturer. If clearances less than 6″ are found, check the manufacturer's instructions to see if the clearance is too small. If more than one side of a unit is obstructed, then the clearance needs to be larger, rising to as much as 10″ to 12″ or more. Clearances above upflow units typically need to be 48″ or more. In no case should an outdoor unit discharge its air anywhere near the intake to another outdoor unit.

Another defective condition to which split systems are susceptible is a contaminated expansion device.[11] Contamination happens when dirt inside the refrigeration system plugs the expansion device. Copper shavings, from when refrigeration tubing is cut, are a common contaminant. Symptoms include low pressures in both the evaporator and the condenser, low power draw, and inadequate capacity. A contaminated expansion device will also cause low energy efficiency. Symptoms are similar to a low-charge condition, and so this deficiency is often misdiagnosed, but adding refrigerant charge will not solve the problem.

Split systems are also vulnerable to air recirculating at the condensing unit, especially if the unit is located in an enclosed or semi-enclosed area such as in a shed, closet, or in an exterior stairwell, which is frequently done in order to conceal the units. The efficiency penalty can be estimated by measuring the air temperature entering the condenser. For example, if the outdoor air temperature is 90 F, but hot air from the condensing unit is recirculating such that the air temperature entering the condenser is 110 F, its efficiency might be 10 EER instead of 14 EER, if we use the example for which a graph of efficiency against outdoor air temperature was provided, above. Research has shown that there is roughly a 1.5 percent decrease in efficiency for each degree of higher air temperature entering the condenser.[12]

Single-Package Systems

A single-package unit has all four components in one unit, typically outdoors, with ductwork carrying conditioned air indoors.

Rooftop single-package units (Figure 7.21) are one of the most widely used cooling systems for commercial buildings in the United States, especially for single-story buildings, common in retail and food service buildings. Single-package units are occasionally located on the ground outdoors as well (Figure 7.22).

It should be noted that while room air conditioners and PTACs in fact also contain all four refrigeration components in one package, they are not formally classified as single-package equipment for purposes of energy ratings.

The rooftop unit most commonly comprises air conditioning along with a natural gas furnace. Other heat sources include electric resistance heat and heat pumps. Rooftop units have ductwork that penetrates the building envelope and transports indoor air to the unit (return air) and heated or cooled air from the unit back to the building (supply air). The orientation of this ductwork can either be to/from the side of the unit (*side discharge;* Figure 7.23) or vertically under the unit (*vertical discharge;* Figure 7.24).

Single-package units typically integrate ventilation, most commonly with an outdoor air intake, identified by a typical triangular-shaped rain hood or, in some cases, vertical dampers or louvers. This ventilation can be used both for fresh air to the

Figure 7.19   Heat pump outdoor unit mounted too close to a wall.

Figure 7.20   Vegetation covering a heat pump outdoor unit.

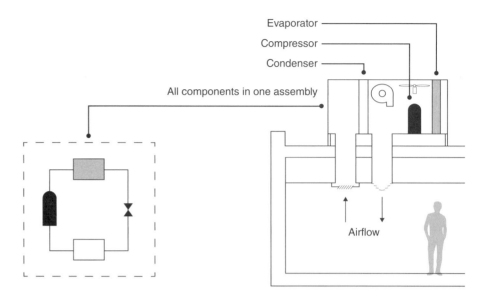

Figure 7.21   Packaged rooftop unit.

Figure 7.22   Ground-mounted single-package unit

building, and for what is called *free cooling*, or *economizer cooling*, by drawing outdoor air into a building when the outdoor air is cool or dry and the building interior is hot or humid.

Some single-package units are dedicated only to bringing in outdoor air and do not recirculate any air from the building. These units are called *makeup air units*, or *100 percent outdoor air* units. They can still integrate heating and sometimes cooling. One can differentiate between a typical heating/cooling single-package unit, and a dedicated makeup air unit, as follows:

- In general, typical heating/cooling single-package units have a condenser (indicates cooling), while makeup air units do not.
- In general, typical heating/cooling single-package units have return air (a second duct connection to the building), while makeup air units do not.
- In general, typical heating/cooling single-package units have thermostats in the space, whereas makeup units have supply air temperature control inside the unit itself.

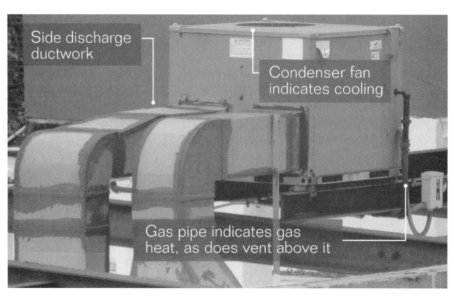

Figure 7.23   Side discharge rooftop unit.

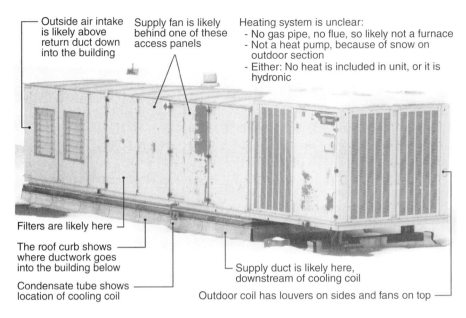

Outside air intake is likely above return duct down into the building

Supply fan is likely behind one of these access panels

Heating system is unclear:
- No gas pipe, no flue, so likely not a furnace
- Not a heat pump, because of snow on outdoor section
- Either: No heat is included in unit, or it is hydronic

Filters are likely here

The roof curb shows where ductwork goes into the building below

Condensate tube shows location of cooling coil

Supply duct is likely here, downstream of cooling coil

Outdoor coil has louvers on sides and fans on top

Figure 7.24   Vertical discharge rooftop unit.

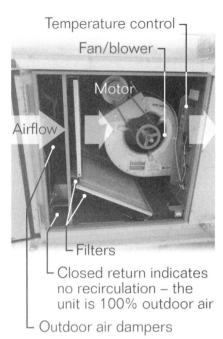

Temperature control

Fan/blower

Motor

Airflow

Filters

Closed return indicates no recirculation – the unit is 100% outdoor air

Outdoor air dampers

Figure 7.25   Return air compartment of rooftop unit.

■ By brand: Typical heating/cooling single-package units—Carrier, York, Trane, Rheem, and McQuay are common brands. Makeup air: common brands include Reznor, Sterling, and Modine.

Single-package units, typically installed on rooftops, span from very small sizes of 2 tons up to large field-assembled systems over 150 tons. Smaller and medium sizes have traditionally been *constant volume* (or *CV*, in other words, constant fan speed), and larger sizes have been *variable air volume* (*VAV*). Recently, variable speed motors have begun being used on smaller size units as well, although the vast majority of smaller systems found in the field are still constant volume.

It is informative to open the return air compartment on a rooftop unit, during field inspection (Figure 7.25). First, this allows seeing if there is a return air duct connection down to the building, in other words if the unit is 100 percent outdoor air (if no return duct) or if it is a typical rooftop unit that recirculates indoor air (if there is a return duct). Second, inspecting the return air compartment frequently allows seeing the size and type of motor. Finally, one can see the type and condition of outdoor air control to see if it is just a fixed opening or if there is a modulating damper and, if so, if it is working and if it seals well when it closes.

The average efficiency of rooftop units (Figure 7.26) in 2003 was 9.2 EER in cooling and 3.1 COP in heat pump mode. In 2007, the U.S. average efficiency for new rooftop units was 10.1 EER in cooling and 3.2 COP in heat pump mode.[13]

Another report shows average efficiencies for single-package units under 65,000 Btu/hr (see Table 7.3).[14]

Federal efficiency requirements for single-package equipment has generally been identical to those for split systems, with a few exceptions. One notable exception is

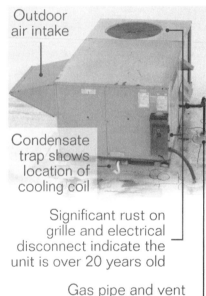

Outdoor air intake

Condensate trap shows location of cooling coil

Significant rust on grille and electrical disconnect indicate the unit is over 20 years old

Gas pipe and vent indicate that it is gas-fired

Figure 7.26   Vertical-discharge, gas-fired rooftop unit.

TABLE 7.3

Average Cooling Efficiencies for Single-Package Units under 65,000 Btu/hr

Year	1975	1985	1995	2005
Average SEER	7.0	8.9	10.7	11.3

Figure 7.27 Through-wall single-package units on a temporary classroom.

Figure 7.28 Indoor through-wall single-package unit.

that the required efficiency for single package equipment smaller than 65,000 Btu/hr is 14 SEER, effective 2015, whereas it remained 13 SEER for split systems in that size range.

A less common form of single-package unit is a through-wall system, as shown in Figure 7.27. Located either on the outside of a building or inside a building (e.g., in a closet; Figure 7.28), these are typically rectangular in shape and oriented vertically. Like other single-package units, the heating can be gas or electric.

Generally inefficient, and frequently used on construction sites as temporary heating and cooling, they are also used on temporary buildings such as classrooms and have a way of becoming permanent.

The DOE has accommodated their low efficiency by requiring relative low minimum efficiency ratings.

Older versions of packaged air conditioners are located indoors and are water-cooled instead of air-cooled, either using a cooling tower, or occasionally using city water that is discarded after use, a practice that is frowned on and, in many localities, is illegal.

Because single-package systems are factory assembled, they are slightly less prone to some of the problems to which split systems are susceptible, such as incorrect refrigerant charge and contamination of the expansion device. However, one study of refrigerant problems in single-package units still found refrigerant problems in over 40 percent of the units.[14] Other problems, such as dirty filters, bent coil fins, and dirty coils also occur with single-package systems and should be inspected for.

Single-package systems have their own vulnerabilities, specifically air leakage from the outdoors into the indoor section, across the interior partition. Unwanted air leakage through ventilation dampers, when they should be closed, is another common problem. The problems are exacerbated because the indoor section is mostly at a negative pressure, as the fan works to draw return air from the building and therefore draws in unwanted outdoor air. Air leakage problems can be identified, diagnosed, and quantified by air mixing measurements (Appendix C). Solutions include air sealing of the cabinet and replacing outdoor air dampers with low-leakage dampers.

A strategy to boost the efficiency of air conditioners has been to delay turning off the indoor fan for a period after the end of an air conditioner or heat pump cycle. Research has found that this strategy should be used only if the air handler is located outside the thermal boundary.[15] Since most single-package units are located outside the thermal boundary, there may be a benefit to keeping the fan on temporarily at the end of a cooling or heat pump heating cycle for these units.

Pacific Northwest National Labs maintains an online energy estimator for rooftop units.[16]

Chillers

An *air-cooled chiller* (Figure 7.29) similarly has all four refrigeration components in one unit, outdoors, but sends chilled water indoors instead of chilled air. The chilled water is used indoors in *fan coils* or *air handlers*.

A *water-cooled chiller* also has all four refrigeration components in one unit, but is typically located indoors. It also sends chilled water to fan coils or air handlers. A *cooling tower* (Figure 7.30), typically located outdoors, rejects heat from the chiller's condenser, and this heat is transported to the cooling tower in pumped water. A water-cooled chiller (Figure 7.31) is fairly synonymous with a chiller that uses a cooling tower. Water-cooled chillers are the most common form of cooling for large commercial buildings.

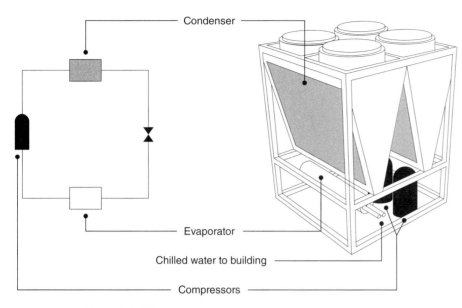

Figure 7.29 Air-cooled chiller.

Figure 7.30 Cooling tower.

Chillers are also classified by the type of compressors they use. Large water-cooled chillers use centrifugal compressors, covering the range of 100 to 6,000 tons. Screw compressors are used in medium-size water-cooled chillers, ranging from 75 to 550 tons. Smaller water-cooled chillers use scroll compressors, in the 15 to 200 ton range (Figure 7.32). Air-cooled chillers do not use centrifugal compressors, but use scroll and screw compressors, ranging from 10 to 550 tons. Reciprocating compressors are no longer typical in new chillers, but are common in older chillers found in the field, particularly in air-cooled chillers.

Chillers use a variety of different refrigerants. Old chillers use refrigerants CFC-11 and CFC-12, which have been banned because of their harmful effect on the ozone layer. CFC-11 and CFC-12 were no longer produced as of January 1, 1996. The shift away from chlorofluorocarbon (CFC) refrigerants, which had been dominant as late as 1990, first took chillers to using hydrochlorofluorocarbon (HCFC) refrigerants, mostly HCFC-123 (used by the Trane Company) and HCFC-22 (by other

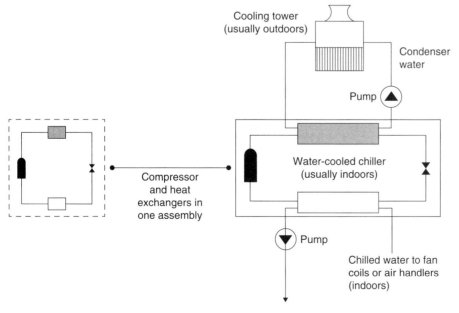

Figure 7.31 Water-cooled chiller

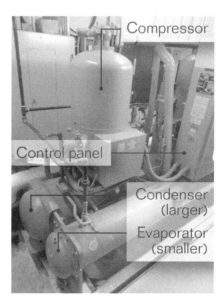

Figure 7.32 Water-cooled chiller.

manufacturers), which became the dominant types of refrigerants in chillers by 1995. By 2000, the refrigerant HFC-134a had just moved to about 50 percent market share in chillers, with the other 50 percent still HCFC-123 and HCFC-22. And by 2006, the market share of refrigerants in chillers had become over 75 percent HFC-134a, with HCFC-22 almost phased out, and HCFC-123 continuing to diminish.[17]

A type of chiller that does not use the traditional vapor compression cycle but rather uses heat and a chemical cycle is the absorption chiller. Absorption chillers make sense only when there is a free source of heat, such as industrial waste heat. Otherwise, significant savings have been shown to convert absorption chillers to electric chillers. For example, a high-rise building in upstate New York found 88 percent overall energy savings (site BTU basis) to convert from an absorption chiller to an electric chiller.[18]

Figure 7.33 Air-cooled chiller.

An air-cooled chiller (Figure 7.33) can be visually distinguished from a single-package unit in a few ways, even though they have similarities, such as a condenser and fans. The single-package unit is ducted into the building, whereas the air-cooled chiller is only connected to the building with two pipes. Most air-cooled chiller manufacturers leave the compressors exposed, and so are easily visible, and locate the condenser and fans above the compressors.

Like air cooled chillers, cooling towers do not have ductwork connecting them to a building, but rather just two pipes. Cooling towers can be visually distinguished from other outdoor cooling equipment, such as air-cooled chillers, because they do not have compressors. They are typically large. Major manufacturers include Baltimore Aircoil (BAC), Marley, and Evapco. Cooling towers are typically located outdoors, on roofs (Figure 7.34) or on the ground, but are occasionally located indoors (Figure 7.35).

Figure 7.34 Roof-mounted cooling tower.

One metric for chiller energy use is kW/ton. Note that this is not an efficiency metric, but the inverse of efficiency, in other words 1/efficiency. For kW/ton, a lower number is more efficient, whereas with a pure efficiency metric, such as EER, higher numbers are more efficient. KW/ton may be converted to EER or COP by:

$$EER = 12/(kW/ton)$$

$$COP = 12/(3.412 \times kW/ton)$$

Likewise, to convert to kW/ton from EER and COP:

$$KW/ton = 12/EER$$

$$KW/ton = 12/(3.412 \times COP)$$

AHRI maintains a rating program for chillers. There are no federal requirements for chiller efficiency, as there are for smaller equipment. There are also no ENERGY STAR requirements. Requirements for chiller efficiency are based on state and local construction codes, or organization-specific requirements, such as the Federal Energy Management Program (FEMP), which covers federal agency purchases. Appendix J shows a variety of past and current chiller efficiency requirements, as well as FEMP requirements, and above-code (high-performance) requirements of ASHRAE Standard 189.

A few years ago, ASHRAE Standard 90 introduced two optional paths for chiller efficiency rating compliance. Path A is for chillers that primarily operate at full load, and Path B is for chillers that primarily operate at part load.

It is important to note that water-cooled chiller ratings do not include pump power for chilled water or condenser water, or fan power for a cooling tower, or fan power for air handlers or fan coils to distribute the chilled water. Air-cooled chillers include fan power in the efficiency rating, but not chilled water pump power or fan power for air handlers or fan coils. For this reason, it is important that energy calculations account for this auxiliary power consumption.

Figure 7.35 Cooling tower located indoors.

The test conditions for water-cooled chiller full-load rating tests are 44 F leaving water temperature, 2.4 GPM/ton chilled water flow rate, 3 GPM/ton condenser flow rate, and 85 F condenser entering water temperature. For air-cooled chillers, the air temperature entering the condenser is 95 F.

Chillers are rated both for full load efficiency, but also for part load efficiency. The integrated part load value (IPLV) is used, in units of either EER or kW/ton. In units of EER, the IPLV is calculated as:

$$IPLV = 0.01A + 0.42B + 0.45C + 0.12D$$

A is the efficiency at 100 percent load, B at 75 percent load, C at 50 percent load, and D at 25 percent load.

For water-cooled chillers, the 75 percent load test is at 75 F condenser entering water temperature, and the 50 percent and 25 percent tests are both at 65 F. For air-cooled chillers, the 75 percent test is at 80 F air temperature entering the condenser, the 50 percent is at 65 F, and the 25 percent test is at 55 F.

For example, a chiller has a 10.5 EER at 100 percent load, a 13 EER at 75 percent load, a 15 EER at 50 percent load, and a 14.75 EER at 25 percent load. The IPLV is:

$$IPLV = 0.01 \times 10.5 + 0.45 \times 13 + 0.45 \times 15 + 0.12 \times 14.75$$

$$= 14.1 \ EER$$

COOLING TOWERS

Cooling towers can be open or closed. Open cooling towers consume water, whereas closed cooling towers do not consume water. Open cooling towers are more efficient because they can cool condenser water down close to the wet bulb temperature of the outdoor air, often 10 or more degrees cooler than the outdoor air temperature. Cooler condenser water means higher chiller efficiency.

Cooling towers are rated according to standards set by the Cooling Technology Institute. The efficiency metric is GPM/hp, where GPM is the rated water flow rate, and hp is the rated fan power, both at full load. Standards such as ASHRAE Standard 90 specify a minimum GPM/hp. For example, ASHRAE 90.1–2004 requires a minimum 38.2 GPM/hp for cooling towers with axial fans, and a minimum 20 GPM/hp for cooling towers with centrifugal fans. ASHRAE 90.1–2010 raised the efficiency requirement to 76.4 GPM/hp, regardless of fan type.

Since chiller rating tests specify 3 GPM/ton for condenser water flow, we can translate the GPM/hp requirement into kW/ton, by dividing the GPM/hp by .7457 (kW/HP), dividing by 3 (GPM/ton), and inverting the resulting ton/kW to obtain kW/ton. Accordingly, 20 GPM/hp is 0.11 kW/ton, 38.2 GPM/hp is 0.06 kW/ton, and 76.4 GPM/hp is 0.03 kW/ton.

For an existing cooling tower, the fan energy use may be estimated by the nameplate fan power, and assuming 3 GPM/ton. For old cooling towers without nameplate data, fan power may be estimated as 0.12 kW/ton for centrifugal fans (blower) and 0.06 kW/ton for propeller fans.

Figure 7.36   Ground-mounted cooling tower.

CHILLER IMPROVEMENTS

Energy improvements for chiller plants include:

■ Replace a low-efficiency chiller with a high-efficiency chiller of the same type. It is important to evaluate the entire system when replacing a chiller. A high-efficiency water-cooled chiller depends on a cooling tower that can deliver low-temperature condenser water. Modeling the entire system also allows optimization of the system set points, at both design and off-design conditions. For example, as the airflow in a cooling tower is increased, the cooling tower fan power increases, but the chiller compressor power decreases. There is an optimum condenser water temperature at which the total power is the lowest.

- Change the type of chiller, if allowed by the size range. For example, centrifugal chillers are higher in efficiency than air-cooled chillers.
- Replace the chiller control. For example, a chiller with a variable speed drive is more efficient than an old chiller with inlet guide vanes for capacity control.
- Replace the cooling tower. Energy improvements include:
  - Lower fan power. For example, a cooling tower with propeller fans typically uses one half the fan power as a cooling tower with centrifugal fans (blowers).
  - Lower condenser water temperature, resulting in higher chiller efficiency. Strategies might include oversizing the cooling tower, or changing the type of cooling of tower.
- Replace low-efficiency pump and fan motors with high-efficiency motors and/or install variable speed drives. Pumps include chilled water and condenser water. Fans include air handler and cooling tower.
- Control modifications, including raising the chilled water temperature, when possible.
- Consider a heat recovery chiller, which has the capability to recover heat from the condenser, if there are simultaneous needs for chilled water and heat.

FEMP maintains simplified online energy calculators for air-cooled chiller replacements.[19]

References

1. Developing Energy Efficiency Rating Matrix for Measures Subject to California and Federal Minimum Appliance Efficiency Standards, SBW Consulting, Inc., Contract No. 7–15–93. Submitted to Base Efficiency Studies Subcommittee, California DSM Measurement Advisory Committee, March 30, 1994, p. 15, for 1976–1992; for years 2000 and 2004, p. 5–26, 2009 Buildings Energy Data Book, Prepared for Buildings Technologies Program, U.S. Department of Energy, by D&R International, Ltd., year 1996 is interpolated.
2. https://www1.eere.energy.gov/buildings/appliance_standards/commercial/pdfs/ptac_pthp_tsd/chapter_10.pdf. Accessed April 8, 2015.
3. ASHRAE Standard 189.1–2009, p. 75.
4. http://www.cooloff.org/. Accessed April 7, 2015.
5. http://www.ahrinet.org/site/1/Home. Accessed April 8, 2015.
6. https://www.ahridirectory.org/ahridirectory/pages/ac/defaultSearch.aspx. Accessed April 7, 2015.
7. Technical Support Document (TSD): Energy Efficient Program for Commercial and Industrial Equipment: Efficiency Standards for Commercial Heating, Air Conditioning, and Water Heating Equipment, U.S. Department of Energy, Office of Energy Efficiency and Renewable Energy, Building Technologies Program, Appliances and Commercial Equipment Standards, Washington, DC, March 2, 2006, p. 49.
8. 98.501B. Prepared by Proctor Engineering Group, Ltd. San Rafael, CA. John Proctor, P.E., Rob deKieffer, Mary O'Drain, PG&E, and Amalia Klinger, PG&E, "Commercial High Efficiency Air Conditioners—Savings Persistence." Prepared for 1999 International Energy Program Evaluation Conference, Final Report, August 15, 1999.
9. http://dms.hvacpartners.com//docs/1005/Public/04/38AP-12PD.pdf, p. 19. Accessed April 9, 2015.

10. Woohyun Kim and James E. Braun, "Impacts of Refrigerant Charge on Air Conditioner and Heat Pump Performance." International Refrigeration and Air Conditioning Conference, 2010. Paper 1122, Fig. 11, p. 5.

11. John Tomczyk, "Restricted TXV Metering Device." *Air Conditioning/Heating/ Refrigeration News*, January 17, 2011. Accessed online April 10, 2015.

12. Ramin Faramarzi, P.E., Technology Test Centers (TTC), Design and Engineering Services, Southern California Edison (SCE), "HVAC—Maintenance and Technologies." *Federal Utility Partnership Working Group Meeting, Providence, RI*, April 15, 2010.

13. U.S. Department of Energy, "Building Energy Databook," 2010, pp. 5–10.

14. Reid Hart, Portland Energy Conservation, Inc.; Dan Morehouse & Will Price, Eugene Water & Electric Board; John Taylor, Consortium for Energy Efficiency; Howard Reichmuth & Mark Cherniack, New Buildings Institute, "Up on the Roof: From the Past to the Future." ACEEE Summer Study on Energy Efficiency in Buildings, 2008, pp. 3–120.

15. W. J. Mulroy, "The Effect of Short Cycling and Fan Delay on the Efficiency of a Modified Residential Heat Pump." *ASHRAE Transactions 92*, Pt. 1, SF-86-17, 1986.

16. http://www.pnnl.gov/uac/costestimator/main.stm. Accessed April 12, 2015.

17. Tony Digmanese, "The Refrigerant Shift in Centrifugal Chillers: From HCFCs to HFCs." *Heating/Piping/AirConditioning*, July 22, 2008. http://hpac.com/fastrack/From-HCFCs-to-HFCs. Accessed April 11, 2015.

18. Taitem Engineering, "Post-Installation Energy Monitoring Analysis." March 2009, Project #07189.

19. http://energy.gov/eere/femp/energy-cost-savings-calculator-air-cooled-electric-chillers. Accessed April 12, 2015.

Chapter 8

Heating and Cooling Distribution

Forced Air Systems

DUCTWORK

Ductwork is a common way to distribute heating, cooling, and ventilation in commercial buildings. Most commonly, ductwork is fabricated from sheet metal and can be either rectangular or round. Round ductwork is frequently prefabricated spiral duct (Figure 8.1).

Ductwork can also be fabricated from rigid insulation board, referred to as *duct-board*, found especially in small commercial installations. Ductwork can also be made of flexible materials, including plastic jackets supported by wire and/or aluminum, commonly referred to as *flex duct* (Figure 8.2).

In older buildings, air is sometimes ducted in the space between joists, most typically return air, enclosed by a piece of sheet metal known as *panning* (Figure 8.3). This type of ducting is also referred to as a *joist bay return*. In older buildings, air is sometimes ducted inside other structural elements, such as in masonry chases.

A duct system is made up of different types of ducts (Figure 8.4). Large ducts connected to air handlers are called *trunks* or *main ducts*. Smaller ducts serving zones or individual spaces are called *takeoff ducts*, *branch ducts* or *runout ducts*. Vertical ducts routed from floor to floor are called *risers*.

Ductwork loses energy through heat conduction and air leakage. Much ductwork is located in unconditioned spaces, or outdoors, where energy losses are more serious. Unconditioned spaces include basements, attics, exterior wall cavities in wood-frame walls, the space above ceilings, and more.

Outdoors, ductwork is subject to physical damage and corrosion. Corrosion is common on the top of horizontal stretches of rectangular ductwork, where rain tends to puddle (Figure 8.5).

Duct locations require attention in the order of the seriousness of duct loss energy impacts. In other words, the priority for duct improvements for ducts serving heated and cooled air should be, from higher to lower priority:

1. Outdoors.
2. In highly vented spaces outside the thermal boundary:
   a. Vented attics.
   b. Vented crawlspaces.
   c. Vented ceiling cavities if the thermal boundary is at the ceiling level rather than at the roof level.
3. In alignment with the thermal boundary (e.g., ductwork in exterior wall cavities).
4. In unconditioned spaces, such as basements and mechanical rooms.
5. Above ceilings, especially if there is no return ductwork (plenum return).
6. In conditioned spaces.

Figure 8.1 Spiral duct.

Figure 8.2 Flex duct.

Figure 8.3 Panning (shown at top).

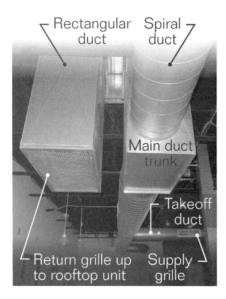

Figure 8.4 Duct system.

Figure 8.5 Corrosion on outdoor ductwork.

Exhaust ductwork deserves separate attention. Insulating exhaust ductwork is not merited. Sealing leaks in exhaust ducts has benefits that relate more to fan power savings, rather than a combination of thermal energy savings and fan power savings, and to delivering balanced exhaust for health and safety reasons. However, duct sealing can also reduce stack effect air pressures in buildings, and so reduce infiltration. Exhaust ducts typically are routed above ceilings and then vertically through chases toward the building exterior. They may also pass through other unconditioned spaces such as mechanical rooms, and vented unconditioned spaces such as attics, before terminating at an exterior wall or at the roof. Exhaust systems are predominantly under negative pressure, with the fan at the termination of the duct (wall or roof). In this case, duct leakage means that more airflow is required in order for the design airflow to be provided back at the original site of exhaust, whether it is a bathroom, kitchen, or lab hood. Sealing the ducts means that the fan speed can be reduced, or the fan can be replaced with a smaller fan or motor. Air leakage can also mean that the design airflow is in fact not being extracted from some or all of the exhaust locations, in which case duct sealing will improve indoor air quality. The negative pressure caused by air leakage into the duct, in the spaces through which the duct is routed, also means that the duct leakage is inducing unwanted infiltration into the building, adding to heating and cooling energy loads and energy usage. See Figure 8.6.

In addition to losses from ductwork, there are also leakage losses from associated components, such as filter housings, and duct connections to heating, ventilating, and air conditioning (HVAC) equipment (Figure 8.7).

Flex duct is typically pre-insulated. Duct board is intrinsically insulated. Large commercial ductwork is frequently insulated on the interior, with *duct lining* (Figure 8.8), and the insulation serves both to reduce energy losses and for noise control. Much ductwork is externally insulated. However, much ductwork in buildings is uninsulated. And, frequently, insulation on ductwork is detached (Figure 8.9).

Sites of duct air leakage include:

■ Transverse joints between duct sections.

■ Longitudinal joints.

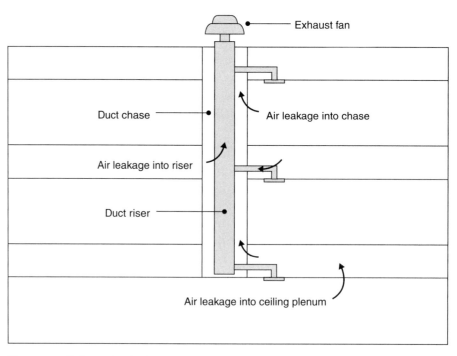

Figure 8.6 Exhaust duct leakage.

- Joints at equipment.
- Joints at registers/grilles.
- Filter boxes.
- Duct penetrations, for instrumentation, test holes that were not sealed, access panels (Figure 8.10), and occasionally structural members or other penetrations.
- Air handler leakage.
- Panning, and leaks in a panned cavity (e.g., between joists and floor in basement panning).
- Sometimes the ends of ductwork are not capped (Figure 8.11), or the caps become detached.
- Catastrophic failures (Figure 8.12): Occasionally, a piece of ductwork has a major failure, with a large hole, or has entirely come away from a piece of equipment or other ductwork.

Figure 8.7   Leakage at duct connection to furnace and at filter housing.

Duct leakage (Figure 8.13) causes energy losses beyond just the leakage of heated or cooled air. Duct leakage can be a path for overall building infiltration losses, when an air handler is not running. This is especially true for ductwork in highly vented spaces such as attics or crawlspaces, exterior ductwork, or ductwork in exterior building wall cavities. Duct leakage also causes increased fan energy use, because air handler fans need to run longer to deliver the added heating or cooling required to compensate for the leakage losses. In some situations, duct leakage can cause a lower thermodynamic efficiency of the heating or cooling plant. For example, duct leakage from a supply duct to a return plenum can cause hot air in heating, or cold air in cooling, to return to a heating or cooling system. This will reduce the efficiency of combustion systems, in heating, or of vapor compressions systems, such as air conditioners or heat pumps, in both heating and cooling.

A survey of 13 large commercial building systems found average duct leakage of 23 percent. A survey of nine large commercial building exhaust systems found leakage ranging from 9 percent to 74 percent, averaging 39 percent.[1] Duct leakage appears to be ubiquitous: It has been found to be a problem in virtually all ducted systems.

Duct leakage can be measured by isolating the ductwork (e.g., covering grilles), pressurizing the ductwork, and measuring the airflow being used to pressurize the ductwork, which equals the leakage rate. See Figure 8.14.

Air leakage tests and their results can be used for a variety of purposes:

- Assessing if ductwork is leaky or tight.
- Identifying the locations of duct air leakage.
- Evaluating potential energy savings.
  - For fan energy savings.
  - For thermal savings in terms of reduced leakage of heated or cooled air
  - For thermal savings in terms of reduced whole-building air leakage, for example, if ductwork is located in a vented space outside the thermal boundary
- Presenting results in equivalent air leakage, or estimated hole size, in order to provide it in terms that an owner or property manager are better able to understand.
- For quality control during ductwork sealing, and at the conclusion of sealing.
- For compliance with codes, standards, and guidelines.

Figure 8.8   Duct lining.

Estimating energy savings by reducing duct leakage depends on the initial duct leakage rate (itself dependent on duct pressure and the size of holes), the effectiveness

Figure 8.9   Detached duct insulation.

Figure 8.10   Duct access panel.

Figure 8.11   Masonry duct shaft, not capped, inadvertently open to ceiling plenum below.

of sealing, the location of the ducts, the type of leakage (into or out of the ducts), and the temperature of leakage air. Air leakage testing is helpful in order to baseline existing conditions, but is still only an estimate because results must be corrected to actual duct pressure, which varies along the length of the ductwork, and the post-retrofit leakage must still be estimated.

For purposes of owners and property managers visualizing duct leakage, we can convert measured air leakage rates into equivalent leakage area[2]:

$$\text{Leakage area in square inches} = \frac{\text{Measured duct system leakage rate in CFM}}{(1.06 \times (\text{Test pressure in Pascals})^{0.5})}$$

A supplementary duct leakage test, performed in conjunction with a whole-building blower door test, can be used to measure duct leakage outside the thermal boundary, in other words, to the outdoors. By pressurizing a building with the blower door to the same pressure as the ductwork, air ostensibly cannot leak from the duct to the building indoors, and so any leakage measured by the duct test must be leakage to the outdoors.

A useful spreadsheet for calculating distribution improvements was developed for residential systems, in accordance with American Society of Heating, Refrigerating, and Air-Conditioning Engineers (ASHRAE) Standard 152.[3] The calculations have been used on a limited basis for small commercial systems and are also informative for how they approach energy savings estimates. The approach taken by ASHRAE Standard 152 is to calculate a distribution system efficiency. The power of this approach is that it can be applied in a manner similar to heating or cooling plant efficiency. For example, if we know the heating fuel usage U, if we estimate the existing heating plant efficiency as Ep, and we estimate the existing distribution system efficiency as Ed (an estimate provided by the spreadsheet), we can calculate the annual building heating losses as L = U − (U × Ep × Ed). We can then estimate the new fuel usage from the building losses (which presumably do not change) and proposed new distribution system efficiency, and so calculate the energy savings.

Another approach is to presume a reduction in air leakage, as a fraction of design airflow in the system.

Whereas the Sheet Metal and Air Conditioning Contractors' National Association's (SMACNA's) test standard provides air leakage at 25 Pascals (Pa) in cubic feet per minute (CFM) per 100 square feet of ductwork, some guidelines specify test air leakage normalized to the building floor area, for example, CFM per square foot of building floor area.

Some old ductwork already has sealed joints. The adhesive tape used for this often contains asbestos and should not be disturbed. Options for asbestos on duct joints include professional removal, encapsulation, and simply leaving it in place with precautions to not disturb it. A common asbestos-containing product was Celotex Carey Duct Adhesive, installed between 1940 and 1955.[4] Asbestos-containing sealants products continued to be used until the 1980's.

Ducts may be sealed by coating joints with sealants (brushed or sprayed on), by a variety of specialty tapes, or by pressurizing ducts with an aerosol. Because most traditional cloth-backed duct tape, and many other construction tapes, will not adhere reliably and permanently, specialty duct-sealing tapes are recommended, or mesh tapes covered in mastic sealant. Some in the industry shy away from even referring to duct tapes, and instead refer to specialty duct sealing tapes as *rolled duct sealants*.

Aerosol-based duct sealing has been developed in recent years. An aerosol is injected into a duct system, and as the aerosol accelerates out of leakage sites, it deposits on the duct surfaces at these holes, builds up, and fills them. Aerosol duct sealing has the advantage of being able to reach air leakage sites that cannot be reached by manual coating, for example, ducts in wall cavities, above hard ceilings, in chases, and where ducts have one side close to a wall or ceiling. The process to accomplish aerosol duct sealing is:

- Plug supply and return locations. This is done by inserting a flexible plug into ducts at supply and return grilles (Figure 8.15).
- Isolate air handlers, so they are not contaminated with the aerosol. This is also done by inserting flexible plugs into ducts, upstream and downstream of air handlers.
- Cover or remove controls in the ductwork that could be contaminated with the aerosol.
- Inject the aerosol.

Much of the labor required for aerosol duct sealing is in the setup and teardown. The aerosol injection itself typically takes under 10 minutes for small systems and under an hour for larger systems.

Because the air handler is not reached by the aerosol, there is no sealing of leaks within air handlers, at filter boxes, or at connections between the air handler and ductwork. Aerosol duct sealing can fill cracks up to $5/8''$. Larger failures need to be repaired mechanically. Aerosol duct sealing typically reduces air leakage by 90 percent or more. See Figure 8.16.

Various codes and standards set forth requirements for insulation and sealing of ducts, based on climate zones, duct location, joint locations, and other factors.[5] These include ASHRAE Standard 90, the International Energy Conservation Code, and the International Mechanical Code. Testing typically references the SMACNA HVAC Air Duct Leakage Test Manual, although is generally required only for high-pressure systems.

Informally, reference is made to low-pressure, medium-pressure, and high-pressure ductwork, with inconsistent definitions either based on pressure or velocity. For example, ductwork is commonly referred to as low-pressure if it has less than $2''$ water gauge static pressure and/or less than 1,500 feet per minute air velocity. Formally, SMACNA defines seven duct classes ($1/2''$, $1''$, $2''$, $3''$, $4''$, $6''$, and $10''$), and three seal classes (A, B, and C). See Table 8.1. Seal class A is $4''$ and larger duct classes, and requires sealing of all joints, seams, and wall penetrations. Seal class B is $3''$ to $4''$ duct classes, and requires sealing of transverse joints and seams. Seal class C includes duct classes $1/2''$ to $2''$ and requires sealing of transverse joints only.

Figure 8.12 Catastrophic duct failure.

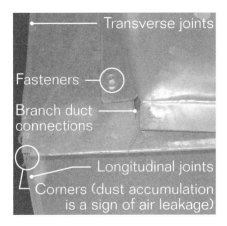

Figure 8.13 Common duct leakage sites.

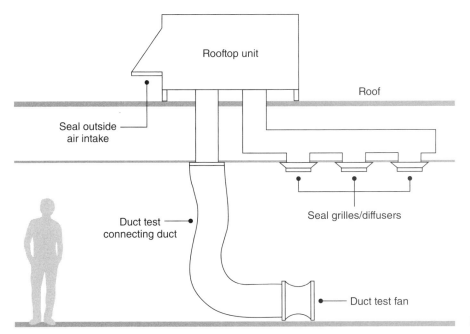

Figure 8.14 Duct leakage test.

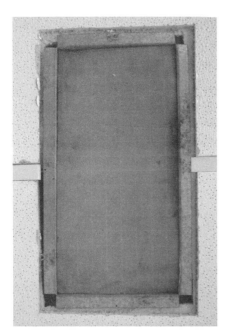

Figure 8.15 Flexible plug inserted into duct at supply grille, to allow aerosol injection.

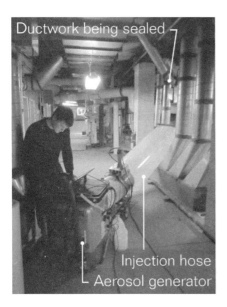

Figure 8.16 Aerosol duct sealing.

TABLE 8.1

Duct Leakage Classification

Duct Class	$1/2''$, $1''$, $2''$ wg	$3''$ wg	$4''$, $6''$, $10''$ wg
Seal Class	C	B	A
Sealing Applicable	Transverse joints only	Transverse joints and seams	Joints, seams, and wall penetrations
	Leakage Class (CL)		
Rectangular Metal	24	12	6
Round Metal	12	6	3

Notes:
Duct class means the maximum pressure. For example, $1/2''$ means $1/2''$ maximum, $1''$ means $0.6''$–$1''$ maximum, etc.
wg: water gauge.
$F = CL \times P^{0.65}$, F—maximum leakage (CFM/100 square feet of duct); P—pressure (inches wg)
Source: SMACNA HVAC Air Duct Leakage Test Manual

Even though duct leakage testing is only required by codes and standards for high-pressure ductwork, testing for energy diagnostics and energy improvement in all ductwork is beginning to occur.

Air is a fluid that knows how to leak. Neither air nor ductwork pay attention to what duct class or seal class they are, and air tends to leak anywhere and everywhere, including not only joints, seams and wall penetrations, but also corners, connections between ductwork, connections to equipment, within equipment, within accessories such as filter boxes, and more.

One can identify if an existing duct is sealed with traditional sealants or tapes by inspecting the duct exterior. See Figure 8.17.

Most duct sealants are white or gray. The absence of duct sealing is evident by air leaking out of supply ducts or into return or exhaust ducts, as shown by smoke tracing or visual inspection. Dust or dirt is also a sign of leakage out of ducts and equipment (Figure 8.18).

Traditionally, most small commercial and multifamily residential ductwork has not been sealed, and only some large commercial ductwork has been sealed.

AIR HANDLERS

Air handlers (Figure 8.19) typically contain a fan, air filters, and heat exchangers (coils) for heating and cooling. Air handlers are also known as *air handling units* (AHUs). Smaller air handlers are referred to as *packaged air handlers*, and larger, customized air handlers are called *central station air handlers*. Air handlers optionally include connections for outdoor air, for ventilation, and, less frequently, a second fan for the purpose of exhausting air. Other optional features include heat recovery, humidification, and specialty dehumidification, beyond the dehumidification that is provided by air conditioning. Provision is also made in air handlers for removal of condensate that has come from dehumidification. In a sense, packaged units, such as rooftop units or even forced air furnaces, are air handlers, but they are typically not referred to as such, and the term is primarily used for units with fans, filters, and coils.

Air handlers can be inspected by first, and always, shutting power to the unit, and then opening the fan section. A power switch is typically within sight of every air handler. Larger air handlers are large enough to walk into; smaller air handlers can be inspected from the fan access panel. If walking into an air handler, it is important to recognize the risks of a high-speed fan in a confined space. It should be treated a potentially hazardous confined space, and added safety precautions taken, such as a second person standing watch outside the air handler. Like rooftop units, the fan section of an air handler reveals much information:

- Fan type (axial (Figure 8.20), centrifugal (Figure 8.21)—also referred to as a blower wheel, etc.) and fan motor nameplate data.
- Type and condition of filters.
- Whether the air handler has recirculating airflow (return air), or outside air, or both.
- Type of outdoor air control (fixed, modulating, interlocked with return air, etc.).

Air handling units (Figures 8.22 through 8.24) can have water or steam coils for heating, chilled water for cooling, or refrigerant (DX) coils for cooling. In some older systems, two heating coils are provided, one for *preheating*, and one after the cooling coil for *reheating*, in a type of dehumidification control strategy that has fallen out of favor due to energy inefficiency.

Energy improvements to air handlers include:

- High-efficiency motors.
- Variable speed drives, possibly as part of a larger conversion from a constant volume (CV) system to a variable air volume (VAV) system.
- Air sealing, if there is evidence of air leakage into or out of the air handler.
- Change in type of coil, as part of a larger fuel conversion. For example, steam coils might be converted to hot water as part of a project to eliminate a steam heating system. As another example, it has been reported that DX coils have been installed in existing air handlers, replacing chilled water coils, as part of a *variable refrigerant volume (VRF)* system.
- Addition of heat recovery (or replacement of air handlers with new air handlers that have heat recovery capability)
- Entire air handler replacement, to take advantage of higher efficiency fan/motor combinations, larger coils, tighter cabinets, etc.
- Air handler elimination altogether, as part of a larger energy conservation strategy. Traditional air handlers are high in motor power use, and present trade-offs due to the integration of ventilation. Higher-efficiency approaches include distributed fan coils and dedicated outside air for ventilation.

VARIABLE AIR VOLUME BOXES

Variable air volume (VAV) systems have *VAV boxes* that are used to control the temperature in individual thermal zones. These VAV boxes can be as simple as a damper,

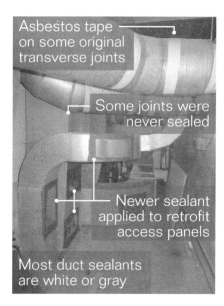

Figure 8.17   Partially sealed ducts.

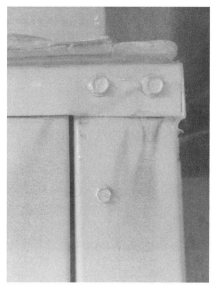

Figure 8.18   Dust or dirt is a sign of air leakage.

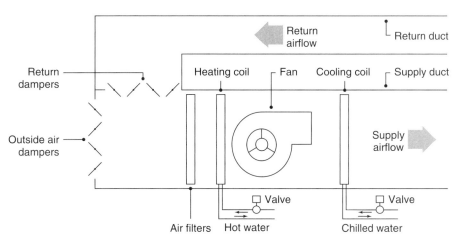

Figure 8.19   Air handler.

Figure 8.20   Axial fan.

Figure 8.21   Centrifugal fan.

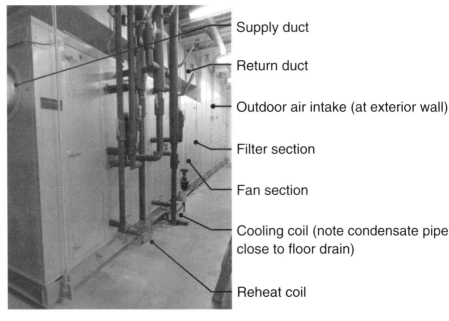

Supply duct

Return duct

Outdoor air intake (at exterior wall)

Filter section

Fan section

Cooling coil (note condensate pipe close to floor drain)

Reheat coil

Figure 8.22   Air handling unit.

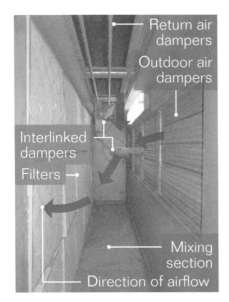

Return air dampers

Outdoor air dampers

Interlinked dampers

Filters

Mixing section

Direction of airflow

Figure 8.23   Air handler—outdoor air section.

Direction of airflow

Filters

Axial fan

Figure 8.24   Air handler—fan section.

which is opened or closed incrementally to increase or reduce hot or cold airflow, in response to a call for either heating or cooling. When these dampers are closed, in response to reduced demand for either heating or cooling, the air pressure in the ductwork increases, and this air pressure signal is used to slow down the VAV air handler fan. VAV boxes can also have fans, in *fan-powered VAV boxes.*

In some VAV systems, the VAV boxes also have some form of heat, referred to as *reheat,* typically either hot water heat exchangers or electric heaters. The strategy is for the air handler to supply cooled air, which satisfies the need for cooling in some thermal zones, while other thermal zones requiring heating are provided with this heat through reheat at the VAV boxes. This obviously means that some thermal zones have air that has been cooled and then reheated. Minimizing the need for reheat becomes a major focus of optimizing central control systems, to reduce energy use.

GRILLES AND DIFFUSERS

Grilles and diffusers are the devices at which air enters or leaves spaces in a building. They are the most visible elements of forced air systems. In general, *diffusers* have blades intended to spread the delivery of supply air, air that is entering a space, and frequently allow the amount of airflow to be adjusted. And, in general, *grilles* serve return airflow, do not have adjustable blades, tend to be larger, and tend to be fewer than the number of diffusers serving the same space. Grilles also serve exhaust air (air leaving the building). It is not unusual for exhaust grilles to allow airflow adjustment, through a set-screw just behind the front face of the grille, near one edge. The terminology for diffusers and grilles in the industry is inconsistent, with diffusers often called grilles, and with the use of *registers* and *vents* also common.

Water and Steam Systems

PIPING AND FITTINGS

Water is used to distribute heating and cooling from boilers and chillers throughout a building. Water piping is typically either cast iron (black) or copper. The large pipes that connect to a boiler or chiller are called *mains*. Pipes that run from mains to radiators and air handlers are called *runouts*. Vertical pipes that run from floor to floor are called *risers*.

Water and steam piping needs to be insulated to reduce energy losses, but also for safety (hot water and steam piping) and to prevent condensation (chilled water piping). See Figures 8.25 through 8.27.

Insulation reduces energy losses in piping and fittings such as valves and flanges. How much energy can be saved depends on the fluid temperature, the length and diameter of piping, the amount of insulation, and whether energy is lost in a space that is heated or cooled, or neither. Insulation savings calculations are typically done by spreadsheet. Online calculators are also available, as well as downloadable programs.[6] When performing savings calculations, it is important to not overestimate the operating hours per year.

Table 8.2 may be used to calculate savings from water piping. Multiply the factors in the table by the length of the piping, and by the difference between the water temperature and the room air temperature, and by the hours of operation per year, to obtain the reduction in heat lost from the surface of the pipe, in Btu/year. Divide this number by the efficiency of the heating plant to obtain fuel saved (Btu/year).

Valves, flanges, and fittings can be insulated using custom-fabricated insulation covers, or flexible blankets, which are typically removable. A standard is available to guide savings calculations for insulating bare valves and flanges.[8] As a rough approximation, a bare valve (Figure 8.28) has the equivalent area of four linear feet of piping of the same diameter as the valve.

A flanged elbow (Figure 8.29) has the equivalent area of approximately two feet of piping of the matching diameter.

Some manufacturers of insulation products for valves, flanges, and specialties provide savings guidance for insulating these fittings. See Figure 8.30.

Table 8.3 provides actual pipe sizes, compared to nominal pipe sizes.

It should be noted that copper piping used for air conditioning and refrigeration (ACR) follows a different pipe diameter convention. For ACR piping, the nominal pipe diameter and the actual outside diameter are the same.

Figure 8.25 Uninsulated piping.

Figure 8.26 Professionally insulated piping—adequate insulation thickness, elbows and fittings insulated, no piping missed.

Figure 8.27 Inadequately insulated piping—inadequate insulation thickness, elbows and fittings not insulated, insulation does not fully cover piping, sections of piping are not insulated.

TABLE 8.2

Energy Savings from Horizontal Pipe Insulation (Btu/hr per linear foot per degree temperature difference between water in pipe and surrounding air)[7]

		\(^1\)/\(_2\)	\(^3\)/\(_4\)	1	1.25	1.5	2	2.5	3	4	5	6	8
		Pipe Size (NPS), inches											
Fiberglass	\(^1\)/\(_2\)″	0.37	0.49	0.59	0.67	0.81	1.03	1.24	1.44	1.84	2.24	2.60	3.30
insulation	1″	0.41	0.51	0.64	0.80	0.91	1.13	1.36	1.61	2.04	2.47	2.90	3.73
thickness,	1.5″	0.44	0.54	0.67	0.84	0.94	1.17	1.43	1.67	2.11	2.57	3.03	3.87
Inches	2″	0.44	0.56	0.69	0.86	0.97	1.19	1.44	1.71	2.16	2.63	3.10	3.94

TABLE 8.3

Pipe Sizes

Nominal pipe size or copper tube size, inches (NPS for iron, CTS for copper)	\(^3\)/\(_4\)	1	1.25	1.5	2	2.5	3	3.5	4	5	6	8
Actual iron pipe outside diameter, inches	1.050	1.315	1.660	1.900	2.375	2.875	3.500	4.000	4.500	5.563	6.625	8.625
Actual copper pipe outside diameter, inches	0.875	1.125	1.375	1.625	2.125	2.625	3.125	3.625	4.125	5.125	6.125	8.125

Figure 8.28   Uninsulated valve.

Figure 8.29   Uninsulated flanged elbow.

Figure 8.30   Insulated valve (left), pump body (center), and elbow (right).

Insulation requirements are established by codes and standards such as the International Energy Conservation Code, ASHRAE Standard 90, and ASHRAE Standard 189 for high-performance installations. For example, for typical hydronic systems (hot water temperatures 141 to 200 F), International Energy Conservation Code (IECC) 2009 requires 1″ thick pipe insulation for nominal pipe or tube sizes less than 4″ diameter, and 1.5″ thick insulation for pipes 4″ diameter and larger. However, the higher-performance ASHRAE Standard 189 requires 1.5″ thick pipe insulation for nominal pipe or tube diameters less than 4″, and 2″ thick insulation for pipes 4″ diameter and larger.

Interestingly, the thickness of the insulation primarily affects percent energy savings, regardless of pipe size, especially for larger pipe sizes (see Figure 8.31). Note how 2″ thick insulation saves approximately 92 percent of heat losses for larger pipes, 1.5″ thick insulation saves 90 percent, 1″ saves 87 percent, and $^1/_2$″ saves 80 percent.

The biggest insulation gains are from insulating bare pipe, even with minimum insulation, and the returns diminish as increased insulation is added. However, it is both easy and well worth evaluating the incremental benefit of adding insulation to already-insulated piping, or of using thicker insulation than required by code. Energy code requirements typically eliminate 85 percent to 90 percent of energy losses. Additional insulation increases savings to the 90 percent to 95 percent range, and the incremental return on investment can be competitive with other energy investments.

At the same time, insulation is widely, and mistakenly, regarded as a low-cost improvement. Properly insulating piping can be a significant cost. For example, insulating 200 feet of 3″ diameter piping to meet the requirements of ASHRAE 189 is estimated to cost almost $5,000.[9]

In addition to insulation, leaks in piping, fittings, and equipment should be sought and eliminated. Leaks should especially be sought in mechanical rooms, where hot water leaks can flow down floor drains without being noticed. Water meters are recommended for boiler feed-water lines, in order allow ongoing assessment of any leaks. A closed hydronic heating system should consume no water.

Further energy savings are possible by reducing the temperature of hot water piping, where possible, and likewise raising the temperature of chilled water.

Some older pipe insulation contains asbestos (Figure 8.32). Asbestos is a dangerous, cancer-causing material when its fibers are disturbed and released into the air. Asbestos is best handled by professionals, to be removed or encapsulated. Otherwise, it should be left undisturbed.

PUMPING

Pumps are used for circulating water, or glycol/water mixtures for systems at risk of freezing. Pumps are most commonly used in hydronic heating, chilled water, condenser water, and other systems such as boiler/tower water loop heat pumps. Smaller pumps can be used in large systems for purposes of zone control, basically as an alternative to zone valves, for example. Pumps are also used in specialty systems, such as solar thermal systems, for water pressure booster systems, for well water systems, and more.

In larger commercial installations, two pumps are frequently installed in parallel, to allow redundancy. One of the two pumps operates when needed, and the other is a spare. In small systems, pumps are referred to as *circulators*.

Pumps are frequently classified by their mounting style. *Base-mounted pumps* (Figure 8.33) are supported by a frame, which itself rests on the floor or on a concrete pad.

Close-coupled pumps can be base-mounted, but the motor and pump are shipped in an assembly, and do not require field alignment. *In-line pumps* (Figure 8.34) are small enough to be supported by the pipes in which they pump water.

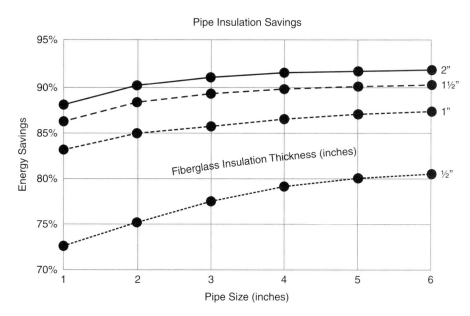

Figure 8.31   Pipe insulation savings for fiberglass insulation on hot water pipe at 150 F, losing heat to 80 F ambient air.

Figure 8.32   Asbestos insulation on water pipe.

In examining a pumped water system, it is helpful to identify the direction of flow. Flow direction may be marked on a pump, or by labels on the piping. If not, the direction of flow may be identified by recognizing that water flows into the middle of the pump (axial direction—in the direction of the pump and motor axis), and leaves from the perimeter of the pump (tangential direction), as shown in Figure 8.35.

Options to save pump motor energy include:

- High efficiency motors.
- Variable speed drives.
  - On larger pump motors, variable speed drives can be retrofit. These are most typically applied to systems in which zone radiation is shut with two-way valves, resulting in an increase in pressure that can be used as a signal to slow the pump speed.
  - On smaller motors, integrated pump/motor/control assemblies have been developed that automatically slow the pump motor speed in response to an increase in pressure. These typically use brushless permanent magnet (BPM) motors.
- Replace a pump with a more efficient pump, including possible downsizing. For example, assess if pump flow has been excessively reduced at balancing valves.
- Controls: turning off pumps if/when not needed. For example, seek hydronic pumps that are inadvertently operating outside of the winter season.
- System redesign/optimization
- Where new piping is being installed, consider using larger pipes. One pipe-size larger than usual significantly reduces pumping power.

Figure 8.33   Base-mounted pump.

HEAT DELIVERY

Heat delivery in hot water and steam systems is most commonly performed by *perimeter radiation*. This can be fin-tube radiation, cast iron radiation, or fan coils.

Figure 8.34   In-line pumps.

Low pressure on left and high pressure on right confirms that flow is from left to right

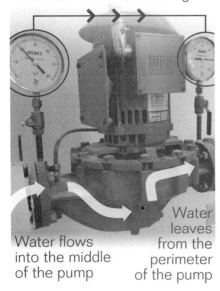

Water flows into the middle of the pump

Water leaves from the perimeter of the pump

Figure 8.35   Pump flow direction.

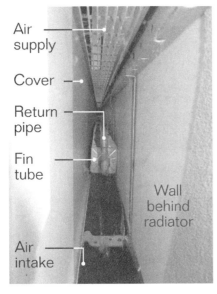

Air supply

Cover

Return pipe

Fin tube

Air intake

Wall behind radiator

Figure 8.36   Perimeter commercial baseboard radiation.

Commercial baseboard (Figure 8.36) one-tier (row of fin tube) high is 8 to 10″ high, using 4.5″ square fin on tube between 1″ and 1 1/4″ in diameter. Two-tier high is typically enclosed in a cabinet 14″ high or taller, and three-tier is typically enclosed in a cabinet at least 28″ high. Residential baseboard radiation (Figure 8.37) is typically enclosed in a cabinet that is 8 to 10″ high, using fin that is 3″ square, or smaller, on a 3/4″ tube.

Fan coils are small enclosures that contain a fan and a coil. They come in many different sizes and shapes, including *cabinet heaters*, PTACs, *unit ventilators* (larger fan coils, typically in classrooms), and *suspended unit heaters* (Figure 8.38). If the fan coil is served by separate piping systems for heating and cooling, it is called a *four-pipe fan coil*, identified by four pipes connecting to the fan coil. If the fan coil is served by a single piping system that serves heating in winter and cooling in summer, it is called a *two-pipe fan coil*.

Refrigerant Systems

Refrigerant is the medium used for transporting heating or cooling from outdoor units to fan coils or air handlers in split systems, and within packaged systems.

Refrigerant is typically carried in refrigerant-grade copper *piping*, also referred to as *tubing*, or *lines*. Unlike copper piping used for water, refrigerant piping has nominal dimensions that are identical to the outside diameter.

Energy codes and standards require insulation for refrigerant piping. For typical air conditioning piping with temperatures between 40 and 60 F, IECC 2009 requires 0.5″ insulation for piping smaller than 1″ diameter, and 1″ insulation for piping 1″ and larger. ASHRAE Standard 189.1–2009 requires 1″ insulation for pipes less than 1.5″ diameter, and 1.5″ insulation for pipes 1.5″ and larger.

There are two refrigerant pipes that connect an outdoor unit with its indoor unit(s): the low-pressure *suction line* and the high-pressure *liquid line*. The suction line is larger in diameter than the liquid line.

By energy codes and standards, only the suction line must typically be insulated, because the liquid line is generally above 60 F. However, standard practice is to insulate both pipes.

Insulation on the suction line is used not only for energy conservation, but also to prevent condensation (also called *sweating*) on the outside of the cold pipe. As such, the insulation also needs to stand up to exposure to moisture.

Elastomeric (rubber) insulation is common for refrigerant piping, because the piping does not need to withstand high temperatures. This insulation is effective; however, it deteriorates when exposed to sunlight (Figure 8.39). Protection can be provided with ultraviolet-resistant paint, or by jacketing the insulation.

The impact of refrigerant pipe insulation on energy use is not high. An uninsulated suction line will result in heating of the cold suction gas, increasing its *superheat* (temperature difference between the suction gas and the evaporator temperature). Increased superheat causes an increase in capacity, but also causes an increase in compressor power, so the impact on efficiency is small.[10]

An approach to boosting efficiency has been to install a heat exchanger between the suction line and liquid line. However, research on this improvement has shown that whereas under some conditions, this increases the efficiency of the system, under other conditions it actually reduces the efficiency.[11]

However, an aspect of refrigerant piping that does strongly impact air conditioner efficiency, as well as that of heat pumps, is refrigerant leaks. Leaks can be identified through lack of system capacity, in other words, a building that cannot stay sufficiently cool during air conditioning, or that cannot stay sufficiently warm during heat pump heating. However, a system low on refrigerant, and low in efficiency, may not always have sufficiently low capacity to not meet comfort temperature set points. In other words, a system may be low on refrigerant, operating inefficiently, but still deliver comfort. Adequate refrigerant can also be evaluated by measuring refrigerant pressures and temperatures. Leaks can also be detected by refrigerant gas leak detectors. Once a condition of low refrigerant is known, the exact location can be determined by placing liquid soap on refrigerant joints, where leaks will bubble.

Other energy losses related to refrigerant piping are unsealed pipe penetrations, through walls or roofs, which are sites for air infiltration.

References

1. Mark Modera, "Remote Duct Sealing in Residential and Commercial Buildings: Saving Money, Saving Energy and Improving Performance." Berkeley, CA: Lawrence Berkeley National Laboratory, Environmental Energy Technologies Division, 2006, pp. 12, 47.
2. Minneapolis Duct Blaster, Operation Manual (Series B Systems), Minneapolis, MN: The Energy Conservatory, p. 34. July 2014.
3. http://energy.gov/eere/buildings/downloads/ashrae-standard-152-spreadsheet. Accessed April 18, 2015.
4. http://www.mesothelioma.com/asbestos-exposure/products/duct-adhesive/. Accessed April 19, 2015.
5. U.S. Department of Energy, "Duct Insulation and Sealing Requirements in Commercial Buildings." Energy Efficiency and Renewable Energy, PNNL-SA-88299, May 2012.
6. http://www.wbdg.org/design/midg_design_echp.php. Accessed April 21, 2015; http://www.engineersedge.com/heat_transfer/heatlossinsulatedpipe/heat_loss_insulated_pipe_equation_and_calculator_13169.htm. Accessed April 21, 2015; http://www.pipeinsulation.org/. Accessed April 21, 2015.
7. Developed from "Piping Insulation, Continuing Education from the American Society of Plumbing Engineers," February 2013, p. 5.
8. ASTM C 1129–12, "Standard Practice for Estimation of Savings by Adding Thermal Insulation to Bare Valves and Flanges." American Society for Testing and Materials.
9. http://www.wbdg.org/design/midg_design_echp.php. Accessed April 22, 2015.
10. A. E. Dabiri and C. K. Rice, "A Compressor Simulation Model with Corrections for the Level of Suction Gas Superheat." *ASHRAE Transactions* 87, Pt. 2 (1981).
11. P. A. Domanski, D. A. Didion, and J. P. Doyle, "Evaluation of Suction Line-Liquid Line Heat Exchange in the Refrigeration Cycle." International Refrigeration and Air Conditioning Conference, Paper 149, 1992.

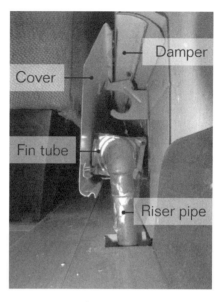

Figure 8.37 Perimeter residential baseboard radiation.

Figure 8.38 Suspended hot water unit heater. It is identifiable as a hot water heater by the insulated hot water pipes, rather than a gas-fired unit heater (there is no gas pipe or vent), or electric heater (there is no large electric power supply to the unit).

Figure 8.39 Deteriorated exterior refrigerant pipe. insulation

Chapter 9

Ventilation

Background

Mechanical ventilation is the movement of air by fans, in or out a building, in order to improve indoor air quality. *Natural ventilation* is the movement of air by natural forces, such as wind and buoyancy.

Indoor air quality is compromised when airborne contaminants accumulate, and their concentrations become objectionable. Contaminants can be gases, particulates, odors, and more. Contaminants can be highly dangerous, such as carbon monoxide, or simply objectionable, such as odors. Airborne water, or *humidity*, is a contaminant that may not be objectionable, but that can cause many problems, such as mold growth and the deterioration of materials.

Approaches to maintain good indoor air quality fall into four categories, presented in order of priority:

1. **Reduce or eliminate the source of contaminants** Great strides have been made using this approach, including reduction of indoor tobacco smoking, and the development of materials such as paints and other finishes that emit fewer harmful contaminants.

2. **Source capture and removal** This is the approach of exhaust fans, such as those used in bathrooms, kitchens, and lab fume hoods.

3. **Source filtration** For contaminants that have not been captured at their source, we try to remove them with filters, typically in air handlers and fan coils. A trend toward higher-efficiency filtration, and even the removal of some gases with chemical filters, such as charcoal filters, has broadened possibilities in source filtration.

4. **Contaminant dilution** By diluting contaminants with outdoor air brought indoors, we seek to reduce the contaminant concentrations below levels at which they are objectionable.

Ventilation serves the second and fourth of these functions. Exhaust fans provide source capture and removal. Outdoor air brought into buildings provides contaminant dilution.

An increase in ventilation in a building dramatically and immediately impacts energy use. Mechanical ventilation requires the use of electricity for fans, and, more significantly, outside air brought into a building increases the use of energy for heating and cooling. Even if a building has only exhaust fans, the exhaust draws in outside air by infiltration. As soon as outside air enters a space that has heating or cooling, it impacts the temperature in the space and becomes an added load that increases heating or cooling energy use.

Providing effective and energy-efficient ventilation in buildings is a challenge. Common problems include underventilation (poor indoor air quality), overventilation (high energy use), discomfort (if ventilation is not heated or cooled correctly), and ventilation equipment failures that go unnoticed. In examining buildings to find ways

Figure 9.1   Upblast exhaust fan.

Figure 9.2   Downblast exhaust fan.

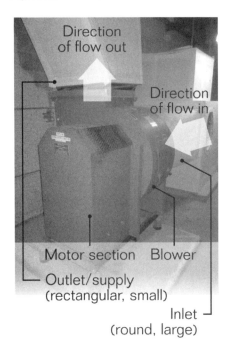

Figure 9.3   Centrifugal fan (blower).

to save energy, we frequently find inadequate ventilation, the solution for which may be increasing energy use, rather than reducing it. The design and operation of ventilation is subject to changing best practices and changing code requirements. When ventilation equipment fails, we frequently do not know that it has failed, unlike other energy loads (heating, cooling, hot water heating, lighting) for which failures are immediately noted and reported by building occupants. In summary, it is difficult to get ventilation right. Nevertheless, ventilation presents a major opportunity for saving energy in buildings, and many effective improvements present themselves.

Identifying Ventilation Equipment

EXHAUST

Exhaust fans use approximately 33 percent of all fan and pump power in buildings, more than all pumps combined, for example.[1] Exhaust fans in commercial buildings are most often located on flat roofs, but can also be wall-mounted or located indoors. Most exhaust fans are *centrifugal fans*, also called *blowers, blower wheels*, or *squirrel-cage wheels*. The blower wheel is cylindrical in shape, air flows into the wheel in the direction of the cylinder's axis, and leaves tangentially. In other words, the air turns 90 degrees. A few exhaust fans are *propeller fans*, or *axial fans*, in which air enters and leaves in the same direction, moving along the fan's axis.

On drawings, roof-mounted exhaust fans are typically identified by the acronym *PRE, for power roof exhausters*. Exhaust fans serving kitchens are usually in an *upblast* configuration (Figure 9.1), to exhaust grease-laden air up and away from the building. Exhaust fans serving bathrooms usually exhaust their air down toward the building, in a *downblast* configuration (Figure 9.2).

When located inside buildings, exhaust fans can be ceiling-mounted with a built-in exhaust grille, or hood-integrated kitchen exhaust fans, or can be any of a variety of in-line fans, such as cabinet fans or in-line axial fines.

It is frequently helpful to tell which way air is flowing in a ducted ventilation system. Blowers have a distinctive shape that allows identifying the direction of flow (Figure 9.3). The smaller discharge duct, often rectangular, is where air leaves the fan.

Exhaust fans can be *belt-driven* (Figure 9.4), in which the motor is connected to the fan shaft through two pulleys and a belt, or *direct-driven*, in which the motor shaft is also the fan shaft. Belt-driven fans allow mechanical adjustment to the fan speed. Direct-drive fans are more efficient because they do not have pulley losses. Over time, the trend seems to be toward direct-drive fans, not only because of efficiency but fewer moving parts (and so fewer parts to fail), and because of the advent of variable speed drives to allow speed adjustment through the motor speed rather than through mechanical adjustment.

GRAVITY VENTILATORS AND GOOSENECKS

Sometimes, what looks like an exhaust fan is actually just a *gravity ventilator* (Figure 9.5), in other words, an opening in the building that allows air to leave (or enter). These openings may either be connected to in-line fans inside the building, or may be simply openings in the building intended to relieve building pressure, or to allow air to enter. Gravity ventilators are slightly lower profile than exhaust fans, and usually have only one cap, rather than two. Other gravity/relief ventilators are *goosenecks*, shaped to provide built-in rain protection, as well as relief dampers built into equipment such as packaged heating/cooling units (Figures 9.6 and 9.7).

TURBINE VENTILATORS

Most exhaust fans are motor driven and therefore consume electricity. However, some exhaust fans are wind driven, and are referred to as *turbine ventilators, rotary turbines,* or *attic vents.* Turbine ventilators (Figure 9.8) can be visually distinguished by their unique shape, with a visible turbine turning in the wind. Turbine ventilators do eliminate the need for fan motor energy. However, they rely on prevailing wind, and so have limited control, and can add to the thermal ventilation loads (heating and cooling) if they run 24/7. They are frequently used to ventilate unconditioned spaces, such as attics, for which they are a good application. Some control can be added to turbine ventilators, by using automatic dampers, which open only when ventilation is needed.

GRILLES

In examining a building, if we do not have the benefit of drawings, we frequently wish to tell if a specific grille (in ceiling, wall, or floor) is for ventilation, or for heating/cooling. Peering in the grille can allow seeing if it is connected to a *riser* (vertical duct), in which case the fan is usually above, on the roof or, occasionally, below, in the basement or first floor mechanical room. Grille location is often a hint of what it is connected to: Grilles in bathrooms and kitchens are typically connected to exhaust fans, whereas grilles in corridors are typically connected to makeup air units, which draw in and deliver outdoor air to the building. The direction of airflow helps to visually distinguish between either exhaust/return, if air is entering the grille, and supply/makeup, if air is coming out of the grille. Bathroom grilles (Figure 9.9) are typically small, and either wall mounted or in the ceiling. Kitchens can be exhausted through either similar wall or ceiling grilles, or through hoods.

HOODS

Hoods are a type of exhaust capture device, located over a source of contaminants.

Commercial Kitchen Hoods
Commercial kitchen hoods (Figure 9.10) are typically several feet long, and are located over commercial cooking equipment. The airflow rates can be estimated from the length of the front edge of the hood[2]:

- Light duty: cooking surface < 450 F : 150 CFM/foot
- Medium duty: cooking surface 450–600 F : 250 CFM/foot
- Heavy duty: cooking surface > 600 F : 300 CFM/foot
- Extra heavy duty: cooking surface > 700 F : 500 CFM/foot

Extra heavy duty hoods are mainly for solid fuels like charcoal, and heavy duty hoods are mainly broilers, so most typical commercial kitchen hoods are light or medium duty. If the cooking surface temperature is not known, 250 CFM/foot is a reasonable starting assumption for airflow rate.[3]

Residential Kitchen Hoods
Residential kitchen hoods are typically two feet wide. If a hood has a grille on the front, it is a recirculating hood, and so is not providing any exhaust ventilation. Nominal airflow rates are typically 100 CFM. The hoods can have built-in fans or can be connected to a remote fan.

Lab/Fume Hoods
In labs and industrial settings, fume hoods are used to capture and remove contaminants. See Figure 9.11.

OUTSIDE AIR

Outside air is air brought into a building for purposes of diluting contaminant concentrations in indoor air, or to replace exhaust air, or both. Outside air is abbreviated as *OA* on drawings. Outside air is also called *makeup air.*

Figure 9.4   Belt-driven exhaust fan.

Figure 9.5   Gravity ventilator.

Figure 9.6   Relief damper built into packaged heating/cooling unit.

Figure 9.7   Gooseneck air intake on a vegetated roof.

Figure 9.8   Turbine ventilator.

Figure 9.9   Bathroom exhaust grille.

Outside air can either be brought in with a dedicated fan, or it can be drawn in by the low pressure at the intake of an air handler fan, typically combined with return airflow. When brought in with a dedicated fan, it is typically referred to as *100 percent outside air*, in what is called a *makeup air unit*, which frequently integrates heating, and sometimes includes cooling. A *heat or energy recovery ventilator* is a form of makeup air unit which also integrates exhaust. When outside air is brought in at the return air stream of an air handler, it is frequently combined with the return air in a *mixing box*, with two sets of dampers, one in the return air stream and one in the outdoor air stream, controlled by a single linkage connected to a motorized actuator. Increasing the outside air decreases the return air and vice versa.

Outside air intake locations can be identified on the outside of buildings by *louvers*, the intent of which is to allow air into a building while keeping rain out (Figure 9.12). Louvers usually include a wire mesh or screen, to also keep birds and other animals out. Outside air connections can also be identified from within air handlers (Figure 9.13) or mechanical rooms by openings to the outdoors, and the characteristic interlocked dampers serving the two airstreams, outside air and return air.

Establishing Existing Ventilation Rates and Electric Power Draw

Three ventilation quantities are of interest in establishing existing ventilation energy use and in developing a baseline against which improvements can be assessed to predict possible energy savings:

- Airflow rates
- Airflow duration
- Fan power draw

AIRFLOW RATES

Airflow is measured in units of cubic feet per minute (CFM).

Design airflow may be obtained from original or as-built drawings. Design airflow may not match actual airflow. Actual peak airflow may be obtained from *test and balance* (TAB) measurement reports. Airflow may also be measured and displayed on energy management systems. Airflow may be obtained from the make and model of fans, although actual airflow depends on many factors, including the size of ductwork and other flow restrictions, so make and model is of limited accuracy in establishing existing airflow rates or power draw.

Airflow may be obtained by direct measurement. Airflow may be measured at grilles or diffusers using a balometer (flow hood), or indirectly measured at a grille, or across a duct, by measuring air velocity and multiplying velocity by the open cross-sectional area of the grille or duct. Measuring airflow is not easy. Airflow can be measured at grilles and summed over all grilles; however, this does not account for air leakage, which is usually substantial. Measuring at the fan itself, typically on a roof or wall, is possible but is also not easy, requiring a shroud to be placed over the fan, but challenged by wind, and by access difficulty. Airflow traverse measurements, measuring air velocity across a duct, and multiplying the average air velocity by the cross-sectional area of the duct, is another option, but is subject to fluctuations.

Outside air brought in at an air handler may be measured indirectly, if its temperature is different than the return air temperature, by using the air mixing method. See Appendix C.

Useful times of day for measuring outside airflow in most commercial buildings is early morning or evening. These are times when ventilation should be at a minimum, due to low occupancy in typical commercial buildings, before or after working hours. If ventilation is high during these times, this can indicate an opportunity for reducing ventilation. In the absence of airflow measurement, even observation of outside air damper positions can be informative at these times, to assess if there is significant outside air being drawn into a building during unoccupied periods. If the outdoor air damper position is closed, it is still worth examining if there is substantial leakage, and if the dampers have gaskets to minimize air leakage.

Measuring carbon dioxide (CO_2) concentration in buildings can be used as an indirect measurement of the combination of ventilation and infiltration. If we assume typical rates of human activity, the combined rate of ventilation and infiltration can be estimated as:

$$V = P \times 10,500 / (Cs - Co)$$

In which:

V = combined ventilation and infiltration rate, CFM
P = number of people
Cs = indoor CO_2 concentration (ppm)
Co = outdoor CO_2 concentration (ppm)

For example, if 100 people are in a building, the indoor CO_2 concentration is 600 ppm, and the outdoor CO_2 concentration is 400 ppm, the combined rate of ventilation and infiltration is $100 \times 10,500 / (600 - 400) = 5,250$ CFM. Carbon dioxide measurements are best taken at a steady condition, after people have been in the space for a few hours, so make sense to measure CO_2 in the midafternoon in a typical commercial building. A good location to take a carbon dioxide measurement is in the return airstream (before outdoor air mixing), as this provides a mechanical average of indoor carbon dioxide concentration.

When we measure or estimate ventilation airflow rates, a single measurement or data point is typically sufficient for constant flow systems, such as typical exhaust fans. However, for variable flow systems, such as the modulating outdoor airflow that is drawn in by many air handlers, or for systems being controlled by variable speed drives, test and balance measurements or design airflow on drawings only provide peak capacity, and not the airflows that vary over time. The varying airflow rates of ventilation present one of the biggest challenges of understanding energy use in buildings, and yet are important as ventilation can represent 20 percent or more of the energy use. The best way to understand such a varying-airflow ventilation system is to track ventilation rates over at least 24 hours, and preferably for a full week. This allows accurate estimates of energy savings through ventilation improvements.

AIRFLOW DURATION

In parallel to the challenge of knowing how much ventilation is occurring is the challenge of knowing when it is happening. This problem is amplified for energy auditors by the short duration of an energy audit site visit. And so this is where, again, the collaboration of the energy manager and the energy auditor can be put to best effect. The energy manager may know the existing ventilation schedules, or is in a good position to find out.

Tables can be used to track and report the ventilation schedule (see Appendices K and L). This information may be obtained from an energy management system. If an energy management system is not available, short-term monitoring of parameters such as outside air temperature, return air temperature, and mixed air temperature may be done, in order to obtain the fraction of outside air, using Appendix C. Or, at a minimum, the position of the outdoor air damper may be observed at hourly intervals over a day. However, caution is advised using this last approach because the position

Figure 9.10   Commercial kitchen exhaust hood.

Figure 9.11   Lab fume hood.

Figure 9.12   Louvers serving an outside air intake.

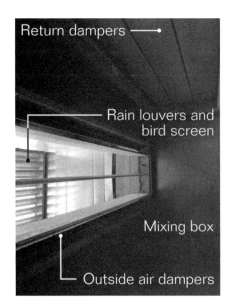

Return dampers

Rain louvers and bird screen

Mixing box

Outside air dampers

Figure 9.13 Outside air intake, seen from inside an air handler.

of the outdoor damper (fraction open) does not relate simply (linearly) to outdoor airflow.

With the duration and magnitude of existing ventilation, the existing occupancy can be compared to ventilation code requirements, in order to assess potential energy savings from matching the ventilation to code requirements.

FAN POWER

Fan power can vary widely, depending on fan efficiency, motor efficiency, and duct design. Research has shown variation in overall airflow efficiency between 0.8 and 5.3 CFM/watt.[4] Fan power is ideally measured, and this measurement is justified when multiple fans are candidates for replacement on a building. In the absence of measurement, fan power can be estimated as the rated horsepower (HP) × 0.75 (in kilowatts).

Codes and Standards

Ventilation requirements are important for energy work. Ventilation requirements are typically captured within the mechanical code. The mechanical code itself typically follows changes to American Society of Heating, Refrigerating, and Air-Conditioning Engineers (ASHRAE) Standard 62, usually with a delay. While ASHRAE 62 is a standard, and not a code, it carries great weight as a best practice and is often adopted as a code requirement.

It is helpful to know how the existing ventilation in a building might have sought to comply with the code requirements at the time the building was designed, and to know what current requirements are, when evaluating improvements to save energy. We also seek to do no harm. We want to improve indoor air quality in buildings, as we seek to reduce energy use.

HISTORY AND MAJOR CHANGES

In 1836, Thomas Tredgold, a Cornish mining engineer, estimated that just over 4 CFM per person was required for human metabolic needs. In the latter half of the 1800s, it was observed that diseases spread more rapidly in army hospitals with poor ventilation, and as a result, the American Society of Heating and Ventilation Engineers (ASHVE) adopted a rate of about 30 CFM per person, in 1914.[5]

ASHRAE's first version of Standard 62 reduced the recommended ventilation rate in response to the energy crisis of 1973, down to 10 CFM per person, and in 1981 this was further reduced down to 5 CFM. In 1989, ASHRAE 62 increased the rate to 20 CFM per person for most commercial spaces (offices, dining areas, etc.), slightly less (typically 15 CFM per person) for a variety of other spaces (assembly, lobbies, etc.), and more for spaces with unusual contaminants, up to 60 CFM per person for smoking lounges. These rates carried through the 1999 version of ASHRAE 62. ASHRAE 62–1989 also introduced a rationale for 15 CFM per person to control carbon dioxide concentrations below 1,000 ppm, which in 1999 was changed to a target of 700 ppm above the outdoor concentration. In 2004, ventilation rates were split into a per-person component and a per-area component. For example, for offices, the standard requires 5 CFM per person, plus 0.06 CFM per square foot. For a default occupant density of 5 people per 1,000 square feet (if the actual occupant density is not known), this represents 17 CFM per person, again, for offices. In 2004, ASHRAE 62 was split into a commercial version (62.1) and a residential/low-rise version (62.2). Ventilation for healthcare facilities are typically covered by health codes, and by ASHRAE Standard 170.

ASHRAE's Standard 62 has long allowed two compliance paths, the Ventilation Rate Procedure, and the Indoor Air Quality Procedure. Most buildings were designed according to the Ventilation Rate Procedure. The Indoor Air Quality Procedure requires the documentation of contaminants of concern, goals for their control (target maximum concentrations), the design approach to control the contaminants, and strategies to confirm that the goals have been met. In 2007, a prescriptive Natural Ventilation Procedure was added, in its own section, allowing the use of such strategies as operable windows, but still requiring backup mechanical ventilation capacity.

The International Mechanical Code (IMC) typically adopts significant portions of ASHRAE Standard 62, although usually with a delay. For example, through 2007, the IMC used ventilation rates from ASHRAE 62–1989. In 2007, the IMC adopted the ventilation rate procedure from ASHRAE 62–2004, and this went into effect in the 2009 version of the IMC. The 2012 IMC referenced ASHRAE 62–2010. State and local jurisdictions usually adopt the IMC, after a delay of their own, but alternately sometimes keep prior ventilation rates from earlier versions of the IMC or ASHRAE 62. So there is merit to knowing one's own state or local ventilation requirements. We cannot assume that the current version of ASHRAE 62 is in effect, even if we look to it as a best practice.

ASHRAE 62 also contains information on a variety of ventilation practices, such as the proximity of exhausts to intakes, filtration, and more.

REDUCING VENTILATION AND SHUTTING VENTILATION

All ventilation codes and standards allow ventilation to be reduced, in response to reduced occupancy. Early versions of ASHRAE Standard 62 allowed ventilation to be shut off during unoccupied periods, as long as a space was ventilated in advance of occupancy.

However, ASHRAE ruled unequivocally in 2012 that under Standard 62–2010 outside air could not be shut off, in other words, that a minimum level of outside air needs to be provided, to meet the area-based ventilation requirement, even when a space is unoccupied.[6] It is likely that, in the future, ASHRAE 62 will again allow outside air to be shut off when a space is unoccupied.[7] Again, state and local ventilation requirements may well allow ventilation to be shut off, even if the current version of ASHRAE Standard 62 does not.

Estimating Ventilation Energy Use

FAN POWER

The ventilation fan power energy consumption (kWh) is the fan power, in kilowatts, multiplied by the duration in hours.

VENTILATION HEATING/COOLING ENERGY USE

For ventilation heating energy use:

$$Vh = C \times (Ti - To) \times 1.08 \,/\, Eh$$

Vh = heating energy use (Btu/hr), C = airflow (CFM), Ti = indoor temperature (F), To = outdoor air temperature (F), and Eh = heating system efficiency (percent).

A challenge arises because both the airflow and the outdoor temperature can change. The energy use is best summed by performing calculations using the *bin method*, in which a calculation is done at each of a series of different outdoor temperatures, or a full *hourly model*, in which a calculation is done for each hour of the year. For ventilation cooling energy use:

$$Vc = C \times 0.0045 \times (hi - ho) / Ec$$

Vc = cooling energy use (kW), C = airflow (CFM), hi = indoor enthalpy (Btu/lb), ho = outdoor enthalpy (Btu/lb), and Ec = cooling system efficiency (EER).

The enthalpy is the total heat content of air, accounting for humidity. We need the enthalpy for cooling calculations because dehumidification occurs and requires energy from the cooling system. Again, a summing, or integration, calculation must be done over varying outdoor conditions, and over varying ventilation rates.

Ventilation Improvements

Ventilation improvements include:

- Reduce overventilation.
- Heat recovery or energy recovery.
- High-efficiency fans.
- High-efficiency motors and drives.

REDUCE OVERVENTILATION

Reducing overventilation is done in several different ways, such as substantially reducing ventilation during unoccupied periods, reducing overventilation to code-compliant levels during occupied periods, increasing ventilation effectiveness, and sealing ductwork while balancing ventilation to limit uncontrolled ventilation of spaces that do not need it.

Reducing ventilation during unoccupied periods means turning off exhaust fans, and reducing outside air to minimum levels, at the end of occupied periods. The reduction of outside air can continue into the beginning of the occupied period, taking credit for the residual fresh air that is in the building at the end of the unoccupied period.

Reducing overventilation during occupied periods means comparing current ventilation rates to rates required by code, typically by examining occupancy schedules and existing ventilation schedules. A table (see Appendix M) can be used to plan the proposed ventilation schedule, for purposes of estimating savings when compared to the baseline existing ventilation schedule. The more detail and the more accuracy on the occupancy schedule, the more potential energy savings by setting ventilation at the correct level, rather than over-ventilating.

Any ducted ventilation system, such as exhaust or makeup risers, is subject to duct air leakage. Air typically leaks into exhaust ductwork, drawn from spaces not requiring exhaust, and causing over-ventilation, and vice versa for makeup air systems. Even if ventilation is enclosed in a chase, leakage occurs into or out of the chase as well, through a variety of paths, such as through structural, electrical and piping penetrations (Figure 9.14). By sealing ductwork, and by balancing the ventilation airflows at grilles to code-compliant values, the airflow can be reduced and over-ventilation eliminated. Aerosol-based duct sealing is particularly effective at sealing concealed ductwork.

Controls play a key role in reducing overventilation. Timer control can be used to turn off exhaust fans during unoccupied periods. *Demand-controlled ventilation* means reducing ventilation in response to a demand signal. The most common demand signal is CO_2, which is used to control general ventilation for occupied spaces, because CO_2 is a good surrogate for occupancy. Examples of other demand signals include humidity for spaces where moisture is the primary contaminant

Figure 9.14 Air leakage through structural, electrical, and piping penetrations into chases.

(e.g., showers), temperature and smoke above commercial stoves for kitchen exhaust, air velocity for lab hoods, and carbon monoxide (and/or nitrogen dioxide) for parking garage ventilation.

A fairly active debate continues about the merits and risks of demand-controlled ventilation using CO_2. CO_2 is a good indicator of occupancy, and a high level of CO_2 correlates reasonably well with high levels of body odor. However, CO_2 does not tell us anything about contaminants other than people who are breathing. And control based on CO_2 is subject to a variety of vulnerabilities: sensor precision, sensor drift over time, questions relating to where to place CO_2 sensors, variability of outdoor CO_2 concentration, and more. The debate about if and how to use CO_2 control for demand-controlled ventilation spills over to discussions about whether codes and standards support its use. CO_2-based demand-controlled ventilation remains fairly widely used. See Figure 9.15.

What are anticipated ventilation rates in response to CO_2-based control? From before:

$$V = P \times 10,500 / (Cs - Co)$$

In which:

V – ventilation rate, CFM
P – number of people
Cs – indoor CO_2 concentration (ppm)
Co – outdoor CO_2 concentration (ppm)

Figure 9.15 Carbon dioxide control mounted on return duct.

This relation holds for adults performing office work, and calculates to 15 CFM/person for a difference of 700 ppm between the indoor and outdoor CO_2 concentrations ($Cs - Co$). For people engaged in different levels of activity, corrections may be made from Table 9.1.

For example, for a space in which light machine work is being done, we expect the ventilation rate to be twice as high as in a space where people are engaged in office work at desks.

For CO_2-based control, what should the CO_2 concentration set point be? The set point is important: The ventilation rate at 750 ppm indoors is twice as high as the ventilation rate at 1,100 ppm indoors, if the outdoor concentration is 400 ppm. So, ventilation energy consumption, using demand control, will be twice as high at 750 ppm indoors as at 1,100 ppm indoors. Historically, 1,000 ppm indoors was viewed as an acceptable target, in older versions of ASHRAE Standard 62. More recent versions of ASHRAE Standard 62 have recommended a 700-ppm rise above the outdoor concentration, which is typically 400 ppm, so this implies a target of 1,100 ppm indoors. A study of demand-controlled ventilation in a sample of buildings in California found set points that varied from 500 to 1,100 ppm, with an average of 860 ppm, and a most commonly used set point of 800 ppm. This means that buildings with demand-controlled ventilation, with these low CO_2 set points, are still overventilating, almost by a factor of two.[9] It also means that demand-controlled ventilation does not necessarily mean that energy savings will be delivered, if the set point is not set high enough, and in fact can increase ventilation rates and increase energy use if the set-point is not high enough. We recommend a set point of 1,100 ppm (700 ppm over the prevalent outdoor average of 400 ppm).

Demand control can also be applied to specialty ventilation systems, such as exhaust hoods (Figure 9.16). For example, a variety of sensors are available for the control of kitchen hood exhaust flow rates, including optical sensors, temperature sensors, and infrared sensors.[10]

For parking garages, in 2009 the International Mechanical Code (IMC) reduced the required ventilation rate from 1.5 CFM per square foot of garage to 0.75 CFM/SF. In 2012, the IMC introduced carbon monoxide (CO, not to be confused with carbon

TABLE 9.1

Ventilation Rate to Maintain 700 ppm Difference between Indoors and Outdoors[8]

Activity	Ventilation Rate (CFM/person)
Sleeping	10
Office work	15
Light machine work	30
Heavy work	53

Figure 9.16 Commercial kitchen exhaust hood.

TABLE 9.2

Ventilation Effectiveness (Ez)[12]

Supply	Return	Supply Air Temperature	Ez
Ceiling	Ceiling or floor	Cool	1.0
Ceiling	Floor	Warm	1.0
Ceiling	Ceiling	Warm ($<$ Tspace + 15 F)	1.0
Ceiling	Ceiling	Hot ($>$ Tspace + 15 F)	0.8
Floor	Ceiling	Cool (underfloor air distribution)	1.0
Floor	Ceiling	Cool (displacement)	1.2
Floor	Ceiling	Warm	0.7
Floor	Floor	Warm	1.0
Makeup air drawn in, return/exhaust at opposite side of room			0.8
Makeup air drawn in, return/exhaust near supply			0.5

dioxide) and nitrogen dioxide as demand-controlled options for ventilation control in parking garages. These two provisions allow for significant reduction of overventilation, while maintaining indoor air quality, both for gasoline-engine vehicles (using carbon monoxide) and diesel-engine vehicles (using nitrogen dioxide), or both.[11]

Another approach to reducing over-ventilation might be called *ventilation zoning*, in which the ventilation system is divided into smaller, separate ventilation systems. A related approach is called *dedicated outdoor air systems (DOAS)*, in which the ventilation is entirely removed from the central heating and cooling system. Whether using smaller, separate ventilation or using a separate dedicated outdoor air system, the goals are twofold:

- To reduce the overventilation that is inherent in using a large air handler to deliver ventilation air to multiple spaces, some of which may not be occupied
- To reduce the high fan power required by a large air handler for the typically smaller ventilation airflow, during periods when the high fan power of the air handler is not required for heating and cooling

Yet another approach to reducing over-ventilation is to increase the *ventilation effectiveness*. Ventilation is effective only to the degree that it reaches the *breathing zone* of building occupants. A ventilation effectiveness of 1 means that the ventilation delivered to a space is expected to all reach the breathing zone of the occupants, whereas a ventilation effectiveness of 0.5 means that only 50 percent of the delivered ventilation reaches the breathing zone of occupants. So a ventilation system with a ventilation effectiveness of 0.5 requires twice as much ventilation airflow, and associated fan energy and thermal energy, as a system with a ventilation effectiveness of 1. ASHRAE 62 provides estimates of effectiveness (see Table 9.2).

For example, a rooftop system typically supplies and returns from the ceiling. During heating, its effectiveness is 0.8. Ventilation could ostensibly be reduced by 20 percent if a separate dedicated ventilation system were installed that delivers a ventilation effectiveness of 1.

HEAT RECOVERY AND ENERGY RECOVERY

Because ventilation ideally involves the simultaneous exhaust of indoor air and the intake of outdoor air, and because the energy required to heat or cool outside air is a significant portion of ventilation-related energy use, we can save energy by using the exhaust air to heat or cool the intake air. This is done by exchanging heat between

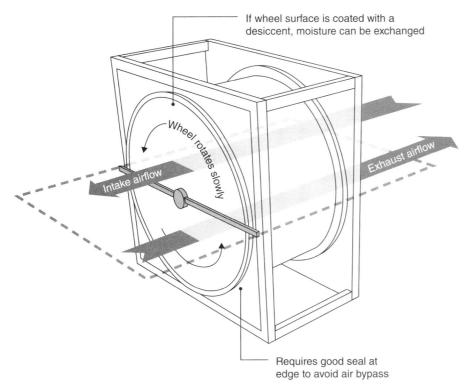

If wheel surface is coated with a desiccent, moisture can be exchanged

Wheel rotates slowly

Intake airflow

Exhaust airflow

Requires good seal at edge to avoid air bypass

Figure 9.17 Wheel heat/energy recovery ventilator.

the two air streams, and is called *heat recovery ventilation*, accomplished by a *heat recovery ventilator* (HRV). Heat recovery ventilators have also been called *air-to-air heat exchangers*.

In some situations, it is also beneficial to exchange moisture between the two airstreams. If both heat and moisture are transferred between incoming and outgoing airstreams, this is accomplished by an *energy recovery ventilator* (ERV). When is it beneficial to also transfer moisture? In winter, it is always desirable to transfer moisture, because humidity is low, and we seek to increase it, indoors. In winter, the outdoor air is dry, the indoor air has more moisture (from people, plants, and other indoor moisture sources), and it is beneficial to retain the indoor moisture indoors, in other words to transfer it from the exhaust airstream to the intake airstream. However, in the summer, the desirability of transferring moisture depends on whether or not the building is air-conditioned. In the summer, humidity is high in general, and we seek to lower humidity levels inside buildings. If a building is air-conditioned, air conditioning also dehumidifies air, and so the indoor humidity level is low, while the outdoor humidity level is high. So, in the case of air-conditioned buildings, using an energy recovery ventilator is desirable: The moisture in the outside airstream is better transferred to the outgoing airstream and sent back outdoors. However, if a building is not air-conditioned, then its summertime humidity level is higher than outdoors, due to indoor moisture sources. In this case, we would prefer not to use an energy recovery ventilator. We do not want to retain the higher level of moisture indoors. We want the higher indoor moisture level to be exhausted outdoors.

In summary, for buildings that are not air-conditioned, we prefer HRVs to ERVs. For buildings that are air-conditioned, we prefer ERV's.

The Air-Conditioning, Heating, and Refrigeration Institute (AHRI) refers to both heat recovery ventilators and energy recovery ventilators under the single term energy recovery ventilators. In other words, heat recovery ventilators are simply a class of energy recovery ventilators that do not transfer moisture.

We will use the term *recovery ventilator* to refer to both heat recovery and energy recovery ventilators, unless one or the other is specifically indicated.

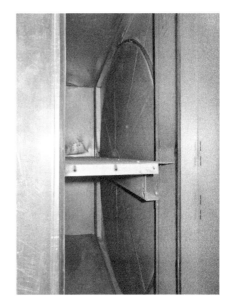

Figure 9.18 Rotating-wheel heat recovery ventilator.

Figure 9.19 Flat plate recovery ventilator—heat exchanger is at the bottom, supply and exhaust fans are at the top.

There are five types of recovery ventilators:

Wheel. The two airstreams each pass through a separate half of a rotating wheel (Figure 9.17). One airstream heats the wheel, and, as the wheel turns, it gives up its heat to the second airstream. If the wheel is coated with a desiccant (a solid that can absorb moisture), it can also serve as an energy recovery ventilator; in other words, it can also transfer moisture. Wheels are compact and efficient, but present a risk of air spilling over from one airstream to the other, and can leak air if their edge seals are not tight (Figure 9.18). Wheels have the benefit of easily allowing zero recovery (no heat transfer), by stopping the wheel, under conditions where recovery is not sought, such as using outdoor air for free cooling (economizer cooling) when outdoor conditions are cool but a building or space needs cooling, such as conditions of high internal heat gain.

Flat plate. The two airstreams each pass through separate passages of a fixed heat exchanger. These heat exchangers can be made to also transfer moisture, and so serve as energy recovery ventilators. There is low leakage with flat plate heat exchangers, and there are no moving parts, other than the fans. See Figure 9.19.

Heat pipe. The two airstreams each pass over a set of heat pipes, in which a fluid that evaporates and condenses serves as an intermediate heat transfer medium. Available heat pipe recovery ventilators only transfer heat, not moisture.

Run-around. This approach involves placing separate air-to-water heat exchangers in each airstream, connected with water piping. A pump circulates the water between the two heat exchangers. A benefit of this approach is that the two airstreams can be remote from each other. Another benefit is the provision of zero heat transfer, by stopping the pump, under conditions where no heat transfer is sought, such as free cooling (economizer cooling).

Heat Pump. A separate heat pump moves heat from one airstream to the other. Like the run-around heat recovery approach, the two airstreams can be remote from each other and economizer cooling can be provided when needed, by stopping the heat pump. A heat pump uses energy to save energy, but in some cases this may make sense.

Heat and energy recovery ventilation are proven technologies. They are particularly cost-effective if ventilation runs for many hours per day.

If kitchen exhaust forms part or all of the exhaust airstream of a heat recovery ventilator, it is important that the heat exchanger be cleanable, because commercial kitchen exhaust carries airborne grease, which contaminates a heat exchanger and reduces its effectiveness.[13]

Heat and energy recovery ventilators form frost in climates in which intake air is cold enough to bring exhaust air to its dew point, condensing water on the heat exchanger, and freezing. Most heat and energy recovery ventilators can sense conditions of frost formation, and periodically go through a defrost cycle to eliminate the frost.

Smaller heat recovery ventilators (below approximately 300 CFM) are rated by the Home Ventilating Institute (HVI). Larger energy recovery ventilators, over 50 CFM and generally up to 5000 CFM, with some provision for even larger devices, are rated by the AHRI.[14] The ability of a recovery ventilator to transfer heat is called the *sensible effectiveness*. The ability of a recovery ventilator to transfer moisture is called the *latent effectiveness*. The ability of a recovery ventilator to transfer both heat and moisture is called the *total effectiveness*. Enthalpy is a thermodynamic property of air that refers to its total energy content.

Heat recovery ventilators are rated for their effectiveness based on the following relation:

$$e = [C_2(x_1 - x_2)] / [C_{min}(x_1 - x_3)]$$

Where:

C = capacity rate for each airstream
  = $\dot{m}c_p$ for sensible effectiveness
  = $\dot{m}h_{fg}$ for latent effectiveness
  = $\dot{m}$ for total effectiveness
C_{min} = The lower of C_2 or C_3
  c_p = Specific heat of dry air, Btu/lb-F
  h_{fg} = heat of vaporization of water, Btu/lb
  $\dot{m}$ = mass flow rate of dry air, lbm/min
  x = dry-bulb temperature, T (for sensible effectiveness), humidity ratio,
    W (for latent effectiveness), or total enthalpy, h (for total effectiveness)
  e = sensible, latent, or total effectiveness

And the measurement locations, as indicated by the equation subscripts, are:

1 = outside air intake
2 = supply air from ventilator to the building (outside air after passing through the ventilator)
3 = exhaust air from the space
4 = exhaust air leaving the building (exhaust air after passing through the ventilator)

For example, the sensible effectiveness for a ventilator with equal exhaust and intake airflows, $C_2 = C_{min}$, the effectiveness is:

$$e = (T_1 - T_2) / (T_1 - T_3)$$

ANSI/AHRI Standard 1060 uses standard test conditions of 35 F entering air with 70 F exhaust indoor air (heating/winter), and 95 F entering air with 75 F exhaust indoor air (cooling/summer).

The sensible effectiveness generally varies little with changes in entering air temperature conditions. This is helpful: The rated effectiveness can then be used for a wide range of entering air conditions, for example to model savings over a whole year. This is also helpful to confirm if a particular recovery ventilator is performing according to its rated effectiveness. To be sure, the performance should be confirmed with performance at off-design conditions, which are often provided on manufacturer curves, or with computer programs provided by the manufacturer. Latent effectiveness is more dependent on entering moist air conditions, and is best confirmed with manufacturer data.

Performance also varies if the exhaust and intake are not the same airflow rates. The ratings assume the same airflow rate for both airstreams. However, in practice, we often might be faced with situations in which the exhaust airflow is different than the intake airflow. Here, the definition of effectiveness is important to understand. The heat transfer we seek is heat into or out of the intake airstream, referred to in the equation as C_2. The equation divides by the maximum possible recoverable energy, represented by the minimum airflow rate. Results can sometimes seem counterintuitive. For example, a 1,500-CFM wheel ERV has a rated sensible effectiveness of 71 percent, at standard winter conditions (1,500 CFM in each airstream, 30 F and 70 F entering air temperatures). If the exhaust airflow is reduced by half to 750 CFM, the *effectiveness* rises to 94 percent. However, the heat transferred is less than for the full-airflow effectiveness of 71 percent.

In this example, for the full-airflow case: e = 71 percent = [1500 × 1.08 × (30 − T2)] / [1,500 × 1.08 × (30 − 70)], for which T2 = 58.4, and the recovered heat is

$1,500 \times 1.08 \times (30 - 58.4) = 46,008$ Btu/hr. But for the half-exhaust airflow case: $e = 94$ percent $= [1,500 \times 1.08 \times (30 - T2)] / [750 \times 1.08 \times (30 - 70)]$, for which $T2 = 48.8$ F, and the recovered heat is $1,500 \times 1.08 \times (30 - 48.8) = 30,456$ Btu/hr. Note how the half-exhaust airflow case recovers less heat than the full-exhaust case, even though the effectiveness is higher. So, just because an effectiveness is higher does not mean that the recovered energy is higher, unless the two airflows are the same. If the airflows are not the same, the full effectiveness equation needs to be used to calculate the recovered heat. And we also conclude that if the exhaust airflow is lower than the supply airflow, the recovered heat will be less than if the two airflows are the same.

In the AHRI directory (summer 2015), the majority of the products are wheels (76 percent), followed by flat plates (21 percent), and heat pipes (3 percent), with over 30 manufacturers listing their recovery ventilator products. Heat pipes appear to be made by just one manufacturer. Ninety-five percent of the listed wheels transfer moisture (in addition to heat), whereas only 29 percent of flat plates transfer moisture, and none of the heat pipes do. Overall, 78 percent of the listed products transfer moisture.

The highest available sensible effectiveness is 82 percent for both wheels and flat plates, but only 53 percent for heat pipes. However, the average wheel sensible effectiveness is 71 percent, somewhat higher than the average flat plate sensible effectiveness of 62 percent.

There is generally little difference between the rated sensible effectiveness of wheels at the test heating condition (70 F and 30 F entering temperatures) and at the test cooling condition (95 F and 75 F entering temperatures).

However, the effectiveness does rise measurably when lower airflow is used in the same ventilator. For example, the average sensible effectiveness of wheels at 100 percent airflow is 70.7 percent, whereas the average sensible effectiveness at 75 percent airflow is 75.6 percent. So, if we can run recovery ventilators at variable speed, during part load, we can modestly boost the recovered energy.

Motor power to turn recovery ventilator wheels is generally small. For example, one manufacturer shows 25 watts to turn a $32''$ wheel, and 90 watts to turn a $48''$ wheel.

There are important comfort issues to be considered when using recovery ventilators. A recovery ventilator can never heat incoming air all the way to the temperature of exhaust air, without supplemental heat. For example, in winter, at 30 F outdoors and 70 F indoors, a typical recovery ventilator will heat the incoming air to perhaps $55 - 60$ F. This is a cold temperature, very similar to the temperature of air-conditioned supply air in summer. This air is best heated to a more neutral 70 F.

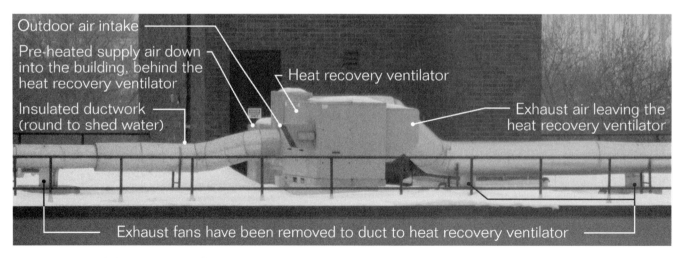

Figure 9.20   Retrofit heat recovery ventilator.

However, supplying heat to a recovery ventilator, which is seeing varying outdoor air temperatures, requires varying the rate at which heat is provided. This is not a problem for a larger system, with automatic controls. It can present challenges for smaller systems that do not have variable capacity control. Although temperature is not as much a challenge for cooling, unconditioned ventilation air can still present comfort issues. In general, attention should be directed to supplemental conditioning (heating and cooling) of makeup air provided by a recovery ventilation system, to provide temperature control, even though the device provides much of the heating and cooling required.

The International Energy Conservation Code (IECC) requires recovery ventilation, with a minimum 50 percent recovery rate (total effectiveness), for larger ventilation systems (varying by climate zone), with provision to permit free cooling (economizer cooling), with a variety of exemptions. See Figure 9.20.

HIGH-EFFICIENCY FANS

Until recently, fan efficiency (sometimes called *efficacy*) has been largely unregulated. Even today, regulation is not widespread, and available fans have a wide range of efficiencies.

For larger fans (shaft power over 1 HP), the Air Movement and Control Association (AMCA) developed Standard 205 in 2010, which was invoked by ASHRAE Standard 90.1 in 2013, with a minimum fan efficiency grade (FEG) of 67. However, FEG does not appear to translate simply into fan power.

ENERGY STAR has a certification for residential fans up to 500 CFM: 2.8 CFM/watt minimum (except bathroom and utility fans 10–80 CFM, for which the requirement is 1.4 CFM/watt minimum).[15]

HVI's residential directory shows the residential ENERGY STAR fans, and the efficiencies reach as high as over 20 CFM/watt.[16] The directory also includes larger fans. For exhaust fans in the 500–1000 CFM range (very typical for commercial and multifamily exhaust fans), there is also a wide range of efficiencies, from 1.5 CFM/watt to 4.6 CFM/watt.

In summary:

- Fan/motor assemblies have a widely varying overall efficiency, on *flow-to-wire* basis (delivered airflow divided by input fan power), varying by over 1,200 percent.

- Regulation of fan/motor assembly efficiency is limited.

- As a result, there appears to be a strong opportunity for energy improvements based on replacing low-efficiency assemblies with ones of high efficiency.

SUPPLYING MAKEUP AIR DIRECTLY TO AN EXHAUST HOOD

Another approach to reduce ventilation energy use, specifically for exhaust hoods such as commercial kitchen exhaust hoods, is to deliver unconditioned makeup air (neither heated nor cooled) directly next to the hood. Here, we are seeking to deliver the makeup air so close to the hood that it can provide its function (to replace exhaust air and to entrain contaminants that are sought to be captured and exhausted) without impacting the building air temperature, and so without becoming a load on the building's heating or cooling system, and without conditioning the makeup air either.

Many hoods do not already have makeup air directly delivered to them. If a roof upblast exhaust fan does not have a makeup air system next to it, it typically does not have direct makeup air. But even if it does have a makeup air system next to the exhaust fan, the makeup air may still be delivered to a location that impacts the heating and cooling load, or have built-in heating and cooling.

To apply this strategy, the positioning of the makeup air is important. It must be located such that the makeup air still sweeps and entrains the source of contaminants. It must be located such that most of the exhaust air is drawn from the makeup air, rather than other room air (which is a load on the heating and cooling system). And it must be located such that if its temperature (the outdoor air temperature) is objectionable, it will not present any comfort problems for people standing next to the hood, for example, cooks in a commercial kitchen. This is especially the case in cold climates, during heating season. One option is to provide the makeup air behind and below the hood. In this way, it can sweep forward and up, entraining contaminants but not flowing over occupants. Another approach is to supply the makeup air through perforations around the perimeter of the hood. Some commercial products are available to deliver makeup air, in these ways, directly to exhaust hoods.

Directly supplying makeup air can deliver significant energy savings for exhaust hoods that operate for several hours each day.

References

1. Detlef Westphalen and Scott Koszalinski, *Energy Consumption Characteristics of Commercial Building HVAC Systems. Volume II: Thermal Distribution, Auxiliary Equipment, and Ventilation*. For: Office of Building Equipment, Office of Building Technology State and Community Programs, U.S. Department of Energy, October 1999.
2. John A. Clark, "Solving Kitchen Ventilation Problems." *ASHRAE Journal*, July 2009.
3. Greenheck Fan Corp., "Kitchen Ventilation Systems: Application and Design Guide," September 2005.
4. I. Shapiro and V. Hayes, "Evaluating Ventilation in Multifamily Buildings." *Home Energy Magazine*, July/August 1994.
5. J. Kuhnl-Kinel, "The History of Ventilation and Air Conditioning: Is CERN Up to Date with the Latest Technological Developments?" Proceedings of the Third ST Workshop CERN. Chamonix, France: European Organization for Nuclear Research (CERN), 2000.
6. Interpretation IC 62.1–2010–4 of ANSI/ASHRAE Standard 62.1–2010, "Ventilation for Acceptable Indoor Air Quality."
7. ASHRAE 62.1–2010, "Ventilation for Acceptable Indoor Air Quality." Presentation by Roger Hedrick, Chair, ASHRAE Standing Standard Project Committee 62.1. Undated.
8. David S. Dougan and Len Damiano, "CO_2-Based Demand Control Ventilation." *ASHRAE Journal*, October 2004.
9. W.J. Fisk, D.P. Sullivan, D. Faulkner, and E. Eliseeva, "CO_2 Monitoring for Demand Controlled Ventilation in Commercial Buildings." Berkeley, CA: Lawrence Berkeley National Laboratory, March 17, 2010, LBNL-3279E.
10. http://www.melinkcorp.com/Products-and-Services/Kitchen-Ventilation-Controls/Melink-Intelli-Hood.aspx. Also: MARVEL, Intelligent Demand Controlled Ventilation System for Professional Galleys, Halton Marine, Pulttikatu 2, FIN-15700 Lahti, Finland. Both accessed June 12, 2015.
11. Tom Nobis, "State of Flux: IMC 2012 Changes Affect Parking Garage Exhaust Systems." International Parking Institute, October 2011, p. 32.
12. Paul Solberg, Dennis Stanke, John Murphy, and Jeanne Harshaw. Trane Engineers Newsletter Live ASHRAE Standard 62.1, 2013.
13. http://www.haltoncompany.com/halton/usa/cms.nsf/pbd/E0A38893E769593FC225774 C004852BE. Accessed August 14, 2015.
14. ANSI/AHRI Standard 1060, "Standard for Performance Rating of Air-to-Air Exchangers for Energy Recovery Ventilation Equipment." Arlington, VA: Air-Conditioning, Heating, and Refrigeration Institute.
15. http://www.energystar.gov/index.cfm?c=vent_fans.pr_crit_vent_fans. Accessed August 14, 2015.
16. http://www.hvi.org/proddirectory/index.cfm. Accessed August 14, 2015.

Chapter 10

Identifying Heating and Cooling Equipment

It is important to be able to recognize heating and cooling equipment, in order to proceed with identifying possible energy improvements. An approach that is helpful is to start from the outside of a building, and then proceed indoors.

OUTDOORS: WALK AROUND THE BUILDING, AND INSPECT THE ROOF

■ *Are there single-package HVAC units, typically on the roof (occasionally on the ground)*? Identify by ductwork into the building. If so, do they provide both cooling and heating? Cooling is identified by a condenser (typically has a round propeller fan on top) and compressor (typically visible by looking past the propeller fan). Heating is typically identified by a gas pipe and a vent (for a gas-heated system), or as a heat pump identifiable by a reversing valve (next to the compressor), or as electric resistance heat identified on the nameplate. Do the units provide ventilation (typically a triangular-shaped rain hood)? If the units provide ventilation, are they 100% outdoor air (no return duct) or not (includes a return duct)?

■ *Are there split system units outdoors, identified by the absence of ductwork, with copper pipes routed into the building, outdoor coil, and compressor inside?* If so, is it a heat pump, identified by a reversing valve (Figure 10.1), typically visible through the fan grille? If an outdoor unit has no reversing valve (Figure 10.2), it is a cooling-only system. If it is a cooling-only unit, then we will focus on finding out what kind of heating is used, when we move indoors.

■ *Are there through-wall or window room air conditioners or PTACs?* If so, proceed indoors to identify what form of heating goes with these units, either integrated into the unit (heat pump, electric, hot water, steam) or separate.

■ *Is there a cooling tower?* If so, then the HVAC system is most likely a water-cooled chiller system, with a smaller chance that it is a boiler/tower heat pump system. Differentiating between these two systems will happen subsequently, indoors.

■ *Is there an air-cooled chiller (no ductwork connected to the building, just water pipes)?* If so, then, again, we can focus on heating when we go into the building.

■ *Are there chimneys, metal vents through the roof, or dual plastic vents?* If vents are seen, we will look for a combustion system when we go indoors, either a furnace or a boiler.

Examining the roof is also helpful, when possible. A roof can tell much about what energy is used in a building. See Figure 10.3.

In the process of walking around a building exterior, information is also obtained on ventilation, including outdoor air intakes, and exhaust fans. See Figure 10.4.

After touring the building outdoors, one can proceed indoors. The mechanical rooms or basement are a typical starting point, indoors.

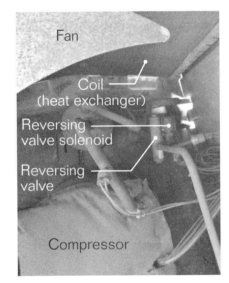

Figure 10.1 Reversing valve indicates that an outdoor unit is a heat pump.

Figure 10.2 Cooling-only outdoor unit (no reversing valve).

Figure 10.3 Rooftop equipment—note upblast exhaust fan, an indicator of a commercial kitchen; also heating and cooling rooftop unit with outdoor supply ductwork, and outdoor air intake.

Figure 10.4 Ventilation outdoor air intake.

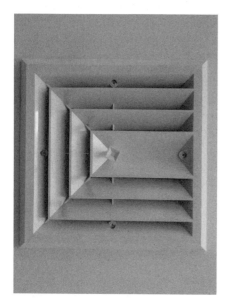

Figure 10.5 Supply diffuser.

MECHANICAL ROOMS, BASEMENTS, AND OTHER UNCONDITIONED SPACES

- *Is there a boiler or furnace?* Both can be identified by *breeching*, the horizontal connection of the boiler or furnace to a chimney or vent. Both can also be identified by the use of combustion fuel, most typically gas piping (black) or fuel oil (an oil burner on the front of the furnace or boiler).

- *If a boiler, is it steam or hot water?* Details on how to make this distinction are provided earlier.

- *Is there a water-cooled chiller?* Water-cooled chillers are most readily identified by two large horizontal cylindrical heat exchangers, and an associated compressor. If we see a cooling tower outdoors (or subsequently indoors), we will be looking for a chiller indoors. If there is no chiller indoors, a cooling tower is a sign of a boiler/tower heat pump system.

- *Are there one or more air handlers?* These can be identified by coil connections, typically hot water and chilled water, or possibly steam piping instead of hot water, or refrigerant piping from an outdoor condensing unit instead of chilled water. Air handlers can also be identified by a blower section, filter section, and typically an outdoor air intake, which meets a return air duct.

CONDITIONED SPACES

- *Examine room air conditioners and packaged terminal air conditioners (PTACs), if any.* Identify the type of heat: electric resistance, heat pump (compressor noise on call for heat), or hot water or steam coil.

- *Examine perimeter radiation, if any: baseboard, cast iron radiators, etc.* If the building has steam heat, the number of pipes connecting to the radiator identifies whether the system is single-pipe or two-pipe.

- *Examine forced air terminals.* If air flows into the room, it is a supply grille/diffuser (Figure 10.5), also referred to simply as a *supply*. Supplies can also be identified because they have louvers that spread the flow of air in the room, and they typically have an adjustment mechanism to increase or decrease the flow of air. *Linear* diffusers (Figure 10.6) are also common. If air leaves the room, it is either a return grille or ventilation exhaust grille. Exhaust grilles are more common in rooms that have a source of odors/contaminants, such as kitchens and bathrooms. Exhaust grilles often have adjustable louvers, with a screw adjustment just behind the louvers. Return grilles (Figure 10.7) are larger than supply grilles, are typically fewer, and typically do not have adjustable louvers.

- *Examine fan coils.* Are they refrigerant, electric, water, or steam fan coils? What is the fan coil mounting: floor mounted, wall mounted, ceiling surface mounted, ceiling recessed, or ducted/concealed?

- *Examine wall-mounted controls, such as thermostats, temperature sensors, and humidity controls.* Do they allow local adjustment, or are they only sensors that provide signals to a central energy management system? If local adjustment is allowed, what are their set points and programmed schedules?

EXAMINE ABOVE CEILINGS, IN ATTICS, AND OTHER CONCEALED SPACES

- *Examine the type and condition of ductwork and piping.* Note the insulation thickness, type, and condition. Is there sealant on ductwork joints, or visible holes and leakage? Are there both supply and return ductwork, or is the space above the ceiling used as return air plenum?

- *Look for the type and condition of variable air volume (VAV) boxes.*

- *Look for water source heat pumps, fan coils, and small air handlers.*

Figure 10.6   Linear diffuser.

EQUIPMENT AGE

When inventorying heating, ventilating, and air conditioning (HVAC) equipment, the approximate age of equipment is helpful to identify for two reasons:

1. Age can provide a rough guide to efficiency.

2. Age is needed to estimate remaining useful life.

Equipment age can be identified in a variety of ways. The owner or facility manager may report that the equipment is original to the building, and so the equipment age is the same as the age of the building.

The year of manufacture of equipment is frequently available from the serial number. For example, for air conditioners[1,2]:

- Bryant: Third character—R–Y is 1964–1971, A–H are 1972–1979; following 1979, the third and fourth characters are the year. Day-Night: Same as Bryant.

- Carrier: 1963–1969 the first digit is the year (e.g., 3 = 1963), starting 1970 the second character is the year (0 = 1970); more recently, third and fourth characters are the year (e.g., 99 = 1999).

- Fedders: Month-year starting with AA for September 1966, through 1977 (letter I not used); for example, CB is November 1967.

- Lennox: Prior to 1973, first two digits are the year; beginning 1974, the third and fourth digits are the year.

- Rheem or Ruud: Four characters indicate week and year; in 1960s and early 1970s, these were the last four characters, more recently in the middle of the serial number.

- Trane: Through the 1970s, year-month was used; for example, 2D was April 1972; since 1980s, date of manufacture stamped on nameplate.

- Whirlpool: Decade-year, for example G4 = 1974.

- York: Third letter, A starts in 1971, skips I, through 2005; since 2004, second and fourth characters are the year of manufacture; for example, X0X4 is 2004.

Figure 10.7   Return grille.

Figure 10.8 Significant rust indicates a unit that is likely over 20 years old.

Figure 10.9 ARI label, which indicates the equipment is older than 2007.

Figure 10.10 AHRI label, which indicates the equipment is newer than 2007.

Some manufacturers have online databases that provide the age of the equipment based on the serial number. Age decoding charts are also available for furnaces.[3] One can also contact the equipment manufacturer with the model and serial numbers, to obtain the age.

A very rough estimate of equipment age may be obtained from the appearance of the equipment, especially outdoor equipment like condensing units and rooftop units. An outdoor unit that is less than 10 years old will show little rust, just dirt. A unit 10 to 20 years old will start to show rust. A unit over 20 years old will show significant rust on components like vinyl-coated wire mesh, sheet metal, screws, and electrical disconnects (Figure 10.8).

For air conditioning equipment rated according to Air-Conditioning, Heating, and Refrigeration Institute (AHRI) standards, if the equipment has an AHRI label, it dates to 2008 or newer. If it has an ARI label, it dates to before the organization changed its name to AHRI, and so the equipment dates to 2007 or older. See Figures 10.9 and 10.10. Similarly, heating equipment with the GAMA label, for Gas Appliance Manufacturer's Association, means that it is 2007 or older. GAMA and ARI merged to form AHRI on January 1, 2008.

References

1. http://www.localinspectioncompany.com/articles/acdata.pdf. Accessed December 20, 2014.
2. http://www.usair-eng.com/pdfs/Serialnumber.pdf. Accessed December 20, 2014.
3. Scott LeMarr, "How to Determine Furnace Age." HHI, LLC, 2012.

Chapter 11

Controls

Heating and cooling controls have many functions, but it may be helpful to recognize that their main function is to control indoor air temperature. With all the complexity of control systems, we sometimes lose sight of this. Additional areas of energy controls include ventilation and lighting, and occasionally other parameters such as humidity. A desired indoor air temperature, or any controlled parameter, is referred to as a *set point*.

Control comprises *measurement*, the *control function* (comparing the measurement to the set point and making control decisions, accordingly), and *actuation* (changing the position of a controlled device to actually effect a change). For example, a 24-volt wall-mounted thermostat measures temperature and compares it to a set point that has been set by a building occupant. If heating is required, the thermostat provides a 24-volt signal to a zone valve actuator and the fan in a fan coil. The signals turn the fan on and opens a valve to allow hot water to flow to the fan coil, which delivers heat. In such simpler systems, measurement and control of temperature are integrated into a stand-alone wall-mounted or equipment-mounted thermostat. In larger systems, these functions are separated, with a temperature sensor measuring temperature in a space and sending a signal to a central control system, which results in a signal for actuation being sent by the central control to an actuator or other physical device.

Central control systems are called *energy management systems* (EMSs), *building management systems* (BMSs), or *building automation systems* (BASs), among other terms. All-electronic systems are called *direct digital control* (DDC) systems.

Thermal Zoning

Controlling indoor air temperatures in multiple spaces in a building is called *thermal zoning*. Thermal zoning reduces loads in two different ways:

- By preventing overheating in spaces that receive heat from other sources, such as lighting, machinery, people, other *internal gains*, or the sun through windows.
- By allowing unoccupied spaces to reduce their air temperature in heating, or raise their temperature in cooling.

There are three levels of thermal zoning:

REQUIRED: **Indoor air temperature control.** Without temperature control in a zone, there is no thermal zoning.

RECOMMENDED: **Temperature control plus *physical separation* between spaces, such as placing doors between thermal zones.** Separating spaces prevents migration of air between spaces, for example by the stack effect, and so increases energy savings from thermal zoning.

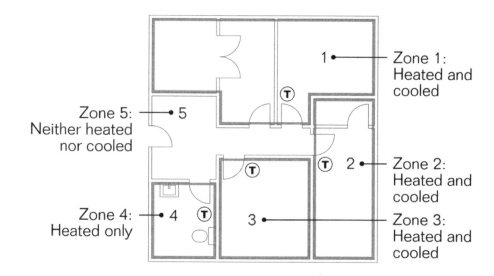

Figure 11.1   Thermal zoning diagram.

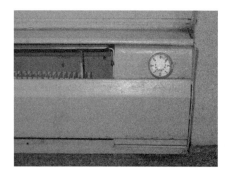

Figure 11.2   Equipment-mounted mechanical thermostat.

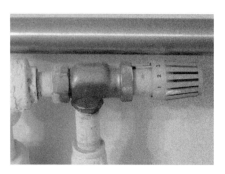

Figure 11.3   Thermostatic radiator valve.

OPTIONAL: **Temperature control, physical separation, and *insulation* between spaces.** This is typically done to separate *conditioned spaces* from *unconditioned spaces*, or where insulation is required anyway for sound isolation, such as in hotels.

A *thermal zoning diagram* (Figure 11.1) is a diagram that the energy auditor or energy manager can prepare to show thermal zones, the location of the temperature control, and whether zones are heated, cooled, both heated and cooled, or are unconditioned (neither heated nor cooled). Thermal zoning diagrams assist in understanding systems, in evaluating adding thermal zones or removing heating/cooling from zones as ways to save energy, and in conveying these issues to building owners.

Stand-Alone Controls

Stand-alone controls for indoor air temperature control are used mainly for smaller systems.

The simplest form of stand-alone control is the equipment-mounted thermostat (Figure 11.2). These include controls on through-wall air conditioners, radiator mounted controls, and controls on fan coils. These controls can be entirely mechanical, like a thermostatic radiator valve (Figure 11.3), can be electromechanical, or can be electronic.

Mechanical controls are typically not calibrated in degrees, but rather in scales such as from 1 to 5. For example, uncalibrated electric heater controls usually have a built-in range of 50 to 90 F. Radiator-mounted thermostatic radiator valves (TRVs) often have an off position, a separate position marked with an asterisk (*) that is for freeze protection, and then a series of numerical marks.

Examples of some set-point temperatures (in F) corresponding to the TRV marks are shown in Table 11.1.

Due to the position of equipment-mounted thermostats (frequently near the floor, and close to the heat source; Figure 11.4), there may not be as much meaning to the value of the set point, as compared to that of a standard wall-mounted thermostat or temperature sensor position. We presume that equipment-mounted thermostats are adjusted to a position of comfort by the occupant. Of more importance is the inability of equipment-mounted thermostats to be programmed for unoccupied periods or to be conveniently adjusted on entering or leaving a space. Remote bulbs do allow more convenient adjustment but no programming.

TABLE 11.1

Thermostatic Radiator Valve Set-Point Temperatures

Mark	Set-Point Temperature (F)			
	Armstrong	**Caleffi**	**Danfoss**	**Honeywell Braukmann**
*	45	45	46	43
1 or I	54	54	54	46
2 or II	61	61	61	54
3 or III	68	68	68	61
4 or IV	75	75	75	68
5	82	82	—	73
6	—	—	—	79

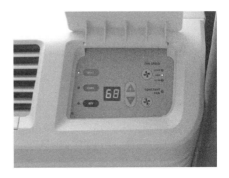

Figure 11.4   Equipment-mounted electronic thermostat.

Figure 11.5   T86 wall-mounted thermostat.

The workhorse of stand-alone controls is the wall-mounted thermostat, typically located 60″ above the finished floor of a space. Older versions of these thermostats use mercury bulbs. Still widely found is the classic Honeywell T86 thermostat (Figure 11.5), designed in the 1940s by the American industrial designer Henry Dreyfuss, and newer derivatives of this thermostat. The original patent for the round thermostat is shown in Figure 11.6. Wall-mounted electromechanical thermostats typically have two scales, one for the set point, and one to read the actual air temperature at the thermostat.

Interior components of an older wall-mounted wall thermostat include a bimetal element that curls and uncurls in response to room air temperature changes, mercury bulbs that make and break electrical contact when moved by the bimetal element, contacts for heating/cooling/fan to be controlled, and an anticipator (a small heater that allows the cycling rate of the heating system to be set). See Figure 11.7.

Wall-mounted thermostats are typically operated with 24-volt AC electricity, with power provided remotely from a transformer. The main exception is for the control of electric resistance heat, where *line voltage* thermostats (Figure 11.8) are most often used. Line voltage typically means 120 or 240 volts.

Conventional thermostats typically have a control range that is limited to 40 to 90 F, and sometimes slightly narrower (42 to 88 F is common, and some are even as narrow as 50 to 80 F). Some thermostats have temperature-limiting functions that allow an energy manager to restrict the range even further, either mechanically with an internal set-screw or electronically.

Thermostats serving systems with a fan typically have two fan positions: auto and on. In the *Auto* position, the fan runs only when there is a call for heating or cooling. In the *On* position, the fan runs continuously. The fan control position should be noted during energy audits.

Thermostats are sometimes kept inside a cover to prevent occupants from tampering with set points (Figure 11.9).

Thermostats are either nonprogrammable, allowing a single temperature to be set, or programmable, allowing different temperatures to be set according to different time schedules.

Electronic programmable thermostats (Figure 11.10) are commonly available in *7-day* versions (separate schedule for each day of the week), *5-2* versions (separate schedules for weekdays and weekends), or *5-1-1* versions (separate schedules for weekdays, Saturdays, and Sundays). A typical programmable thermostat allows different settings for four periods each day, which are generally used for two occupied and two unoccupied periods.

The readout of an electronic thermostat usually provides the actual room temperature, until a set-point button (up or down arrow) is pressed, at which point the current

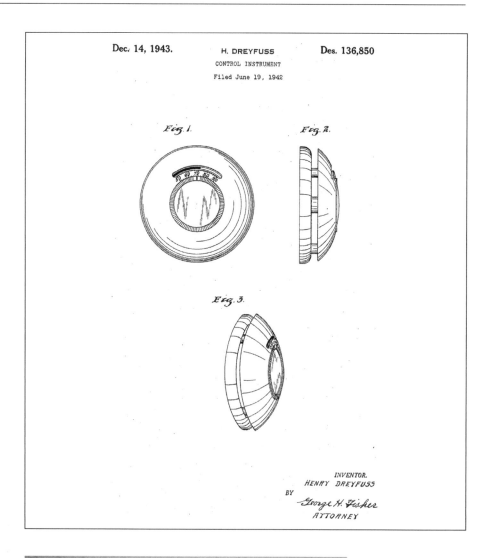

Dec. 14, 1943.

H. DREYFUSS

CONTROL INSTRUMENT

Filed June 19, 1942

Des. 136,850

Fig. 1.

Fig. 2.

Fig. 3.

INVENTOR.
HENRY DREYFUSS
BY
George H. Fisher
ATTORNEY

Figure 11.6 Original Dreyfuss patent for round thermostat.

Bimetal element

Set-point lever

W - heating
R - power to thermostat
O - heat pump reversing valve
Y - cooling
G - fan
X - backup heat

Mercury bulbs   Anticipator

Figure 11.7 Components of an older thermostat.

set point is displayed. Set points at other times in the schedule can be obtained by retrieving the programmed schedule, typically by pressing a Program or Menu button. Older programmable thermostats are electromechanical, with pins used to set occupied and unoccupied periods.

It is instructive to *call for heating* (set the set point higher), or *call for cooling* (set the set point lower), to see the current set point and to see how the heating or cooling system responds to these calls.

TABLE 11.2

Example Record of Programmable Thermostat Set-points

Day	Time	Set Point (F)		Notes
		Heating	**Cooling**	
Mon–Fri	6:00 A.M.	70	74	Workday setup
	8:30 A.M.	70	74	
	5:00 P.M.	55	85	End of work
	10:00 P.M.	55	85	
Sat–Sun	6:00 A.M.	55	85	Weekend setback
	8:30 A.M.	70	74	Weekend morning
	11:30 A.M.	55	85	
	10:00 P.M.	55	85	
Thermostat Location: Office				Fan position (auto/on): On

It is important to record set points for each thermostat. This helps with documenting existing conditions, defining improvements to the set points, and estimating energy savings. An example datasheet format for a typical 5-2 heating/cooling programmable thermostat that allows four settings per day is shown in Table 11.2.

Ductless air conditioners and heat pumps, mostly developed overseas, typically come with electronic remote stand-alone controls (Figure 11.11). Features and functions tend to be different than U.S.-made stand-alone controls.

ZONE DAMPER SYSTEMS

A version of stand-alone controls that has seen use in ducted small commercial installations are *zone damper systems*, also called *variable volume/variable temperature (VVT)* systems. Primarily used on constant volume systems (furnaces, heat pumps, and rooftop units), zone dampers control airflow to thermal zones. Each zone damper is controlled by a thermostat. In some systems, the thermostats can communicate with each other and can be remotely controlled over the Internet, which provides control capability that approaches that of a central control system. When installed on a constant volume system, zone dampers can present a problem of high air velocities in zones calling for heating or cooling, when zone dampers close in zones not calling for heating or cooling. This problem is addressed by providing a bypass damper at the air handler.

Zone damper systems can provide good comfort. However, when installed on constant volume air handlers, there is a significant energy penalty associated with air bypass, and also an energy penalty resulting from lower airflow at part-load conditions, which offset the benefits of thermal zoning for these installations. Bypassing heated or cooled supply air into the return air stream substantially reduces the efficiency of heat pumps, air conditioners, and furnaces. An additional energy penalty results from the units cycling between heating and cooling, to serve zones that are separately in heating or cooling mode, as heat exchangers are alternately heated and then cooled. Fan power use, per unit of airflow, has also been found to be high in zone damper systems.[1] Zone damper systems have been found to use as much as 35 percent more energy than a single-thermostat system. Reliability has also been found to be poor, with a median expected life of only 12 years shown in American Society of Heating, Refrigerating, and Air-Conditioning Engineers' (ASHRAE's) expected life database.

Zone damper systems with bypass dampers (Figure 11.12) should not be installed on constant volume air handlers as an energy improvement. Existing zone damper installations may even be candidates for energy conservation, by retrofit replacement with high-efficiency zoned systems such as ductless heat pumps or four-pipe fan coil systems.

Figure 11.8 Line voltage thermostat.

Figure 11.9 Thermostat inside cover.

Figure 11.10 Electronic programmable thermostat.

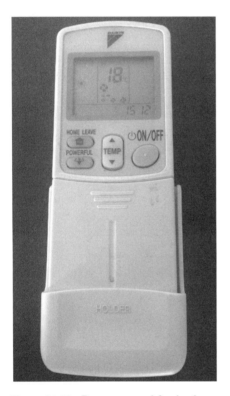

Figure 11.11 Remote control for ductless heat pump.

Central Controls

Central control systems separate the functions of sensing, control decisions, and actuation of devices in response to control decisions. The entire system has become computer based, with features such as advanced user interfaces, storage of data for trending, alarms, management of peak demand, management of operation and maintenance, and remote access over the Internet.

The main functions of central controls are to control heating and cooling, either turning systems on and off, or modulating output capacities, with a focus on controlling temperature in thermal zones. Central controls also increase or decrease ventilation. Central controls are sometimes also used to control lighting. The heating and cooling that is controlled can be any of the many types of heating and cooling, from large chillers and boilers to their associated support equipment (pumps, fans, cooling towers, air handlers, fan coils, valves, dampers, etc.), to small fan coils or packaged terminal air conditioners (PTACs).

Older systems are frequently pneumatic, using compressed-air pressures in the 3- to 13-psi range to control devices, such as to open and close valves and dampers. Newer systems are entirely electronic. Some systems retain a mix of pneumatic and electronic components, for example with newer controls sending electronic control signals, such as 0 to 10 volts DC, to electronic-to-pneumatic (E-to-P) transducers, which convert the electronic signal to a 3- to 13-psi pressure signal for use by older actuators.

Common electronic signals are 0 to 10 volts DC and 4 to 20 mA.

The main functions of central controls are to control indoor air temperature, turn equipment on and off, reduce pump and fan motor speeds when possible, and control outside air, either for ventilation or for free cooling. Accordingly, energy savings through central controls derive from changing the indoor air temperature, turning equipment off or reducing motor speeds when possible, reducing ventilation when not required, and increasing outdoor air for free cooling when possible. Secondary functions, such as trending, alarms, peak demand management, and user interface, are all important but do not necessarily deliver significant energy savings.

Central controls do not guarantee energy efficiency. When a control system is well designed and implemented, it gives a building owner a unique lens into the operation and performance of their building, and empowers them to make improvements that reduce energy and increase comfort. If poorly designed and implemented,

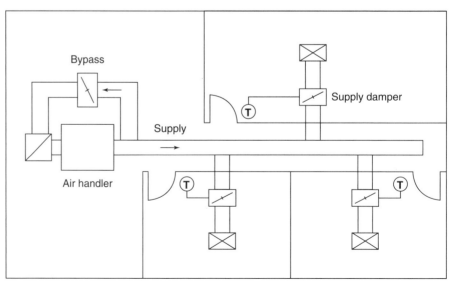

Figure 11.12 Zone damper system with bypass.

a control system does not save energy, and can even inadvertently commit a building to using more energy than necessary.

Because of the strong relationship between central controls and energy use in buildings, some of the largest energy service companies (ESCOs) are manufacturers of controls.

Central controls can give a mistaken sense that energy savings flow from added information, whereas energy savings can only flow from physical changes to a building or its systems. An energy management system can also draw too much attention to energy-consuming equipment under its control, specifically heating/cooling and lighting, at the expense of attention directed to energy loads, such as the envelope, hot water consumption, and uncontrolled machines and appliances such as transformers and plug loads. And, frequently, attention is directed only to the controls themselves, rather than to the energy equipment. We can mistakenly seek to save energy through controls adjustments alone, rather than examining the efficiency of the heating/cooling plant. Or we can mistakenly seek to save energy only through the control of lighting, rather than examining the number or type of light fixtures.

Because central controls involve programmed algorithms, they can be obscure to the energy manager, building owner, property manager, or facilities manager, and can result in control of the building being turned over to specialists. Sometimes, this is a good thing, but at other times this can result in sole sourcing (services provided by a single vendor) or outsourcing of critical decisions involving a building's energy use and performance.

It is important to bring a sense of inquiry and vigilance to central controls in a building. What do they do? How can they work for us? How can they reduce energy use and improve building performance, rather than obscure what we need to know and understand and improve?

SENSORS

The most common sensors are temperature sensors, frequently wall mounted in the same locations as traditional thermostats, 60″ above the floor. Most sensors simply send a temperature signal to the central controller. Some sensors allow limited local control of temperature by occupants. See Figure 11.13.

Other sensors include humidity, carbon dioxide for ventilation control, occupancy sensing, and more. See Figure 11.14.

ACTUATORS AND CONTROL DEVICES

Control signals are sent from the central control system to actuators and control devices. Common types of actuators are those that open and close valves or duct dampers, to control water or steam flow (valves) or air flow (dampers). Control signals are also sent to *variable speed drives*, which control the speed of motors. See Figures 11.15 through 11.17.

EXAMPLE CONTROLS IN AN OLDER COMMERCIAL BUILDING

We offer an example of replacing the control system in a large older commercial office building with a central chiller and boiler that serve two-pipe fan coils in each space.

Without central controls, the system operates as follows: Each fan coil operates on its own, with equipment-mounted thermostats that call for heating by turning on the fan and/or opening a water valve located inside or near the fan coil. A central circulating pump in the mechanical room runs throughout the winter. The boiler fires in response to a temperature control in the supply water pipe, maintaining 180 F water

Figure 11.13 Temperature sensor with limited local control.

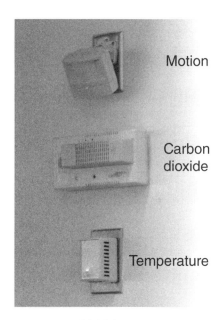

Motion

Carbon dioxide

Temperature

Figure 11.14 Multiple sensors.

Figure 11.15 Motorized valve.

Figure 11.16 Duct damper motor actuator.

Figure 11.17 Outdoor air damper and motor actuator.

throughout the winter. Because it is a two-pipe system, the energy manager makes the decision when to shift from heating to cooling, and manually turns off the boiler and provides power to the chiller, at some point in the spring. Like the boiler, the chiller maintains a constant supply water temperature, perhaps 45 F, and cycles on and off to maintain this temperature, or modulates inlet guide vanes to do so. The same main circulation pump runs throughout the cooling season. A cooling tower serves the chiller system, with its own controls, and delivers 85 F water temperature back to the condenser. There is no ventilation in the building, other than operable windows and bathroom exhaust fans that run continuously.

With central controls, temperature sensors in each space provide signals to the central controls. The controls have programmed occupancy schedules, for example with occupancy Monday through Friday 8 A.M. to 6 P.M., and accordingly has two different indoor air temperature schedules for the fan coils, maintaining 70 F in winter and 77 F in summer during occupied mode, and 60 F winter and 85 F in summer during unoccupied mode. There are no longer equipment-mounted thermostats on the fan coils. Instead, the fan and/or valves are controlled by signals from the central controls. The boiler and chiller are now controlled by the central controls, instead of by stand-alone controls, and the central controls might reset the supply water set points from the boiler and chiller in order to save energy, raising the water temperature by a few degrees in cooling, when outdoor air temperatures in summer are cooler, and vice versa lowering the water temperature in heating, when outdoor air temperatures in winter are warmer. The main circulating pump speed may be reduced, by the central controls, when full water flow is not needed, based on measurement of water pressure downstream of the pump. The water pressure will increase when valves close at the fan coils, providing an indication that less water flow is needed. Variable speed controls may also be applied to the cooling tower pump, and the cooling tower fans, reducing pump and fan motor speeds when full capacity is not required, and lowering the water temperature supplied from the cooling tower to the condenser during off-design conditions. The chilled water and condenser water (cooling tower) pumps may also be fully turned off at times by the central controls, for example during unoccupied mode (nights and weekends) during the cooling season. The central controls may, similarly, turn off the bathroom exhaust fans, during unoccupied mode.

In this example, applying central controls to a system without such controls might facilitate energy savings in several ways: Unoccupied setback of space temperatures in the building, reset of chilled water and boiler water temperatures, variable speed control of pump and fan motors, and by turning off pumps and fans at times when they are not needed. It should be noted that many, if not all, of these energy-saving controls functions could have been applied to the system without adding central controls. For example, programmable thermostats could be added to each fan coil, stand-alone reset controls could be added to reset water temperatures, and timer controls could be installed to control exhaust fans. The central controls simply allow control to be done, and easily changed, from a central system, with added functions such as alarms and trending.

EXAMINING EXISTING CENTRAL CONTROL SYSTEMS

Much information can be gathered from the user interface of an existing central control system.

First and foremost, temperature set points should be obtained, for each thermal zone, for a full weekly schedule. Separately, the set points should be mapped to the spaces served by each thermal zone. This mapping will be helpful in examining the potential for energy savings through changed set points, or through adding thermal zones, or by removing conditioning (heating or cooling) from spaces that do not require it. We direct our attention to spaces that might be overheated or overcooled at temperatures not required for their function: Storage spaces, utility rooms, entry vestibules, corridors, stairwells, and the like.

As-found conditions can be helpful to note: Actual indoor air temperatures in each thermal zone, supply water temperatures (hot water or chilled water), return water temperatures, water temperatures entering condensers, and outdoor air temperature. These might allow determination of whether outdoor reset is being used and, if not, energy savings potential by implementing reset control.

As-found air handler conditions can also be informative: For example, return air temperature, mixed air temperature, and outdoor temperature can be used to estimate the fraction of outdoor air being drawn into the system. The target outdoor air fraction can be compared to this actual outdoor air fraction. The temperature of supply air, also called discharge air, can be useful to assess potential for supply air reset control.

Look particularly for signs of unusual or catastrophic problems, specifically ones that will contribute to high energy use. For example, a cooling system that is unable to meet its set point on a mild summer day might be a sign that the system is low on refrigerant, or is low on airflow, or has some other major problem. Or air temperatures might show an unusual amount of outdoor air, even though the outdoor air damper is supposed to be closed, providing an indication of an outdoor air damper that is stuck open. Or ventilation dampers might be mistakenly set in an open position 24/7.

One of the best areas for reducing energy use is to reduce over-ventilation. Observation of ventilation airflow rate or ventilation damper position (% open) can provide indicators of overventilation. Is ventilation airflow high even when the building is unoccupied? Is ventilation airflow high even when the building is partially occupied or does not need full ventilation, for example in the early morning in an office building? Do bathroom and kitchen exhaust fans run 24/7? Is the ventilation airflow rate higher than required for the current building peak occupancy; use 15 CFM per person as a guide for the maximum ventilation airflow rate. Spot measurements of carbon dioxide, during periods of partial occupancy and peak occupancy, can also be telling: CO_2 concentration measurements below 1,100 ppm during peak occupancy is a sign of overventilation. Measurements can be taken in occupied spaces, or in the return air stream to an air handler (before mixing with outdoor air), which provides a mechanical average of the concentration in the space served by the air handler.

Often, the challenge of assessing potential energy improvements of an existing central control system comes from not knowing its current configuration, control sequences, set points, and control algorithms. Potential savings cannot be accurately estimated without knowing what a control system is doing. The challenge is exacerbated when an energy auditor might have only a few hours to spend in a building, to obtain information about the control system. This is where an energy manager can increase the quality of an energy audit, by obtaining set points and schedules in advance of an energy audit.

Possible sources of control information include:

- Original plans and specifications.
- As-built drawings.
- Control system manuals.
- Control system vendors or service contractors.

If information cannot be found in any of these sources, existing set points and control settings can be obtained from the user interface, and by observing the operation of equipment. Answers to the following key questions are sought:

- What is the position of ventilation dampers (% open) during full building occupancy?
- What is the position of ventilation dampers (% open) when the building is entirely unoccupied?

- What is the position of ventilation dampers (% open) when the building is partially occupied?
- Do large pump or fan motors (over one horsepower) have variable speed controls?
- Do boilers have supply water temperature reset control, based on outdoor air temperature, for both primary and secondary loops?
- Do chillers have supply water temperature reset control, based on outdoor air temperatures?
- Are nonessential fans shut off during unoccupied periods (for example, bathroom exhaust fans)?

Short-term monitoring can also be used to assess potential control strategies to save energy, if trending capability is not provided by existing central controls. Short-term monitoring might include:

- Measurement of carbon dioxide in the return airflow to air handlers. Readings below 1,100 ppm, during peak occupancy, are an indication of overventilation, possibly combined with infiltration, either or both of which can be reduced.
- Space temperatures. Of particular interest are whether space temperatures are set back during unoccupied periods (raised in cooling, reduced in heating), whether temperatures recover too early from setback (or too late), and whether space temperatures in unoccupied spaces can be reduced in heating and raised in cooling.

The function of central controls in saving energy for lighting is covered separately, in the chapter on lighting.

ESTIMATING SAVINGS

Estimating savings by installing or changing a central control system is best accomplished by examining the individual physical changes proposed to the building. Savings should not be based on assuming a flat percentage reduction in energy use, simply because a central control system was installed or upgraded.

Indoor air temperature setback should be estimated based on measured indoor air temperatures, and proposed setback temperatures. The setback temperatures should not be assumed to be reached instantaneously. Either a model that accounts for thermal mass should be used, or time-average setback temperatures should be estimated for the indoor air temperature during setback. Savings of 10 to 20 percent are possible, for buildings where there is currently no setback, and smaller savings are possible in buildings where there is already some setback. See Appendix N.

Control Improvements

SETUP

Setup refers to occupied indoor air temperature set-points when a building is occupied. We seek setup set points that are appropriate for each thermal zone in a building, not too high in heating, and not too low in cooling.

During occupied mode, reasonable starting set-points for frequently-occupied spaces are 68 F for heating, and 78 F for cooling.[2]

Firm recommendations for occupied heating and cooling set-points are not easy to find. The U.S. General Services Administration (GSA) did a survey of summer workplace temperatures, and found more than 60 percent of occupants reported being

TABLE 11.3

Indoor Design Conditions for Spaces Not Regularly Occupied

Space Type	Winter	Summer
Locker rooms	70	78
Electrical closets, elevator machine rooms	55	78
Mechanical rooms, electrical switchgear	55	95
Stairwells	65	None
Storage rooms	65	85

uncomfortably cool. The GSA accordingly recommends summer set points as high as 78 F.[3] The International Energy Conservation Code requires design conditions of 72 F maximum for heating and 75 F minimum for cooling.[4] However, these are design guidelines, meant to size heating and cooling equipment, not operational guidelines, and are likely too warm (heating) and cool (air conditioning) as set points. ASHRAE Standard 55 is a primary reference for comfort in buildings, and correctly points out that comfortable indoor air temperatures depend on activity level, clothing level, air speed, humidity, and the temperature of surfaces in the building. Standard 55 does not recommend or require specific set points, and acceptable comfort is shown to span a range from the upper 60s to the lower 80s, and again, depends on many variables. In general, we seek more than 80 percent of the occupants of a building to be comfortable. In buildings with vulnerable occupants, such as hospitals and nursing homes, health standards may govern temperature set points for certain occupied spaces.

For spaces that are not regularly occupied, the federal government provides recommended design temperatures, shown in Table 11.3, which may be used as a starting point for set points.[5] Over time, we might expect more clear standardization of starting set-points for both occupied and unoccupied modes, for a broader variety of space types.

SETBACK

Setback refers to reducing indoor air temperature set points in heating, and raising them in cooling, during periods when a thermal zone is unoccupied. *Setback* is frequently used to refer to the unoccupied set point. *Recovery* refers to the period during which a system changes from setback to setup.

Setback saves energy because envelope and infiltration losses are reduced during the time that indoor temperatures are set back. Setback also saves energy by reducing infiltration due to stack effect pressures, which are pressures due to the differences in air density between indoors and outdoors. In addition to reducing loads, setback can further save energy in some types of equipment, by raising the thermodynamic efficiency of the heating or cooling plant. This is particularly true for direct expansion cooling systems and heat pumps, as well as for furnaces. It can also be achieved in hydronic and chilled water systems, but only if the water temperatures are adjusted accordingly.

There is a widespread myth that temperature setback does not save energy because a system has to exert more energy when it brings the indoor air temperature back to an occupied set point. This is incorrect. Setback will always reduce load, and will often also increase heating and cooling plant efficiency, both of which provide energy savings.

A challenge with setback is that, in midwinter, the largest heating loads occur before dawn, just when a heating system is typically trying to recover from setback, and when there are the fewest benefits from internal gains (e.g., lights are off) or sun through windows. As a result, a heating system will take longer to recover from setback on the coldest days, and the space may not reach its set point at the desired time, resulting in a space that feels cold for a period of time. This, in turn, frequently results in setback being eliminated; occupants do not like to be cold.

Steps to counter the pattern of people eliminating setback due to discomfort in midwinter include:

- Educate users to program recovery from setback earlier in midwinter.
- Program recovery to start earlier on colder days in winter on the basis of an outdoor reset. In other words, start earlier on colder days, and later on milder days.
- Use temperature controls with *adaptive features*, which automatically begin recovery earlier on colder days in winter, and hotter days in summer. Adaptive control is also called *optimum start/stop*.

Recovery from setback takes more time in buildings with poor thermal boundaries (inadequate insulation and/or high infiltration rates). Recovery from setback is faster in buildings with better thermal boundaries. Recovery takes more time in buildings with more thermal mass, and vice versa: Light frame buildings recover faster from setback.

Recovery from setback is primarily a challenge in heating. In cooling, the indoor-outdoor temperature differences are not as large as in heating. Recovery from setback is typically from the night, when many cooling loads (solar gains, occupancy, lights and machines, outdoor air temperature and humidity) are not at their peak.

A simplified model of recovery from setback is provided in Appendix N. Using the model, one can examine the impact of different variables: Heating system capacity, building heat loss rate, setback temperature, duration of recovery, and indoor temperature. As an example, a building with a 25 percent oversized heating system is able to recover from a 60 F setback up to an occupied set point of 70 F in just under three hours, when the outdoor temperature is 40 F. However, when the outdoor temperature is 10 F, it takes over seven hours to recover to 70 F. If the recovery had been started three hours before the building is occupied, the building would be at only 65 F at the beginning of occupancy.

The following rules can be used to estimate the time to recover from setback, by measuring the time a building takes to cool one quarter of the way to the outdoor temperature, when the heat is turned off:

- The time to recover from a 15 F setback at 40 F outdoors is approximately equal to the time it takes to cool one quarter of the way to the outdoor temperature.
- The time to recover from a 15 F setback at 15 F outdoors is approximately twice the time it takes to cool one quarter of the way to the outdoor temperature.

The rules assume that the heating delivered to the spaces is oversized by 25 percent at design conditions. If the heating is oversized by more than 25 percent, recovery is faster, and vice versa.

For example, a building at 70 F goes into setback at 5 P.M., when the outdoor temperature is 30 F. As it enters setback, the heat is off. It takes two hours to cool to 60 F; in other words, it has cooled by 10 F, or one quarter of the way to the outdoor temperature of 30 F. Based on this measurement, if the outdoor temperature is 40 F, it will take about two hours to recover from a 15 F setback (e.g., from 55 F back to 70 F). If the outdoor temperature is 15 F, it will take about four hours to recover from the same setback (from 55 F to 70 F).

THERMAL ZONING

Adding more thermal zones is an energy improvement, delivering energy savings through reduced overheating and overcooling, and through setback control for more unoccupied spaces for more of the time. Adding thermal zones may not be easy, especially in buildings with systems designed around providing limited thermal zoning, such as a rooftop system serving multiple separate areas of a building with a single thermostat, or even a VAV system with VAV boxes which each serve multiple separate areas. Thermal zoning is best provided by separate heating and cooling in each thermal zone, such as with fan coils, rather than by inefficient control of a central system, such as a bypass/zone damper system, or reheat on VAV systems.

Savings can also be obtained by removing heating and cooling from spaces that do not require it. In other words, turn conditioned spaces into unconditioned spaces, wherever possible. This might apply to stairwells, entrance vestibules, corridors, mechanical and electrical rooms, and more.

ELIMINATING SIMULTANEOUS HEATING AND COOLING

More often than we may be aware, spaces are simultaneously heated and cooled. This may occur because the heating and cooling systems serving a single space are controlled separately, or because of traditional central system reheat that is not being optimally controlled. We actively look for situations in which a single space has separate temperature controls for heating and cooling, for example, a space with a room air conditioner and an electric baseboard heater. And we actively look for systems with reheat, and then note the control set points. Improvements might include interlocking separate heating and cooling controls, training building occupants to avoid simultaneous heating and cooling, or optimizing set points to minimize reheat on central systems.

REDUCING OVERHEATING

In buildings with temperature control problems, it is important to measure indoor air temperatures to assess overheating or overcooling, and to evaluate the energy savings potential from bringing temperatures under control. An easy way to measure indoor air temperature is with an infrared thermometer pointed at an indoor surface like a wall, away from heat sources such as heating, cooling, lighting, and solar gains. Measurements should be taken at about the height of a typical thermostat or temperature sensor, approximately five feet above the floor. Random sample measurements should be taken in different locations and on different floors. It is important to avoid the mistake of just measuring air temperatures in problem areas, such as warm upper floors, because this leads to overestimation of savings from improvements such as thermal zone control. It can also be informative to survey tenants and facility managers about temperature control and comfort, in order to identify problem areas and how widespread the problems are. However, it is better to rely on randomly sampled air temperature measurement data for improvement calculations.

It has been observed that buildings that have bad overheating in some areas, for example, indoor temperatures well over 80 F in areas such as upper floors, and that as a result give a strong impression of being overheated, are *on average* overheated by only 1 or 2 F, when all areas of the building are accounted for. Therefore, assuming that it is possible to deliver a 10 F average reduction in indoor air temperature is unrealistic. For overheated buildings, a 1 to 2 F typical average building-wide temperature reduction might be possible. To achieve any temperature reduction, a physical change in space-by-space temperature control is required, such as adding zone controls to a building that currently lacks such controls. Minor or token changes to controls, such as adding temperature sensors and averaging them, which do not have an associated physical change to the delivery of heating or cooling, generally do not deliver measurable energy savings.

Temperature-limiting controls can also be used to prevent overheating, in thermal zones where occupants might be setting the temperature too high. For example, some thermostatic radiator valves (TRVs) allow temperature limiting with the use of internal pins that can be adjusted.

RESET CONTROL

Reset control refers generally to adjusting water or air temperatures, typically in response to outdoor air temperature, to maximize energy efficiency of heating and cooling systems.

One of the most common applications of reset controls is boiler water temperature reduction. Energy savings from temperature reduction comes in two different ways:

- Water temperature losses are lower. Temperature losses occur from pipes, valves, fittings, pumps, and the boilers themselves.
- Boilers operate at higher combustion efficiency when the water temperature returning to the boiler is lower. See Figure 11.18.

Traditionally, hydronic systems operate at 180 F supply water temperature, with a 20 F temperature rise through the boiler, and a 20 F temperature drop through the distribution system, for a 160 F return water temperature at design (midwinter) conditions. In these traditional systems, at off-design conditions, the 180 F temperature is not varied, regardless of the load, in other words, regardless of the outdoor air temperature. See Figure 11.19. Note how the return water temperature actually rises, as the load decreases as the outdoor temperature increases. In other words, the combustion efficiency drops at part load.

The main approach to water temperature reset for boiler systems is to reduce the water temperature in response to outdoor air temperature. We refer to this as *outdoor reset control*. For example, if the design supply water temperature is 180 F at a design outdoor temperature of 0 F, we might draw a straight line between this point and 60 F supply water temperature at 60 F outdoor air temperature. See Figure 11.20.

Reducing the supply water temperature all the way to 60 F works well for boilers that can accept a low return water temperature, like condensing boilers. The *reset ratio* is defined as the slope of the supply water temperature curve. For example, in the example shown, the reset ratio is (180 − 60) / (60 − 10) = 2.4.

However, if we want to apply outdoor reset control to a non-condensing boiler, we cannot provide return water temperatures below 140 F because of a risk of flue gas condensation, for which noncondensing boilers are not designed. So we limit

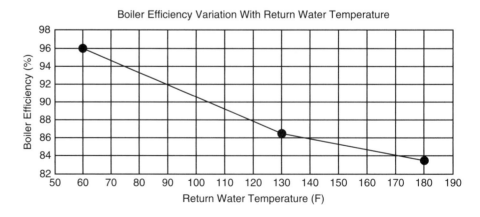

Figure 11.18 Boilers operate at higher efficiency when the water temperature is lower.

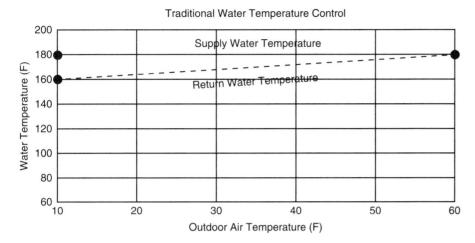

Figure 11.19 With traditional control, supply water temperature is not varied.

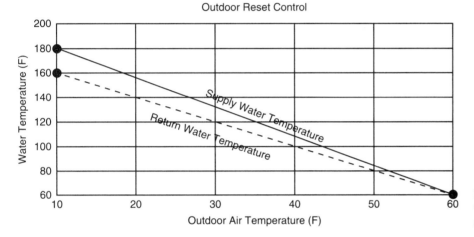

Figure 11.20 With outdoor reset control, water temperature is reduced when the outdoor temperature is warmer.

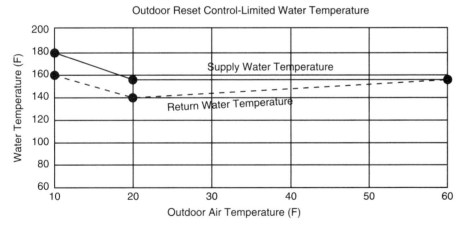

Figure 11.21 Noncondensing boilers cannot accept low water temperatures, so outdoor reset can only be done down to about 160 F supply water temperature, and no savings are seen beyond this point.

the supply water temperature, and the return water temperature is similarly limited. The curve might look like the one in Figure 11.21. Therefore, for noncondensing boilers, the gain in efficiency from an outdoor reset control is more modest than for condensing boilers.

A major challenge for outdoor reset is that a single comfort complaint causes facility managers to eliminate outdoor reset, and raise the water temperature up to where there is no chance of comfort complaints. Comfort complaints should first be resolved by checking to see that there is not a problem with balancing; with radiation malfunctions (such as motor failure at a fan coil); with a night setback that is too deep;

or with recovery from night setback that does not start early enough. If the complaint does relate to the water temperature being too low, small adjustments to the water temperature should first be made, rather than simply eliminating outdoor reset. The reset curve can be raised, but this should be done in small increments.

It is also sometimes possible to simply reduce the water temperature over the full range of outdoor temperatures. Even with outdoor reset, the temperature of hot water can still be too high. Overheating of the supply water temperature can be seen if radiator water temperatures are fluctuating too much. Place a temperature monitor on a radiator to see if the temperature fluctuates significantly and, if so, adjust the supply water temperature curve downward. For example, by lowering the design supply water temperature in a hydronic system from 180 F to 170 F, with the outdoor reset adjusted accordingly, heating savings of 1 percent are possible. Be cautious not to lower it too much, to avoid not meeting the heating load and making spaces cold.

Additionally, it is possible that the supply water temperature is being unnecessarily kept too high if a single thermal zone has an undersized distribution system. For example, if water flow to a specific thermal zone is too low, or if the perimeter radiators are too small, that zone could be always cold in winter. An inefficient solution to this issue is to keep the supply water temperature high, just to satisfy this one problematic zone. The problem is better handled by fixing the distribution system, for example by adding radiation to the problematic zone, so that the supply water temperature can be reduced, and so that water temperature can be effectively reduced over the full range of outdoor air temperatures. Identifying a problematic thermal zone can be as simple as asking occupants if their spaces are consistently uncomfortable. Examining trends on an energy management system can also be used to identify outlier thermal zones that have inadequate heating or cooling distribution or delivery. This issue also occurs with chilled-water cooling systems.

Another strategy to increase energy savings is to reduce pump flow rates. This reduces return water temperature, which increases boiler efficiency and reduces pipe losses.

Chilled water reset uses a similar strategy as reset for hydronic heating systems. At part-load cooling conditions, the chilled water temperature is increased, which increases the return water temperature, and consequently increases the efficiency of the chiller. Increases in chilled water temperature also need to be done in small increments, in order to avoid comfort complaints. Higher water temperatures also place dehumidification at risk, so indoor humidity levels should be monitored as chilled water reset is implemented.

Condenser water reset is another strategy, in which the condenser water temperature is reduced, at part load. This also increases the chiller efficiency.

Supply air reset refers to increasing supply air temperatures in cooling, at part load, another strategy to save energy.

CONTROL OPTIMIZATION

Central control offers the possibility of *control optimization*, adjusting operating settings for various pieces of equipment in a complex system in order to reduce overall power consumption. For example, water flow rates and airflow rates in a cooling tower for a chilled water system can both be adjusted in order to minimize the combined power consumption of the cooling tower fans, condenser water pumps, and chiller. Similarly, optimization can be applied to combinations of chillers, chilled water pump flow rates, and air handler operation, as well as heating systems. Modest savings have been reported as possible, in the 2 to 14 percent range, for systems such as chiller/tower optimization.[6,7] Optimum control strategies are best established by detailed energy modeling. Control optimization can also be achieved through trial and error, if done carefully and methodically, by adjusting controls on equipment in real time, and observing total power use.

ECONOMIZERS

Economizers deliver cooling when outdoor temperatures are low, but when internal gains or solar gains place a net cooling load on a space. Economizers make sense for buildings with significant internal gains, such as office buildings, which might see a need for cooling at milder outdoor temperatures. When optimized, savings for economizers have been estimated to be 30 to 50 percent of cooling usage.[8] We identify if a building has potential for economizer savings if we see significant cooling usage in spring and fall.

Economizers in small commercial buildings are built into equipment such as rooftop units, and combine with the ventilation function. Likewise, the economizer function is typically part of an air handler outside air intake. Economizers that are integrated with ventilation outside air intake, in packaged units or air handlers, are called *air-side economizers*. By contrast, economizers that circulate water to an outdoor heat exchanger, and use this cooled water in air handlers or fan coils indoors, are called *water-side economizers*. Water-side economizers are more common in larger buildings.

Potential savings for economizers are lower as lighting and plug load equipment efficiencies become greater, in other words, as internal gains decrease.

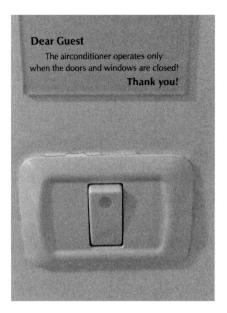

Dear Guest
The airconditioner operates only when the doors and windows are closed!
Thank you!

Figure 11.22   Override control.

OTHER

Innovative control strategies are being adopted to further reduce energy use in buildings. For example, some hotels require insertion of the room card key into a control device inside the room in order to energize power to the room. Some hotels and dormitories use proximity sensors on windows and doors, which cut power to heating and cooling if any door or window is left open. See Figure 11.22. Opportunities abound for creative control strategies.

References

1. John Proctor, PE, "*Zoning Ducted Air Conditioners, Heat Pumps, and Furnaces: Summary of a Case Study Prepared for the California Statewide Utility Codes & Standards Program.*" Prepared for San Diego Gas & Electric Co., Southern California Gas, Southern California Edison, Pacific Gas & Electric Co., November 2011.
2. http://energy.gov/energysaver/articles/thermostats. Accessed August 14, 2015.
3. U.S. General Services Administration, "Energy Savings and Performance Gains in GSA Buildings—Seven Cost-Effective Strategies." GSA Public Buildings Service, 2009.
4. Energy Conservation Construction Code of New York State, section 302.1, 2010.
5. Facilities Standards for the Public Buildings Service. Revised March 2005, PBS-P100, Chapter 5—Mechanical Engineering, p. 122.
6. James W. Furlong and Frank T. Morrison, "Optimization of Water-Cooled Chiller–Cooling Tower Combinations." Baltimore Aircoil Company, *CTI Journal 26*, no. 1, p. 17.
7. "Condenser Water System Savings Optimizing Flow Rates and Control." *Trane Engineers Newsletter* 41, no. 3 (2012), p. 3.
8. Reid Hart, "Use EMS to Improve Simulation of Outside Air Economizer and Fan Control for Unitary Air Conditioners." Pacific Northwest National Laboratories. PNNL-SA-100271, January 2014.

Chapter 12

Water

Hot water energy use can be one of the largest energy loads in some types of commercial buildings, such as hotels, multifamily buildings, laundry buildings, and restaurants. On the other hand, it can be a relatively small energy load in other types of buildings, such as offices, retail, municipal, and many not-for-profit buildings. Conserving hot water saves both energy and water. Cold water conservation is often a separate but important part of sustainability efforts.

Water Service

Water service to a building is either from a municipal system or from a well. Municipal systems typically have a water meter (Figure 12.1), whereas well systems do not.

Water systems serving buildings with fire sprinklers are split after entering the building (Figure 12.2). A smaller pipe goes one way, through a water meter, to serve the building's metered consumable hot and cold water needs. A larger pipe goes another way, and is not metered, to serve its fire sprinkler system. In energy work, we typically ignore the fire sprinkler system. Consumable water piping is then further split from the metered service, with one pipe going to serve the building's cold water loads, and the other going to the hot water heater.

Domestic Hot Water Heating

Domestic hot water, abbreviated *DHW* and sometimes referred to as *service hot water*, is used to describe hot water for consumable loads, such as washing hands, showers, clothes washing, dishwashing, and the like. It is different than closed hot water systems used for space heating, which are referred to as hydronic systems, although the heating of the two uses is sometimes integrated.

BILLING ANALYSIS

In buildings where a fossil fuel is used only for space heating and domestic hot water heating, the domestic hot water energy use can often be disaggregated by examining the summertime use of the fossil fuel. This is common in buildings without gas cooking appliances or clothes dryers, such as offices and retail buildings. This summertime usage, per month, represents a slightly low annual-average monthly energy use for hot water, because wintertime hot water energy use rises with lower entering cold water temperatures. But it is still a useful start in estimating annual energy use for hot water.

In buildings where there are multiple uses for the fossil fuel, or in which the hot water is electrically heated, energy use for hot water cannot be as easily extracted from utility bills. In these cases, we rely on estimates or on monitoring, or on a combination of these. For example, in a multifamily building in which gas is used for space heating,

Figure 12.1 Water meter.

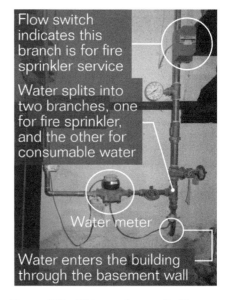

Figure 12.2 Water service, small office building with sprinkler.

Figure 12.3 Natural gas commercial storage water heater.

Figure 12.4 Natural gas water heater.

domestic hot water, and clothes drying, an estimate of gas used for clothes drying may be made on the basis of clothes dryer coin receipts, and this may be subtracted from summertime monthly billing gas use to estimate monthly domestic hot water use.

TYPES OF WATER HEATERS

Domestic hot water heaters are either stand-alone, or are integrated with a space heating system.

Stand-alone hot water heaters can be classified by the type of fuel, including most commonly natural gas (Figure 12.3) or electricity, and less commonly such fuels as propane and fuel oil. Electrically heated water heaters most commonly use electric resistance heaters, but more recently include heat pump water heaters as well.

Stand-alone water heaters can also be classified by whether or not they store water. *Storage water heaters* by definition store water, whereas *instantaneous water heaters* are conversely those that do not store water, defined by the Air-Conditioning, Heating, and Refrigeration Institute (AHRI) as less than 2 gallons for residential gas water heaters (50,000–200,000 Btu/hr), and less than 10 gallons for commercial gas water heaters (over 200,000 Btu/hr input). For electric water heaters, instantaneous is defined as having an input rating over 12 kW.[1] Instantaneous water heaters are sometimes referred to as *tankless heaters* or *on-demand heaters*.

The year of manufacture of many water heaters can be obtained from the serial number. For example, the first two digits represent the year of manufacture for American brand water heaters (also branded as AquaTemp, Aquatherm, Best, Champion, Craftmaster, Envirotemp, Mor-Flo, Revere, Shamrock, Standard, US Craftmaster, Whirlpool). For Bradford-White (also Energy Saver, Golden Knight, Jetglas, and Lochinvar), the first letter is the year of manufacture, starting with 1964, and restarting every 20 years (1984, 2004, etc.), and skipping the letter I. For Rheem (also Ruud, Richmond, ABS, Cimarrom, Citation, Coast to Coast, Energy Master, General Electric, Intertherm, Miller, Lowe's, Mainstream, Montgomery Ward, Professional, Servi-Star, True Value, and Vanguard), the third and fourth characters are the year of manufacture. AO Smith follows a variety of patterns, but if the serial number starts with two letters, then the next two numbers are the year of manufacture, and if the serial number starts with two numbers, those are the year of manufacture, among other patterns.[2]

Water heating can also be provided by boilers, defined by AHRI as conforming to safety requirements of the American Society of Mechanical Engineers' (ASME) Boiler and Pressure Vessel Code, sections I or IV. Water heating by boilers is dominant in larger institutions (colleges and universities, hospitals, correctional facilities), and water heating by stand-alone storage tank systems is dominant in smaller commercial buildings.

Water heaters that are integrated with space heating systems can have a hot water coil (heat exchanger) embedded in the boiler, or can have an external storage tank. Water heating with external storage tanks are often referred to as *indirect water heaters*, or informally as *sidearm water heaters*.

Typical gas or propane storage water heaters have a flue on top, a gas pipe, a visible gas valve, and a burner at the bottom. See Figure 12.4. Oil heaters look similar, but with an oil pipe instead of a gas pipe, and an oil burner instead of a gas valve and gas burner.

Typical electric water heaters do not have a flue on top or a gas pipe or gas valve, but instead have electrical wiring to them. See Figure 12.5.

The nameplate will also typically identify the type of water heater as electric, gas, oil, or otherwise.

Heat pump water heaters (Figure 12.6) do not have a flue on top, or a gas pipe or gas valve. Like electric water heaters, they have electric wiring to them, but they also have a condensate pipe, and have a heat pump section, typically mounted on top of the storage tank, with an air filter, a compressor, and a fan. It makes an audible noise when heating water, which sounds perhaps like a room air conditioner.

Instantaneous water heaters (Figure 12.7) are typically small, and are often wall mounted. Again, if gas fired, a flue and gas pipe are visible; however, gas valves are typically internal and so cannot be seen.

WATER HEATER ENERGY IMPROVEMENTS

A primary improvement for water heating is to install a higher-efficiency water heater. A similar improvement is to replace one type of water heater with a different type, either with a different (more efficient) fuel, or, for example, removing water heating from a steam boiler and providing it as a stand-alone device.

If the replacement water heater is gas or propane, it is preferable that the new water heater not have a standing pilot (see Appendix O). Integral flue dampers also make storage gas water heaters more efficient. The flue damper closes when the hot water heater is not firing and prevents heat from escaping from the heat exchanger inside the hot water heater, either up the vent/chimney or into the space in which the water heater is located. Integral flue dampers are part of the water heater, and should not be confused with vent dampers, that are in the vent, separate from the water heater. A vent damper can prevent stack effect flow up the vent or chimney but will not prevent heat from inside the water heater from escaping into the room. Both the pilot gas energy use and the flue damper benefit are captured in the standby loss rating of a commercial gas water heater and do not specifically need to be evaluated separately. However, as is our practice, we are equally interested in the energy use of an existing water heater, which may well not have ratings readily available. How do we account for either/both a standing pilot and the absence of a flue damper, when examining an existing water heater in a building? Estimating the pilot light energy use is easier: A reasonable estimate for the losses of a water heater pilot is 850 Btu/hr. The losses from a gas or oil water heater flue that does not have a flue damper are harder. Research in this area has been inconsistent, with savings being estimated as high as several thousand Btu/hr. The AHRI directory is itself inconclusive with, for example, the same water heater showing no significant difference in standby losses, with and without a flue damper. So we fall back on an analytical estimate. A 4″ flue in a 60″ tall water heater, with 140 F water and 70 F ambient air, would be expected to lose 350 Btu/hr up the flue, when evaluated on a basis of the required input energy, at 80 percent efficiency. At 120 F water temperature, the losses are approximately 250 Btu/hr.

Separately, we are also interested in jacket losses, through the sides, top, and bottom of the water heater. Old water heaters typically had 1″ insulation. For 1″ insulation, an 80-gallon tank with water at 140 F is estimated to lose approximately 1,000 Btu/hr out of the sides, top, and bottom of the water heater, to ambient air at 70 F. This estimate is input-based, in other words, 1,000 Btu/hr is the input energy required to make up for these jacket losses. These losses vary in proportion to the exposed tank area, but it should be noted that this area varies little with tank size over the range of common tank sizes. At 120 F, the losses are 700 Btu/hr.

So, in summary, an old 80-gallon gas water heater at 140 F water temperature can be assigned 350 Btu/hr in internal flue standby losses, an additional 850 Btu/hr in losses if there is a standing pilot, and an additional 1,000 Btu/hr for exterior jacket losses, for total standby losses of 2,200 Btu/hr. Losses are only slightly higher for larger water heaters (100 gallon, 120 gallon), and only slightly smaller for smaller water heaters. Of more significance is the water temperature and the temperature of air around the water heater: jacket and flue losses are proportional to the difference between the water temperature and the air temperature around the water heater.

When looking to see if an existing water heater has a flue damper, most flue dampers are on the exterior, at the top of the water heater, and have a visible motor,

Figure 12.5   Electric water heater.

Figure 12.6   Heat pump water heater.

Figure 12.7 Instantaneous water heater—note plastic vent and combustion air intake pipes (top), indicating a high-efficiency water heater.

but some are internal to the water heater and can only be seen with a mirror. Also, flue dampers tend to break, so check if the flue damper is working, by turning up the water heater temperature set point to get it to fire, and examining the damper shaft. It should turn when the heater fires.

A highly effective water heater improvement is to correctly size a water heating system, in other words to eliminate oversized water heaters, specifically for gas, oil, and heat pump storage water heaters. An oversized gas, oil, and heat pump water heater operates inefficiently because it operates close to its maximum temperature, which is the lowest thermodynamic efficiency. A correctly sized water heater will have much of its capacity used during hot water draws, filling up with cold water during the process, and therefore will be heated at a lower water temperature at a higher efficiency. Accordingly, an important correction to the rated efficiency is a correction based on use. For example, research has shown that conventional residential water heaters with rated energy factors in the 0.6 range should be corrected to an energy factor of 0.45 at 20 gallons per day, 0.52 at 40 gallons per day, 0.63 at 80 gallons per day, and 0.66 at 100 gallons per day. High efficiency water heaters rated at energy factors of 0.8 should be corrected to 0.75 at 20 gallons per day, 0.78 at 40 gallons per day, 0.83 at 80 gallons per day, and 0.85 at 100 gallons per day.[3]

An old water heater improvement, dating to the early days of the 1970s energy crisis, is to add insulation to the exterior of a water heater. Also known as a water heater *wrap* or *blanket*, the improvement has become less popular as water heaters have become more insulated. Indeed, insulating a water heater that has an exterior surface temperature less than 10 F warmer than the ambient air around the water heater does not make economic sense. However, many old water heaters and water storage tanks are still inadequately insulated. The key is to measure the surface temperature. Online calculators are available to estimate savings, or savings can be estimated by spreadsheet. The calculation is similar to that for pipe insulation, but should model a vertical surface rather than a horizontal one.[4]

Commercial Water Heaters

Commercial water heaters are tested for efficiency to American national standard ANSI Z21.10, which requires a 30-minute test, starting with a tank of water at 70 F, and with the water heated to a maximum temperature of 140 F.[5] They are rated in a *thermal efficiency*, rather than the *energy factor* used for residential water heaters, and are also subject to a *standby loss* limit. Most storage commercial water heaters are in the 80- to 120-gallon range of capacity. For larger applications, hot water heating is performed with boilers and indirect water storage tanks, an approach that is less frequently applied to small and medium size water heating installations.

Gas storage water heaters are slightly more popular than *electric storage water heaters*. Approximately 80,000 to 100,000 gas storage water heaters are installed each year, compared to approximately 60,000 to 75,000 electric storage water heaters. However, electric storage water heaters have seen a steeper increase in installations over the past two decades. Older water heaters are far more predominantly gas heated.

Current commercial water heater ratings are maintained in an online database by AHRI. Almost 2,000 models are listed, under many brand names, but representing only a few distinct manufacturers. Three manufacturers are reported to hold over 90 percent of the U.S. market share of water heaters. The average commercial water heater storage tank size is about 80 gallons. For new gas-fired storage water heaters, the range of rated thermal efficiencies is 80 to 99 percent, with an average of 83 percent, the tank sizes range from 32 to 130 gallons, and the standby losses range from 320 Btu/hr to 1900 Btu/hr with an average of 1120 Btu/hr.

It is important to examine both the thermal efficiency and the standby losses. For example, some high-efficiency water heaters have higher standby losses than lower-efficiency water heaters of the same capacity (water storage volume). New water heaters should be chosen that have both high thermal efficiency and low standby losses. The AHRI directory typically lists the standby losses in Btu/hr, for gas and oil water heaters. It lists the standby losses as a percentage for electric water heaters.

Standby losses in Btu/hr are very helpful. We simply multiply the standby losses by total hours per year (8,760) to obtain the annual standby energy use. This calculation is a slight overestimation because it does not account for lower losses when the tank is at lower temperatures (e.g., after a water draw) or other minor factors, but it is still a reasonable approximation. The standby losses are provided directly in fuel consumption, rather than needing to be divided by the thermal efficiency. The rated standby losses are based on a test using 70 F temperature difference between the water and the ambient air around the water heater. If the temperature difference is significantly different, the standby loss should be adjusted accordingly. For example, if water is stored at 150 F, in a basement that is typically at 60 F, the actual temperature difference is 90 F (from 150 to 60), compared to the rating temperature difference of 70, so the actual expected standby losses would be more accurately estimated as the rated standby losses × 90/70.

The first ENERGY STAR requirements for commercial water heaters went into effect in 2013, at 94 percent thermal efficiency, with an associated maximum standby loss dependent on the water heater size and input rate.

For the baseline thermal efficiency of existing water heaters, if unknown, the following assumptions are prudent: Pre-1990—76 percent, 1990 to 2005—78 percent, after 2005—80 percent. Baseline standby losses were previously estimated as 2,200 Btu/hr for an 80-gallon gas water heater with a standing pilot, with water at 140 F, and 1,000 Btu/hr for an electric water heater at 140 F.

In addition to high-efficiency like-for-like water heater replacements, other options for energy savings include:

■ Replace a storage tank water heater with a high-efficiency instantaneous water heater. Standby losses are lower, and efficiencies can be higher (largely because the entering water temperature is cold). Both of these characteristics are captured in the energy ratings. Instantaneous water heaters are more subject to mineral deposits, and so require occasional cleaning, and are less advisable for hard water areas or well-water systems.

■ Remove domestic hot water heating out of a steam boiler system, and instead generate domestic hot water on a stand-alone basis with a high-efficiency water heater. In large buildings, this is an extremely cost-effective improvement. The high standby losses of an existing steam boiler will be evident by high summertime fossil fuel use.

■ Switch fuels. If DHW is oil heated, propane heated, or electric heated, consider a gas water heater or a heat pump.

■ If DHW loads are small, and distances in the building are large, such as a large office building, consider point-of-use water heaters.

■ Heating water with a variety of free heat sources: solar thermal, drain-heat recovery, refrigerant heat recovery from air conditioning and refrigeration systems (also called desuperheaters), and combined heat and power systems. Such systems typically preheat domestic hot water in a separate preheat tank, from which the water flows to a hot water heater for supplementary/final heating.

The heat pump water heater is an emerging technology that is well worth tracking. These heat pumps typically draw heat from the air around the water heater, using the mechanical vapor compression heat pump cycle, to heat the hot water. They are well suited to basements and other large areas, from which they draw heat. They should not be installed in closets or small rooms, from which they cannot draw adequate heat. Heat pump output capacity tends to be low. Hybrid heat pumps have been developed that allow optional backup use of electric resistance heat, but are still far more efficient than water heaters with electric resistance heat only. Heat pump water heaters work most efficiently when the water temperature is low.

Over time, we expect to see domestic hot water being generated with split system air-source or water-source (geothermal) heat pumps. This will eliminate the concern about putting DHW heat pumps in small rooms or mechanical closets.

A worthwhile investment is to add a water meter to the water pipe feeding the hot water heater. Knowing how much hot water is being used, in contrast with cold water consumption, is helpful in planning both hot water energy conservation and cold water conservation.

Residential Water Heaters

Residential water heaters are used in many small commercial buildings, and also in some multifamily buildings in which a separate water heater is used in each apartment.

Residential water heaters have different energy ratings than commercial water heaters. An *energy factor (EF)* is used as a single metric that accounts for both heating efficiency and standby losses. The energy factor is derived from a 24-hour test: 64.3 gallons are divided into six equal draws (10.7 gallons each), an hour apart, at the beginning of the test. The rest of the 24 hours are in standby. The test conditions are: Ambient temperature (air around the heater) is 67.5 F, the inlet water temperature is 58 F, and the storage tank temperature is 135 F. The energy factor is the delivered energy (presumed to be 64.3 gallons at a 77 F temperature rise) divided by the total input energy over the duration of the test.

A separate *recovery efficiency* is based on the first 10.7 gallons in the energy factor test, dividing delivered energy by the energy consumed. Since the standby energy usage is a smaller component of the recovery efficiency test, it is more closely comparable to the thermal efficiency rating of the test for commercial water heaters.

Finally, a *first-hour rating* measures how much hot water the heater can deliver in an hour. Three gallons per minute (3 GPM) are drawn until the water temperature drops by 25 F from its peak, and then the flow is stopped. The flow is started again when the temperature set point has been satisfied, in other words, when the heater automatically shuts itself off. The cycle is repeated until one hour is up. If a draw is in process, the draw is allowed to finish. The test conditions are the same as the energy factor test.

The intent of the energy factor is to allow a single metric with which to calculate energy usage. For residential installations, the Department of Energy (DOE) recommends assuming a daily load of 64.3 gallons, presumably sufficient for a household of three people, heated from 58 F to 135 F, for a total load of 41045 Btu/day. Annual energy use is then $41,045 \times 365 / EF$, in Btu/year. For residential installations, this approach is a useful estimate, if one prorates the usage per water heater against the presumed occupancy of three people.

However, the energy factor approach is not as useful if a water heater is not seeing a residential per-person-per-day load, for example, if a residential water heater is used in a small office or other commercial application. In these cases, it is helpful to translate the residential ratings into ratings like commercial water heaters, to separate out a thermal efficiency from standby losses. Since the recovery efficiency is similar to the thermal efficiency, we just need to calculate the standby losses. It can be shown that:

$$SL = [(41045/EF) - (41045/RE)]/24$$

SL is the standby loss, in Btu/hr; EF is the rated energy factor, in decimal units; and RE is the rated recovery efficiency, also in decimal units. The recovery efficiency rating is typically provided in percent, but it is important to use decimal units, consistent with the energy factor. So, for example, an 82 percent recovery efficiency should be calculated as 0.82 in this equation.

For example, a 50-gallon residential standing-pilot water heater, dating from 1970, is used in a church. From billing analysis, the summertime gas use is seen to be 100 therms per month, which is extrapolated to 1,200 therms per year. A high-efficiency replacement is being considered, rated at 0.71 EF and 87 percent RE. For the baseline, we presume standby losses of 2,200 Btu/hr (see prior guidance on standby losses) and a thermal efficiency of 76 percent (see prior guidance on thermal efficiency based on water heater age). From the ratings of the new proposed residential water heater, we calculate standby losses of 443 Btu/hr, using the equation above. The existing water heater standby usage is $2200 \times 24 \times 365 / 100000 = 193$ therms per year. So the existing usage that is not standby is $1200 - 193 = 1007$ therms/year. The non-standby load is $1007 * 76$ percent $= 765$ therms. The new non-standby usage is $765 / 87$ percent $= 879$ therms. The new standby usage is $443 \times 24 \times 365 / 100,000 = 39$ therms. And the new total usage is the non-standby usage of 879 therms plus the standby usage of 39 therms $= 918$ therms. The savings are $1,200 - 918 = 282$ therms, or 23.5 percent. If utility bills were not available, we could do the calculations based on estimated load, or perhaps on measured hot water consumption, but still separating out standby losses from the efficiency calculation applied to the hot water load. Note that the load always includes piping losses. It just does not include standby losses.

Hot Water Temperature

Domestic hot water systems can save energy by reducing the hot water temperature. This can be done by reducing the hot water temperature in general, and by further reducing it during unoccupied periods. Energy savings accrue in three ways:

1. Lower standby losses.

2. Lower pipe distribution losses.

3. Higher thermodynamic efficiency of combustion systems and heat pumps systems at lower water temperatures. (Electric water heaters are the only heaters that do not see higher thermodynamic efficiency at lower water temperatures.)

Figure 12.8 Measuring water temperature.

One study that compared operating costs of a 50-gallon electric water heater found 25 percent savings when operated at 120 F, as compared to 140 F.[6] Savings would be expected to be even higher for gas or heat pump water heaters.

Water heater temperature can be controlled with a timer, to reduce the temperature during unoccupied periods. A variety of timer products are available, including controls for gas water heaters. One study of timer savings for gas water heaters found approximately 8 percent savings for a setback from 130 F to 85 F for 85 hours/week (50 percent of the time), and 37 percent savings for setback of 147 hours/week (88 percent of the time).[7] Savings with electric water heaters are anticipated to be less, due to lower standby losses and constant thermodynamic efficiency. One study found only a 0.5 percent increase in energy factor when power to an insulated water heater is turned off during the 18 hour standby period of the 24-hour DOE rating test.[8] Another study found 5.8 percent savings for 56 hours of setback per week.[6]

Existing water temperatures can be measured on pipe surfaces during a hot water draw, or where hot water is delivered. See Figure 12.8.

Residential-style water heater settings typically do not have the actual temperature printed on the control, but rather a series of numbers or letters, with a Low or Warm setting at one end and a Hot or High setting at the other end of the range. The Low or Warm setting is typically between 80 and 110 F, and the High or Hot setting is 140–160 F.[9] If the control has a Vacation setting, this is intended to save energy while preventing freezing of water in winter, and typically corresponds to 50 to 60 F. See Figure 12.9.

Figure 12.9 Residential water heater temperature control.

Figure 12.10　Commercial water heater temperature control.

Figure 12.11　Mixing valve—hot water enters from left (marked H), cold water enters from right (marked C), mixed water leaves at bottom.

Commercial-style water heaters are more likely to have controls that are calibrated in degrees, and their range is slightly higher, typically going up to 180 F, in order to be able to serve commercial kitchens, and also higher at the low end, typically starting at a minimum of 120 F. See Figure 12.10.

What is the right temperature set point for hot water heaters? For energy savings, we want the temperature as low as possible, and 120 F is a widely recommended set point, including by the DOE, unless users have a suppressed immune system or chronic respiratory disease, for which a setting of 140 F is recommended.[10] To prevent scalding, we also want it as low as possible, and 120 F has also been recommended for this reason by anti-scalding safety advocates.[11] The International Plumbing Code recommends that tempered (automatically mixed) water for nonresidential purposes be delivered at no more than 110 F.

On the contrary, concerns about *Legionella*, a group of bacteria, have given rise to recommendations for higher temperatures. The World Health Organization (WHO) reports that above 158 F *Legionella* dies almost instantly, at 140 F 90 percent of *Legionella* die in 2 minutes, at 122 F 90 percent die in 80 to 124 minutes depending on the strain, at 118 to 122 F *Legionella* can survive but do not multiply, and between 90 to 108 F is the ideal growth range for *Legionella*.[12] One study found *Legionella* in 40 percent of residential electric water heaters, but in none of the gas or oil water heaters, and so recommends 140 F for electric heaters and 120 F for gas or oil water heaters.[11] WHO recommends that the water in a water heater should fully reach 140 F at least once a day.

Some systems store water at higher temperatures but deliver it at lower temperatures, using a mixing valve (Figure 12.11).

Domestic Hot Water Distribution

Pipe insulation reduces energy losses. How much energy can be saved depends on the water temperature, whether water is recirculating, how often water is used, the length and diameter of piping, the amount of insulation, and whether energy is lost in a space that is heated or not. Insulation savings calculations are typically done by spreadsheet. Online calculators are also available, as well as free downloadable programs.[13–15] When performing savings calculations, it is important to not overestimate the operating hours per year. Table 12.1 provides savings per 100′ of water pipe at 120 F, for 1″ of fiberglass insulation.

Water heaters in larger buildings use recirculating hot water flow, in order for hot water to be readily available at all points of use. See Figure 12.12.

TABLE 12.1

Energy Savings from 1″ Fiberglass Pipe Insulation (120 F water, 70 F ambient air)

Actual Pipe Diameter (inches)	Operating Hours/Year			
	2,000	4,000	6,000	8,000
	kBtu/year savings per 100 ft of pipe			
1	7,823	15,647	23,470	31,294
1.5	10,322	20,644	30,966	41,288
2	12,570	25,140	37,710	50,280

Note: Savings are provided in terms of reduced heat loss, not reduced fuel use. Divided by efficiency of heat generation in order to obtain reduced fuel use.

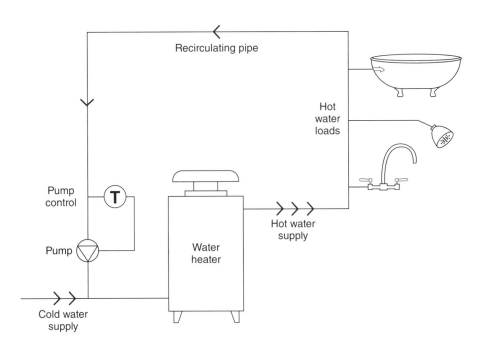

Figure 12.12 Recirculating domestic hot water.

For recirculating systems, a small pipe is routed back to the water heater from the points of use that are farthest from the water heater. A pump is located at the end of this pipe, near the water heater, typically near the entering cold water pipe. The pump may not be controlled, in other words, may run 24/7. Or it may be controlled by a thermostat. If a thermostat is not installed, one should be installed, which will limit the runtime of the pump by limiting the pump operation to when the temperature is too low. The temperature should be set at least 10 F lower than the set point in a storage water heater. If a pump is already controlled by a thermostat, its temperature setting should be checked, to make sure it is not too high. If the temperature setting is too high, the pump will run continuously. See Figure 12.13.

Water Loads

Water conservation saves both the energy required to heat hot water, for hot water loads, and water use itself, for both hot and cold water loads.

Common hot water loads include showers, bathroom and kitchen faucets, clothes washers, bath spouts, utility sinks, and dishwashers. Common cold water loads include water closets, urinals, water fountains, and irrigation. We examine hot water loads first, then cold water loads.

Loads are reduced by reducing either the flow rate or the duration of flow, or both.

Helpful benchmarks are federal requirements, many of which were established in 1992 with the Energy Policy Act, commonly known as EPAct, and newer better-than-code requirements such as those suggested by EPA's WaterSense program, as shown in Table 12.2.

As water conservation products have been developed in recent years, a variety of performance problems and consumer acceptance issues have arisen. Many of these have been overcome, but some have not.[16] Energy auditors and property managers interested to push the bounds of water and energy conservation should stay apprised of current developments, maintaining the demand for reliable new developments in conservation, while avoiding known problems with new products.

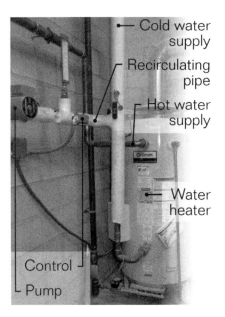

Figure 12.13 Recirculating hot water system.

TABLE 12.2

Water Usage Efficiency Requirements

	Federal Requirement	EPA Water Sense Requirement
Showerheads	2.5 GPM	2.0 GPM
Urinals	1 GPF	0.5 GPF
Residential toilets	1.6 GPF	1.28 GPF
Commercial faucets (private lavatories)	2.2 GPM	1.5 GPM
Residential bathroom faucets	2.2 GPM	1.5 GPM

GPM: gallons per minute; GPF: gallons per flush.

FAUCETS

In addition to the federal requirements for residential and private commercial faucets, since the mid-1990's most model codes limit the flow of public faucets to 0.5 GPM. Public faucets are defined as all faucets other than those in residential buildings, hotel/motel rooms, and in private rooms in hospitals. Adherence to this requirement has been poor, and presents a significant energy conservation opportunity.

When doing fieldwork, a best practice is to measure the faucet flow. Flows vary with water pressure, and with faucet design. Calibrated flow bags (Figure 12.14) are available for measuring faucet flow, which fit easily under a faucet even if the clearance below the faucet to the sink is small. Beakers, gallon bottles, or buckets can also be used but will sometimes not fit under a faucet. Whatever measuring container is used, a stopwatch is also needed, now available on most smartphones. Calibrated flow-measuring cups, or weir cups, are also available and do not require a stopwatch.

If flow measurement is not possible, visual observations can provide a guide to the flow rate or installed faucet flow device. Often, the rated flow of a faucet is printed on its side (Figure 12.15). If the flow rate is stamped in liters/minute (l/min), multiply by 0.2642 to obtain GPM.

If the flow is so high that the sink begins to fill, this is an indication that the flow is over 2 GPM (Figure 12.16).

Aerators with flows of approximately 1.5 GPM typically have multiple holes or a screen, again, most easily observed with a photograph from below, and the water flow looks like a solid aerated flow. See Figure 12.17.

Very low-flow faucets, nominally either 0.35 GPM or 0.5 GPM, typically have a single set of holes around the perimeter, and this is visible as a cylindrical flow pattern of small individual jets, sometimes referred to as *needle flow*. See Figure 12.18.

In other words, just because the flow is aerated does not mean that it is indeed low flow, such as 0.35 or 0.5 GPM. Many aerated-looking faucet flows are actually high flow rates. The term *low-flow faucet insert* is preferable to faucet aerator, especially when referring to inserts that are 0.5 GPM and lower in flow.

When specifying low-flow faucet inserts, it is important to not simply require "a faucet aerator," but rather to specify its required maximum/rated flow. Consumer acceptance of 1.5 GPM low-flow faucet inserts, appears good.[17] Anecdotal reports of 0.35 and 0.5 GPM faucet flow rates and effectiveness are also positive, indicating adequate flow for hand washing, and no increase in the time required to wash hands due to the lower flow. The lower flows are less effective for the scrubbing action required to wash dirt off objects.

In residential settings, faucets see approximately 8 minutes of flow per person per day.[18] The duration of flow for hand washing is fairly short. One study found the average duration in public bathrooms to be 3.7 seconds for women, and 5.2 seconds for men.

Figure 12.14   Calibrated bag for faucet or shower water flow measurement.

Figure 12.15   Rated faucet flow.

For kitchen sinks, water is used not only for hand washing but also for dishwashing (including the need for adequate water flow to clean food off plates and utensils), and for filling containers such as pots and kettles. So, for kitchens, lower flow may not be acceptable. For example, the state of California requires 1.8 GPM for kitchen faucets. At the present time, for energy conservation recommendations, it might be prudent to require 1.8 GPM for kitchen faucets, even if low flows such as 0.35 to 0.5 GPM are used for bathroom faucets.

In estimating savings, we must recognize that people do not always open a faucet to its maximum flow, especially if the flow is high. To be conservative, it might be prudent to cap the pre-retrofit flow at 3 GPM, even if a faucet has a measured flow over 3 GPM.

Also, when calculating energy savings, we note that the reduction in hot water is less than the reduction in total water flow. The ratio of hot water flow to total water flow depends on the temperature of the mixed flow, the temperature of the hot water, and the temperature of the cold water. This ratio (hot water to total water) can vary from 40 percent to 80 percent. For a mixed temperature of 100 F, entering hot water temperature of 130 F, and entering cold water of 55 F, the hot water is 60 percent of the total flow.

In other words, if an existing faucet has a flow of 3 GPM, its hot water flow is approximately 60 percent × 3 = 1.8 GPM. If it is replaced with a 0.5 GPM faucet aerator, the new hot water flow is approximately 60 percent × 0.5 = 0.3 GPM. The hot water savings are 1.8 − 0.3 = 1.5 GPM. Note that the 1.5 GPM hot water savings are less than the 2.5 GPM total water savings.

Metering faucets (Figure 12.19), also known as self-closing or push-button faucets, dispense a fixed amount of water after a button or handle is pushed.

Metering faucets have a federal standard of 0.25 gallons/cycle, at a test pressure of 60 psi.[19] The amount of water dispensed by metering faucets is typically adjustable, by adjusting the cycle time, done by making a mechanical adjustment to internal nuts or washers. The factory preset cycle time is typically 10 to 15 seconds. Metering faucets can also have varying flow rates. When evaluating existing metering faucets, measure both the flow rate and the cycle time. If the flow rate is more than 0.5 GPM and the cycle time is over 10 seconds, evaluate savings to reduce flow to 0.5 GPM (or 0.35 GPM) and to reduce the cycle time to 10 seconds or less. Note that a flow of 0.5 GPM with a cycle time of 9 seconds delivers 0.075 gallons of water, 75 percent less than the federal standard of 0.25 gallons. Metering faucets typically are observed to continue operating for several seconds beyond the end of use, and so conservation beyond factory-default settings are plausible and should be pursued.

Automatic faucets (Figure 12.20) use motion sensors to detect the presence of hands under the faucet, and so to start and end flow. More costly than mechanical metering faucets, they also have higher energy savings potential, because the time between removal of hands and the end of the water flow is shorter. This off-delay is typically factory set at 2 seconds. Some models allow the off-delay to be varied.

It should be noted that some research has shown that metering and automatic faucets do not save water, and in fact use more water than conventional handle faucets.[20]

SHOWERHEADS

Older showerheads deliver flow over 3 GPM. Newer showerheads must meet a 2.5 GPM federal requirement. EPA's Water Sense program has a limit of 2.0 GPM. Several manufacturers make 1.5 GPM showerheads, and showerheads are available as low as 0.625 GPM. See Figure 12.21.

A 2011 California survey of available showerheads found that the majority sold were still 2.5 GPM.[21] Consumer acceptance of 2 GPM showerheads appears to be high, in at least two studies.[21] Anecdotal reports of 1.5 GPM showers indicate that acceptance is increasing with newer designs.

Figure 12.16   High faucet flow over 2 GPM typically will fill a sink even if the drain plug is open.

Figure 12.17   Faucet aerator.

Figure 12.18   Low-flow faucet insert.

Figure 12.19 Metering faucet.

Figure 12.20 Automatic faucet.

Figure 12.21 Showerhead, 2.1 GPM.

As with faucets, showers deliver a mixed flow of hot and cold water. In calculating hot water savings, we must account for the lower hot water flow rate than the total flow of the shower.

A 1999 study found the average duration of low-flow showers (below 2.5 GPM) was 8 minutes 30 seconds, 25 percent longer than the average flow of non-low-flow showers (above 2.5 GPM), which was 6 minutes 48 seconds.[18] This raises the possibility that low-flow showerheads result in longer showers. To be conservative, one might discount estimated savings by 25 percent.

PAUSE VALVES

Pause valves, also called *temporary shutoff valves* or *trickle valves*, can be used to almost stop flow when flow is not needed in showers and faucets, while leaving a trickle to retain a hot or mixed temperature.

Two types of pause valves are available to reduce water use at showers and faucets. One device is mechanically actuated by the user, and allows users to pause the flow while in use, for example to allow lathering. See Figure 12.22.

A newer device is intended primarily for showers and stops the flow automatically, using a temperature activated valve, when hot water reaches the shower (Figure 12.23). This is intended to eliminate waste before someone enters the shower, as they wait for the hot water, and are possibly distracted. The flow is mechanically started again by the user, for example, by pulling a cord.

Authoritative data is not available on energy savings from pause valves. One researcher found that 3.5 gallons of hot water is wasted at the beginning of a typical shower draw in single-family homes.[22] This represents 20 percent of the total water of the average 17.2 gallon shower, but 30 percent of the *hot* water of the average shower. Some of this waste can be reduced through the use of an automatic pause valve. However, some of these 3.5 gallons represents hot water from a prior shower that has become cold. So the net savings for an automatic shower pause valve is likely to average 1 to 2 gallons of hot water per shower in a single family home, and likely less in a multifamily building or hotel, in which recirculating hot water reduces the wait time for hot water at the showerhead.

Because pause valves typically allow a small flow during the pause, and this flow is not enough to keep an instantaneous water heater firing, pause valves are generally discouraged for use with instantaneous water heaters.

PRERINSE SPRAY VALVES

Prerinse spray valves (Figure 12.24) are used in commercial kitchens to clean food off dishes before washing.

They have been found to use 14 percent of the water in California's commercial kitchens.[23] In the United States, prerinse spray uses 51 billion gallons of water per year. These valves can use as much as 3 to 5 gallons per minute.[24]

Federal requirements (EPAct 2005) call for a maximum of 1.6 gpm, the EPA's Water Sense program requires 1.28 gpm,[25] ASHRAE's 189.1 Standard requires 1.3 gpm, and the Federal Energy Management Program requires 1.25 GPM.[24] A field study of prerinse spray valves found that the average flowrate for noncompliant prerinse spray valves was 3.6 GPM.[24] The same study found the temperature range of prerinse spray flow was between 75 F and 129 F, with an average of 106 F, and an average use time of 84 minutes per day. The estimated time to clean a single plate is 15 to 25 seconds. The study found that "use time tends to remain relatively constant regardless of the ... operating flow rate," and that lower-flow valves used less energy and water. However, the study identified some user concern with the performance of prerinse spray valves with flows less than 1 gpm.

DISHWASHERS

Dishwashers see modest residential use (single and multifamily buildings), and occasional use in settings such as office kitchenettes, but are large energy and water users in food service kitchens, such as in restaurants, bars, nursing homes, churches, schools, universities, hospitals, and prisons.

Residential dishwashers are subject to federal efficiency standards, and commercial dishwashers are subject to efficiency requirements for procurement by federal agencies, through the Federal Energy Management Program (FEMP). Residential and commercial dishwashers also both have ENERGY STAR programs that certify products using less energy and water.

Commercial Dishwashers

Commercial dishwashers can be classified as under counter, door type, conveyor, or flight type.[26] Under counter dishwashers are similar in style to residential dishwashers. Door-type dishwashers have doors that slide open, and are loaded with dish racks that typically are slid in and out from adjacent tables (Figure 12.25). Conveyor dishwashers carry dishes on a conveyor system. Flight type are a kind of conveyor dishwasher; however, they do not use racks. They are sometimes called rackless conveyor dishwashers because the dishes are loaded directly onto the conveyor instead of onto racks. These are typically custom fabricated, and are used in facilities that serve over 2,000 meals per hour.

Commercial dishwashers are sometimes referred to as warewashers or dish machines. Under-counter and door-type machines are often classified as stationary rack dishwashers.

Commercial dishwashers are typically available as low-temperature or high-temperature. Low-temperature dishwashers use water around 140 F and use chemical sanitizers. These are often referred to as chemical dishwashers. High-temperature dishwashers use water at 180 F for sanitizing.

Most commercial dishwasher racks are a standard size of 19.75″ × 19.75″ × 4″, commonly referred to as a 20 × 20 rack. Half-size racks are 9 7/8″ in one dimension, nominally 10 × 20. The rack forms a convenient unit of measure for energy and water use.

To determine existing water use, many existing commercial dishwashers are in a database at NSF International's web site, which includes rated water usage.[27] Manufacturers typically also list their rated water usage in their catalogs.

For rules of thumb, FEMP lists "less efficient" models of commercial dishwashers as using 0.85 kW idle energy and 1.44 gallons per rack.[28] The Alliance for Water Efficiency reports that a typical commercial dishwasher uses approximately 4 gallons per rack,[29] but that inefficient models can use over 20 gallons per rack.

Models that comply with FEMP's efficiency requirements use 0.70 kW idle energy and 0.89 gallons per rack. FEMP lists the "best available model" as using 0.16 kW idle energy and 0.55 gallons per rack.

Energy and water savings options include converting to higher efficiency dishwashers and converting to gas-fired booster heaters, in addition to high-efficiency primary water heaters.

EPA's ENERGY STAR program began listing high-efficiency commercial dishwashers in 2007. The requirements vary based on the type of dishwasher. For example, an under counter high-temperature dishwasher must have an idle energy rate less than 0.5 kW and water consumption less than 0.86 gallons per rack.[30]

One study of replacing a conventional commercial dishwasher using an electric booster heater with a gas-fired booster heater found source energy savings of 73 percent, and a simple payback of approximately three years.[31]

ENERGY STAR has a spreadsheet calculator to calculate savings for high-efficiency commercial dishwashers.[32] The calculator can also be used to estimate

Figure 12.22 Pause valve.

Figure 12.23   Temperature activated pause valve.

Figure 12.24   Prerinse spray valve.

savings for gas booster heaters, although the calculation needs to be done iteratively because the spreadsheet does not allow changing both to a high-efficiency dishwasher and to a gas booster heat at the same time.

The most savings come from a combination of improvements:

1. Using a high-efficiency dishwasher.
2. Using a high-efficiency primary water heater.
3. Using a gas booster heater.

This combination maximizes energy savings, water savings, and carbon emission reductions.

Residential Dishwashers

Residential dishwashers, also referred to as consumer dishwashers, are widely used, not only in homes, but also in multifamily buildings, and in commercial buildings such as office kitchenettes, where commercial dishwashers are not specifically required.

The first federal residential dishwasher energy standard was introduced in 1988, and the second in 1994. Between 1981 and 2008, the average energy consumption of residential dishwashers in the United States was cut in half.[33] Much of this reduction is due to reduced hot water use.

Effective May 30, 2013, residential dishwashers must use less than 5 gallons per cycle and 307 kWh/year, and compact dishwashers must use less than 3.5 gallons per cycle and 222 kWh/year.[34] Effective in 2012, new ENERGY STAR standard-size consumer dishwashers must be rated to use less than 4.25 gallons per cycle, and less than 295 kWh/year electricity. New ENERGY STAR compact-size residential dishwashers must meet the same requirements as the federal standard.[35]

A dishwasher energy factor (EF) is defined as: $EF = 1 / (M + W)$, where M is the kWh/cycle and W is the heat required to heat the water, also in units of kWh/cycle. The rated annual energy use (based on DOE's standard 215 cycles per year) can be calculated from the EF as 215/EF kWh/year. For example, a dishwasher with an EF of 1.14 has a rated annual electricity usage of $215/1.14 = 189$ kWh/year.

If a building client kept the Federal Trade Commission (FTC) EnergyGuide label from when the dishwasher was purchased, it can be used to estimate the dishwasher's actual energy usage. The FTC EnergyGuide label does not show the EF, but does show the estimated annual energy usage. The EF can be estimated from the EnergyGuide label, as 215 / (annual kWh rating). For example, if the label shows an annual rating of 649 kWh/year, the rated EF is $215 / 649 = 0.33$. Similarly, the estimated annual usage based on the estimated actual cycles/year can be prorated against the rating and 215 cycles. For example, if the same 649 kWh/year-rated dishwasher is expected to be used 110 cycles per year (see estimated usage for a family of three, below), the estimated annual energy usage is $110/215 \times 649 = 332$ kWh/year.

Unfortunately, many manufacturers do not prominently show any of the dishwasher ratings (EF, kWh/year, or gallons/cycle) in their literature. Fortunately, replacement dishwashers for energy improvements are typically ENERGY STAR rated, and so the EPA ENERGY STAR database can be used to select dishwashers, and identify all three ratings.

A 1999 study found that dishwasher water use averaged 1 gallon per person per day in the U.S., comprising 1.4 percent of residential energy water use, and averaging 0.1 cycle per person per day.[18] This implies 10 gallons per cycle. Again, this is for average dishwashers in operation in the late 1990s.

In predicting savings, we need to know how many cycles a typical dishwasher runs per year, in residential settings such as multifamily buildings. A commonly used statistic is 215 cycles per year. However, this number originates in the test standard for dishwashers and does not appear to be based on rigorous research. A more reliable number is the 0.1 cycle per person per day, previously identified in research.

This translates into 37 cycles/year for one person, 73 cycles/year for two people, 110 cycles/year for three people, and 146 cycles/year for four people, all far lower than the ratings-based 215 cycles per year.

It is also necessary to know current dishwasher energy and water use, when considering replacements. If replacing a large number of identical dishwashers, for example, in a multifamily building, it may be worthwhile to measure dishwasher energy and water use. Many dishwashers are plug-in, so a plug-in watt meter can be used to measure energy consumption. Water consumption can be measured at the drain. It is important to note that 100 percent of the water consumed in a dishwasher is hot water. It is also important to note that measured electricity use only represents the dishwasher's own usage, and not the energy required to heat the water before reaching the dishwasher, and so the consumed electricity should not be compared directly to energy ratings.

Manufacturers may have records of rated energy and water usage. Again, note that ratings are based on 215 cycles per year, so ratings should be adjusted to actual expected cycles per year.

A 2008 study indicated that average dishwasher electricity use in 1993 averaged 2.6 kWh/cycle, with hot water use of 10 gallons/cycle, and by 2004 these numbers had dropped to 1.8 kWh/cycle and 6 gallons of hot water.[36]

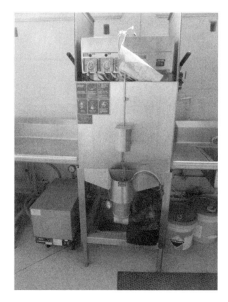

Figure 12.25   Door-type commercial dishwasher.

Based on this report, and on a 2010 MIT study, in the absence of measured or rated data, dishwashers may be assumed to have rated energy usage of 650 kWh/year for years 1993 and prior, 500 kWh/year for years 1994 to 2000, and 400 kWh/year for years 2001 to 2005. If using the dishwasher age in this way to estimate the energy and water usage, the year of manufacture may be obtained from online services.[37]

EPA's ENERGY STAR web site has a spreadsheet calculator for energy and water savings. It uses default baseline values of 355 kWh/year and 6.5 gallons per cycle, representative perhaps of a standard 2004 dishwasher.

For those interested to program their own spreadsheets, it is helpful to disaggregate dishwasher machine energy use from the energy used to heat the hot water, in order to examine the impacts of dishwasher replacements on hot water energy use, and vice versa. DOE's dishwasher rating rules specify a 70 F temperature rise for the hot water. So the water consumption per cycle can be multiplied by $215 \times 8.33 \times 1 \times 70 / 3{,}412$ to obtain the water-heating component of the rating in kWh/year. Subtracting this from the rated annual electricity usage gives the estimated annual dishwasher electricity consumption without water heating. The ENERGY STAR spreadsheet can also be used as the basis for custom calculations, as its equations are transparent, although it appears to assume a constant and presumed fraction of water heating energy use for both the pre-retrofit and post-retrofit scenario.

CLOTHES WASHERS

ENERGY STAR clothes washers are frequently horizontal-axis machines, although some vertical axis ENERGY STAR machines do exist.

Commercial Clothes Washers

Commercial clothes washers (Figure 12.26) are defined by federal energy statute as being designed for use in more than one household, such as in the common areas of multifamily buildings and in coin laundries. The energy regulations focus on front loaders smaller than 3.5 cubic feet, and top loaders smaller than 4 cubic feet. The energy regulations are also limited to soft-mount (free-standing) machines, rather than hard-mounted machines that are typically bolted to the floor. In this book, we will limit our discussion to these single-load soft-mount commercial clothes washers. Larger multiload and tunnel (continuous batch) machines are available for use in larger laundry operations, but will not be discussed here; the reader is referred to other publications for coverage of these.[23, 26, 38] It should be noted that one study found that multiload clothes washers use more water per pound of laundry than smaller front

Figure 12.26   Commercial clothes washers.

load washers.[38] Separately, hard-mounted machines are reportedly less efficient than soft-mount machines, because they do not spin as fast, and so leave more residual water for the dryer than do soft-mount machines.

Commercial clothes washers were first covered by federal efficiency standards in 2007, following earlier standards developed for consumer/residential clothes washers. The efficiency standards were increased in 2013, to the minimum modified energy factors (MEF) and maximum water factors (WF) in Table 12.3.

These requirements will change in 2018 to a minimum 1.35 MEF for top-loading washers and to a minimum 2.0 MEF for front-loading washers. The WF is being changed to an integrated water factor, with minimums of 8.8 and 4.1 for top loaders and front loaders, respectively.[39]

The original 2007 requirements that went into effect in 2010 called for a minimum 1.26 MEF and maximum 9.5 WF.

ENERGY STAR, on the other hand, does not differentiate between front loaders and top loaders. Its requirements for commercial clothes washers as of February 2013 are a minimum MEF of 2 and WF less than 6.

The MEF is $= C / (M + E + D)$, in units of cubic feet per kWh per cycle, where C is the capacity of the clothes washer in cubic feet, M is the machine electrical consumption in kWh/cycle, E is the hot water energy consumption in kWh/cycle, and D is the energy required to remove the moisture remaining in the clothes after the washing is complete, also in kWh/cycle.

The MEF is obtained from lab tests that follow a standard, which specifies multiple tests based on water temperature (hot, warm, or cold), clothing type and weight,

TABLE 12.3

2013 Commercial Clothes Washer Efficiency Requirements

Equipment Class	Modified Energy Factor, cu. ft./kWh/cycle	Water Factor, gal./cu. ft./cycle
Top loading	1.60	8.5
Front loading	2.00	5.5

Source: Federal Register. 10 CFR Ch. II (1–1–11 Edition). Subpart I—Commercial Clothes Washers.

and more. The tests are such that they cannot easily be field replicated, for example, if one wanted to measure the MEF of a specific clothes washer installed in a building.

Prior to the development of the MEF, clothes washers were rated with an EF that did not account for the moisture remaining in the clothes after the washing is complete. Due to the impact of clothes spinning on dryer energy use, the moisture content is now accounted for in the MEF ("D" above).

The Water Factor is the water used per cycle, in gallons, divided by the volume of the clothes washer, in cubic feet. This is primarily useful for estimating the annual water consumption of the clothes washer, and the savings by replacing an older low-efficiency clothes washer with a higher-efficiency model. Multiply the WF by the volume (cubic feet, either rated or measured, see below) of the clothes washer to obtain the gallons per cycle, and then multiply again by the estimated cycles per year to obtain the estimated annual water usage in gallons per year.

Manufacturers always publish the volume in cubic feet (C), and sometimes publish the MEF and the WF. If not published by the manufacturer, the ratings are available for currently available ENERGY STAR machines at the ENERGY STAR web site. As usual, this will likely take care of the proposed replacement high-efficiency clothes washer. If currently manufactured, the ratings may be obtained from the DOE web site, although this database includes only MEF and WF.[40]

Existing in-building clothes washer MEF may be obtained by contacting the manufacturer, or estimating based on the age of the machine. Age-based estimates of MEFs are provided in Table 12.4.

If using the age of the clothes washer to estimate the energy usage, the year of manufacture may be obtained from online services,[41] using the manufacturer, model number, and serial number.

The FTC EnergyGuide label can unfortunately not be used to obtain the MEF. The label shows annual energy use (based on 392 cycles per year), but this energy use does not include the dryer energy use that is included in the MEF. The MEF for current models is also available at the DOE web site.[42]

TABLE 12.4

Estimates of Modified Energy Factor for Commercial Clothes Washers

Year	Top-Loading	Front-Loading	Notes
Pre-2002	1.00	1.40	(1)
2002	1.10	1.50	(2)
2003	1.13	1.53	(2)
2004	1.13	1.53	(2)
2005	1.21	1.61	(2)
2006	1.22	1.62	(3)
2007	1.23	1.63	(3)
2008	1.24	1.64	(3)
2009	1.25	1.65	(3)
2010	1.26	1.66	(4)
2011	1.30	1.70	(5)
2012	1.35	1.75	(5)
2013	1.60	2.00	(4)

(1) Presumed lower than 2002.
(2) From appendix 5a to 2010 Rule. Presumed top-loading clothes washers MEF is 0.2 less than overall average, and front-loading is 0.2 higher than overall average.
(3) Interpolating between 2005 and 2010.
(4) Federal standard. Assume that clothes washers from this year that are sought to be replaced are minimum efficiency, not ENERGY STAR or otherwise high-efficiency.
(5) Presume minor improvement over prior year.

With the MEF, annual estimated energy usage under building-specific conditions can be estimated by multiplying the actual expected cycles per year times the energy use per cycle. The energy use per cycle, for electrically heated water, can be obtained by dividing the clothes washer capacity (in cubic feet) by the MEF. The energy use per cycle for gas-heated water can be obtained by using the ENERGYSTAR calculator. It should be noted that the annual energy use obtained this way, using the MEF, *includes* the estimated separate dryer energy use. If the estimated annual energy use of the clothes washer *exclusive* of dryer energy use is sought, one of the following methods can be used:

■ Use the EnergyGuide label annual energy use (kWh/year), divide by 392 cycles/year to obtain kWh/cycle, and multiply by the estimated building-specific usage (cycles/year).

■ Obtain the per-cycle energy use or annual energy use (exclusive of dryer energy use) from an online database. For example:

   ■ The ENERGY STAR database (above) has a column titled Annual Energy Use, which is washer-only (exclusive of estimated dryer energy use).

   ■ The California Energy Commission maintains a clothes washer database within its appliance database.[43] Select "Advanced Search," select "Cooking and Washing Products," select "Clothes Washers," select "Neither" and select "Select All" for fields. The field "Power Consumption" in this database is the per-cycle power consumption (kWh/cycle). This can be multiplied by annual washer use (cycles/year) to obtain annual electricity consumption, exclusive of dryer use. Another advantage of this database is that it includes non–ENERGY STAR models, and it includes some clothes washers that are no longer manufactured.

   ■ California also maintains a database of historical clothes washer ratings.[44]

   ■ Canada maintains a database of current models available in Canada.[45]

Correcting for anticipated cycles/year for a specific clothes washer in a specific building is especially important for commercial clothes washers, where the number of users per machine varies significantly, and so where the number of cycles/year may significantly impact the savings and cost-effectiveness, per machine. The number of cycles/year used for the EnergyGuide rating is typically far lower than is commonly the case in real buildings, for commercial clothes washers. Using the EnergyGuide rating without correcting for cycles/year will likely significantly underestimate actual savings. The EnergyGuide rating uses 392 cycles per year. The FTC label indicates 8 cycles per week, but this is a rounding of 392 cycles per year divided by 52 weeks.

The capacity of a clothes washer can be obtained from the interior measurements: Capacity (in cubic feet) $= 3.14 \times r \times r \times h / 1728$, where r is the radius of the tub in inches, and h is the height of the tub in inches. This may be necessary when choosing a replacement clothes washer. This can also be used to calculate the energy use per cycle for a clothes washer of unknown volume and MEF rating. This volume is divided by the MEF rating (see Table 12.4) to obtain the energy use per cycle (kWh/cycle, *inclusive* of estimated dryer energy).

When doing fieldwork, it is frequently possible to obtain commercial clothes washer usage (cycles per year) from coin receipts, from the building owner or from the laundry vendor if the owner has a contract with a vendor. Most commercial laundry machines also have factory-installed counters, from which usage data can be obtained, again typically from a laundry vendor. In the absence of such data, usage can be estimated from DOE estimates of averages for multifamily buildings (1,074 cycles per year) and laundromats (1,483 cycles per year).[39] Another source provides slightly higher estimates of 1,241 cycles/year for multifamily buildings and 2,190 cycles/year for coin-operated laundries,[26] although this references 2008 DOE rule making.

If the number of people being served by a group of machines is known (for example in a multifamily building), the number of cycles per year can be calculated from a research finding of 0.37 cycles per person per day.[18] For example, for a multifamily building housing 100 people, and four commercial clothes washers, each machine serves 25 people on average, and so each machine will run $25 \times 0.37 \times 365 = 3,376$ cycles/year.

ENERGY STAR's appliance calculator can be used to estimate annual energy usage, for either electrically heated or gas-heated water.[46]

The main improvement for commercial clothes washers is to replace existing clothes washers with high-efficiency replacements. Increasing the efficiency of domestic hot water supply is another improvement. Behavioral/education improvements include encouraging the use of low temperature water. Multifamily building owners can also encourage tenants to use the common laundry, where higher-efficiency machines can be more cost-effectively installed and maintained, and will deliver more savings per machine, with higher use. Drain heat recovery is another option (see below).

In seeking cost-effectiveness of high-efficiency laundry replacement, it is worthwhile to determine whether a facility has too many washers. Replacing too many washers with the right number of washers is more cost-effective than replacing washers one-for-one: The installed cost of fewer machines is lower, and the savings per machine is higher. Recommendations for the number of machines for multifamily buildings is provided in Table 12.5.

Residential Clothes Washers

As with commercial clothes washers, residential clothes washers impact energy use in three ways:

- Consume electricity to run the machine, primarily the motor that turns the tub, but also standby energy in controls.
- Use hot water.
- By leaving water in the clothes that needs to be removed by clothes dryers.

Conversely, high-efficiency clothes washers consume less electricity, use less water, and leave less water in the clothes.

U.S. federal regulations governing residential clothes washers use similar metrics and test procedures. Definitions of MEF and WF have been the same, but in 2015 a new "Integrated Modified Energy Factor" was introduced for residential clothes washers.

TABLE 12.5

Recommended Laundry Machines for Different Types of Housing

Predominant Resident Profile	One Pair Washer/Dryer Per
Families	8–12 units
Younger working adults	10–15 units
Older working adults	15–20 units
Students	25–40 students
Senior citizens	25–40 units

Source: Multi-housing Laundry Room Planning and Design Guide, WASH Multifamily Laundry Systems, 2011.

TABLE 12.6

Residential Clothes Washers—Estimates for Modified Energy Factor Based on Machine Age

Year	Modified Energy Factor		Notes
	Top-Loading	Front-Loading	
Pre-1994	0.70	1.20	(1), (2)
1994–2003	0.82	1.32	(1), (2), (3)
2004–2006	1.04	1.54	(2), (3)
2007–2014	1.26	1.76	(2), (3)
2015	1.29	1.84	(4)

(1) Richard Bole, "Life-Cycle Optimization of Residential Clothes Washer Replacement." Report No. CSS06–03, April 21, 2006. Center for Sustainable Systems, University of Michigan. MEF is an estimate based on the old EF rating. Presume this data is primarily top loaders.
(2) Presume front loader is 0.5 MEF higher.
(3) Presume top loader meets federal standard.
(4) Presume meets federal standard.

Estimates for MEF are provided in Table 12.6.

Although the FTC EnergyGuide label assumes 8 loads per week for its annual rating (416 cycles/year), and various DOE rules mention 392 cycles/year, a better estimate of usage is the 0.1 cycle per person per day previously identified in field research on a significant number of residential installations. This translates into 37 cycles/year for one person, 73 cycles/year for two people, 110 cycles/year for three people, and 146 cycles/year for four people. We note that these are all lower than the 416 cycles/year used in the rating standard. In general, residential clothes washers are probably used less than indicated by the EnergyGuide or DOE cycles/year, unlike commercial clothes washers, which are probably used more than indicated by EnergyGuide or DOE estimates.

The EnergyGuide label that comes with the machine typically does not show the MEF, but rather shows the annual estimated energy use if the water is electrically heated, and the estimated operation cost for either electrically heated water or gas-heated water. If the EnergyGuide label is available, but the MEF is not, the MEF may be calculated from the annual estimated energy use as follows: Divide the annual electric energy use on the label (kWh/year) by 416 (cycles/year) to obtain rated usage per cycle (kWh/cycle). Then divide the capacity of the clothes washer (cubic feet) by the usage per cycle (kWh/cycle), to obtain the MEF.

If the clothes washer was manufactured prior to 2004, the rating was an EF, and not an MEF. The EF can be obtained from the EnergyGuide label in the same way (divide rated annual usage in kWh/year by 416 to obtain kWh/cycle, and then divide capacity in cubic feet by kWh/cycle). An estimate of the MEF can then be obtained by multiplying the EF by 0.692.[47] The MEF can then be used to estimate building-specific annual usage, on an apples-to-apples basis with current high-efficiency clothes washers.

When the MEF changes to an integrated modified energy factor (IMEF) in 2018, a new set of adjustments will have to be made to correct either EF or MEF for existing clothes washers to IMEF, to compare to future new machines that are rated in IMEF. Early mention of the relationship is that IMEF is approximately MEF minus 0.43.

For those seeking to evaluate the hot water-related energy impacts if the water is heated with gas or another type of water heater (e.g., heat pump water heaters), a few approaches are available. The following strategies are for gas-heated water heating, and can be modified for other types of water heaters.

- If the EnergyGuide label is available, gas usage is provided on an annual basis. Divide by annual cycles per year to obtain rated gas usage per cycle. Or

- Use the MEF to obtain rated kWh per cycle (divide capacity in cubic feet by MEF). Pull out the water heating component of usage per cycle: One estimate is 45.8 percent.[47] Convert kWh/cycle to Btu/cycle by multiplying by 3412. Divide by the efficiency of generating hot water with gas (commonly 75 percent for ratings), divide by 100,000 to obtain therms/cycle, and multiply by cycles per year to obtain therms/year. Or

- Use the WF to obtain gallons per cycle, and multiply by the research finding that 13 percent of clothes washer water use is hot.[48] Multiply gallons per cycle by 7.58 pounds per gallon, multiply by the specific heat of water 1 Btu/lb-F, multiply by a typical water temperature rise (70 F is common for ratings), divide by the efficiency of generating hot water (75 percent is common for ratings), multiply by cycles per year, and divide by 100,000 Btu/therm, to obtain therms of gas usage per year.

Clothes washer ratings are not perfect, in that they do not account for behavioral issues, such as water temperature choices, types and quantities of clothes washed, and other such factors. As such, the ratings only reflect typical user behavior. However, they still provide a useful basis for comparison, and a rough basis for annual energy estimates of both baseline and high-efficiency replacements.

Clothes Washers—Closing Remarks

Are single-load commercial clothes washers more or less efficient than residential clothes washers? The U.S. does not maintain a public database of all clothes washers, but Canada does, presumably with a similar range of models. The Canadian database shows that the average MEF for residential clothes washers is about 9 percent higher than the average for commercial clothes washers. This is not a big difference, but for applications where either type of washer might be eligible, there might be merit to look at residential clothes washers, for efficiency. Of perhaps more significance is the question, "Are larger clothes washers more efficient?" Examining the same database, the answer seems to be fairly conclusively yes. See Figure 12.27. Note how, for the

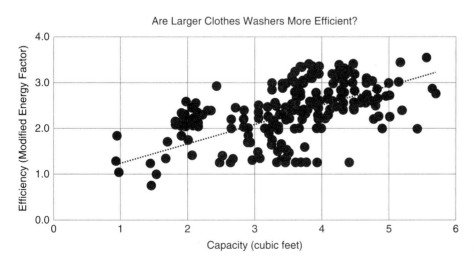

Figure 12.27 Larger clothes washers are more efficient.

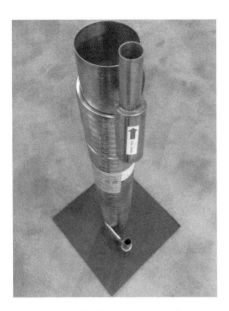

Figure 12.28   Drain heat recovery heat exchanger.

smallest capacities (under 1 CF), there is no machine available with an MEF over 2, and for the largest machines (over 5 CF), there is no machine available with an MEF less than 2.

So we might consider avoiding the smallest clothes washers, and to possibly consider larger high-efficiency machines (over 3 cubic feet), or at least a mix of large and medium-size high-efficiency machines, for example in a laundry room with multiple machines, so users can choose the right size high-efficiency machine that fits the clothes they need to wash.

DRAIN HEAT RECOVERY

Several manufacturers have developed water-to-water heat exchangers (Figure 12.28) that may be used for heat recovery from drain water. These typically comprise a copper drainpipe, around which is wound a smaller copper pipe, through which incoming cold water is routed. Drain heat recovery devices have no moving parts. They take advantage of a physical phenomenon called the falling film effect, in which flow through a large pipe clings to the walls of the pipe. This ensures that drain water is in contact with the drainpipe wall, and readily gives up heat to the incoming cold water flow.

Drain heat recovery works best with simultaneous flow of hot water down a drain and incoming cold water. It does not work as easily in situations where incoming cold water is heated and temporarily stored, and then is drained when there is no incoming cold water, such as in bathtubs, hot tubs, or other batch uses. They work well with loads like showers and faucet water use. In large commercial applications, such as commercial laundries, heat recovery can be designed to accommodate nonsimultaneous hot water drain and incoming water flow, using heat storage and an engineered pump and control system.

The cold water can either be the incoming building cold water pipe, before being routed to the hot water heater, or it can be the cold water pipe serving a specific hot water load or group of loads. For larger buildings, attention must be directed to the pressure drop through the cold water pipe.

Drain heat recovery works more efficiently where there is sufficient space for the heat exchanger to be oriented vertically, and some manufacturers do not recommend horizontal installation. The devices are typically long, in the 30 to 72″ range, with a few added inches of length clearance required for connections. Multiple units can be connected in series for added efficiency, subject to limitations on pressure drop for the incoming water. The longer the heat exchanger, the more efficient it is. For example, for one manufacturer's 4″ drain model, the Canadian government shows rated efficiencies of 40 percent for a 30″ long model, 53 percent for 48″ long, 58 percent for 60″ long, and 63 percent for 72″ long.[49]

The diameter of the heat exchangers is slightly larger than the diameter of the drainpipe, typically 1.25 to 1.5″ larger. For example, a model that fits a 3″ drainpipe might be 4.25 to 4.5″ in overall diameter. Very little clearance is required to nearby materials or construction. There are no temperature hazards and the device does not need to be insulated.

The Canadian government maintains a database of drain heat recovery manufacturers and efficiencies.[50]

Drain heat recovery devices have been independently tested and their performance validated. Most manufacturers provide calculation spreadsheets to support energy calculations for building-specific applications. They are most cost-effective when multiple loads, such as several showers in a hotel or multifamily building, can be connected to a single heat exchanger.

If applied correctly, drain heat recovery devices can save much energy, with no moving parts, and with an expected life that is likely as long as the life of the building.

HEAT RECOVERY FROM REFRIGERATION AND AIR CONDITIONING

Domestic hot water can be provided as heat recovery from refrigeration systems, including air conditioners and commercial refrigeration. These systems are commonly referred to as *desuperheaters*.

TOILETS AND URINALS

Until the 1970s, toilets were commonly 5 to 7 gallons per flush (GPF), with 5.5 gallons reportedly standard, in a two-piece design, with an elevated tank separate from the toilet bowl. Single-piece toilets reportedly had flush volumes over 10 GPF. In the 1970s, a 3.5 GPF toilet was introduced, with many manufacturers shifting to this size around 1977, and California mandating it in 1978.[51] In 1984, a 1.6 GPF toilet was introduced in the United States, and in 1992, the EPAct mandated 1.6 GPF, which is the current national standard.

Field measurement of toilet water use is best done through measurement, by measuring the length and width of water stored in the tank, measuring the depth that the water level in the tank falls during flushing, and dividing the cubic inches by 231 to obtain gallons per flush. For newer toilets, and some older toilets, the rated water use is marked near the toilet seat hinge.

For commercial flush valves (frequently referred to by the brand names Flushometer, made by Sloan, and Aquaflush, made by Zurn), used on toilets without tanks and on urinals, water usage can be estimated at half gallon per second of flushing. So two seconds of flushing on a urinal means a code-compliant 1 GPF, three seconds on a toilet means a code-compliant 1.6 GPF, and longer means noncompliant flow. In a building where all other water uses can be stopped (e.g., when the building is unoccupied), the water use of toilets and urinals with flush valves might be measurable using the building's water meter.

Flush valves (Figure 12.29) frequently are set for 3 to 4 GPF, or more, and the water volume may be reduced without compromising the effectiveness of the flush.[52, 53] The water volume of flush valves can be adjusted inside the device, typically by turning a screw in the control stop, also called the angle stop. Internal valve components can also be replaced to adjust the water volume. If a flush valve is leaking continuously, a frequent repair is to remove debris from a small hole in the large diaphragm in the main body of the valve.

Toilets with lower flow have continued to be developed, with EPA's WaterSense program requiring 1.28 GPF, commonly referred to as a high-efficiency toilet. Dual-flush toilets (typically 1.6 GPF for solids, and 0.8 to 1.0 GPF for liquids; Figure 12.30), and lower-flow single-flush toilets, are increasingly common.

Dual-flush retrofit kits are available for existing toilets, involving replacement of the toilet hardware. Dual-flush valves (Figure 12.31) are also available for toilets without tanks.

For jurisdictions where allowed, composting toilets allow elimination of toilet water use other than for cleaning.

In residential settings, residents flush the toilet approximately 5 times per day.[18]

A device that reduces water use associated with hand washing after use of the toilet is the toilet lid sink (Figure 12.32). After the toilet is flushed, clean water is

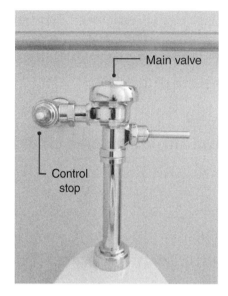

Figure 12.29   Flush valve.

Figure 12.30   Dual-flush toilet.

Figure 12.31   Dual-flush valve.

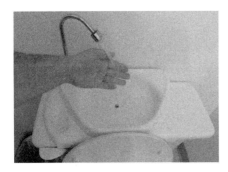

Figure 12.32   Toilet lid sink.

Figure 12.33   Waterless urinal.

first routed through a spout, into a sink above the toilet, for use to wash hands. Even though the flow is equal to a tank of water (ostensibly 0.8 to 1.6 gallons), the savings should be estimated based on typical hand-washing use, in the range of 0.1 to 0.2 gallons per use. Anecdotal reports are that these devices work and are reliable.

Older urinals can use over 3 GPF. The 1992 U.S. federal EPAct legislation required less than 1 GPF, went into effect January 1, 1974, and is still in effect. A 2002 survey in California found that approximately half of the urinals used less than 1 GPF, but over half still used more than 1 GPF. EPA's Water Sense program requires less than 0.5 GPF, and this requirement is in effect in California. Many manufacturers make efficient urinals with less than 0.125 GPF, and waterless urinals (also known as nonwater urinals, requiring no water at all; Figure 12.33) have also become common.[54]

As described above, urinal water flow can be estimated by the duration of flushing with the flush valve, assuming 0.5 gallons per second of flush, or by using the building's water meter if all other water uses can be stopped. Urinal usage has been estimated to be four uses per male per day.

Urinals can either be replaced in their entirety, or flush valves can be retrofit with low-flush spring/diaphragm assemblies. However, it is recommended that a single urinal be retrofit and tested for acceptability before large-scale replacements are undertaken.[55]

IRRIGATION/LANDSCAPING

Water-efficient irrigation and landscaping can reduce water use substantially. Improvement options include replacing water-dependent vegetation with hardier vegetation or nonvegetated natural landscaping, and drip irrigation.

WATER LEAKS

Water leaks may be regarded as an abnormal condition, but they are more common than may be recognized. A significant study of residential water use found leaks to comprise almost 14 percent of indoor water use.[18]

Leaks may be identified in a variety of ways. Fixtures such as faucets, showerheads, bath spouts, and commercial flush valves may be visually inspected for leaks, when they are closed. It can be instructive to look under sinks in bathrooms and kitchens for evidence of leaks. Leaks are also often evident in basements or mechanical rooms, for example near boilers. Building owners and occupants should be surveyed as to possible leaks, as they may be aware of leaks in locations that are less evident, through stained ceilings or other indicators.

Steam heating systems are particularly prone to leaks, as leaked steam is not evident. One study found buildings with steam heating systems to use almost twice as much water as buildings that do not heat with steam.[56] Steam leaks are sometimes visible, through bubbling at joints, and are sometimes audible, such as a steam vent on a radiator that hisses even when the radiator is already hot (when the vent should have closed). There are also tools to evaluate steam leaks, for more in-depth fieldwork.

Toilet leaks are sometimes visible or audible. Dye tablets are also available which, when placed in the toilet tank, will show dye in the toilet bowl if there is a leak. Leaks can be repaired through replacement toilet hardware. A toilet fill valve is also available with a mechanism that prevents the toilet from flushing if there is a leak.

Shower diverters route water from a bath spout up to the showerhead. A study of shower diverters found that 34 percent leaked more than 0.1 GPM when the shower is in use, that the average leak rate was 0.8 GPM, and when fixed over 70 percent of the eliminated leak is pure savings (in other words, does not increase flow to the showerhead). See Figure 12.34. Positive pressure diverter mechanisms, with a lever at the end of the tub spout instead of as a lift device, were found to more reliably prevent leaks.[57] Evaluating the savings potential of fixing shower diverter leaks requires measuring the flow through the tub spout while the shower is running. Multiply the leak by 0.7 to conservatively account for the 30 percent flow that is typically forced up to the showerhead when the diverter is fixed. Further, multiply by 0.6 to account for the fraction of the flow that is hot water.

Large leaks can occasionally be identified by benchmarking water consumption, in other words by comparing a building's water to similar buildings, or to its own historic usage, if leaks are suspected to be recent.

The utility water meter can be used to identify leaks, by examining the leak indicator. See Figure 12.35.

The leak indicator turns even for low flows. Shut off all water loads in the building, and see if the leak detector turns. If it does turn, shut off water inside the building, to assess if the leak is an outdoor load. The flow rate of the leak can be measured by leaving all water loads off, and measuring the leak at the water meter by examining the sweep hand (large hand). One revolution represents one unit of measure, typically 1 cubic foot (sometimes 1 gallon). The unit of measure is typically marked on the water meter. For the common Badger Recordall meters, the unit of measure is indicated by the color of the leak indicator: red is gallons, blue is cubic feet, black is cubic meters.

Additional water meters can be installed to assess leaks, for example for specific tenants in a larger building, or for large single loads that are suspected of having leaks, such as feedwater for boilers.

Figure 12.34   Leaking shower diverter.

References

1. Air Conditioning, Heating, and Refrigeration Institute (AHRI), "Operations Manual—Residential Water Heater Certification Program, AHR RWH OM," December 2013; and "Operations Manual—Commercial Water Heater Certification Program, AHR CWH OM," May 2014.

2. http://inspectapedia.com/plumbing/Water_Heater_age.pdf. Maintained by Honest Home Inspections LLC, by Scott LeMarr. Accessed December 20, 2014.

3. Katrina Leni-Konig, Emanuel G. D'Albora, and Robert A. Davis, Pacific Gas and Electric Company, "Laboratory Testing of Residential Gas Water Heaters," December 2008.

4. http://www.wbdg.org/design/midg_design_ece.php. Accessed August 8, 2015.

5. U.S. Department of Energy, 10 CFR Part 431, Docket Number EERE–2014–BT–TP–0008, RIN 1904–AD18, Energy Conservation Program for Certain Commercial and Industrial Equipment: Test Procedure for Commercial Water Heating Equipment. *Federal Register* 79, no. 39 (February 27, 2014).

6. Chong Hock K. Goh and Jay Apt, "Consumer Strategies for Controlling Electric Water Heaters under Dynamic Pricing." *Carnegie Mellon Electricity Industry Center Working Paper* CEIC-04–02, 2004.

7. "Water Heater Study Reveals Significant Fuel Savings from New Programmable Setback Control." *Plumbing Engineering*, November 2010, p. 50.

8. A. H. Fanney and B. P. Dougherty, "Thermal Performance of Residential Electric Water Heaters Subjected to Various Off-Peak Schedules." *Journal of Solar Energy Engineering* 188 (May 1996): 76.

9. http://www.water-heater-repair-guide.com/waterheatertemperature.html; and, http://waterheatertimer.org/How-to-adjust-water-heater-temperature.html. Accessed August 14, 2015.

10. http://energy.gov/energysaver/projects/savings-project-lower-water-heating-temperature. Accessed August 10, 2015.

11. Benoît Lévesque, Michel Lavoie, and Jean Joly, "Residential Water Heater Temperature: 49 or 60 Degrees Celsius?". *Canadian Journal of Infectious Diseases and Medical Microbiology* 15, no. 1 (January/February 2004). Pages 11-12.

Figure 12.35   Water meter leak indicator.

12. World Health Organization (WHO). "*Legionella* and the Prevention of Legionellosis," 2007.
13. http://www.wbdg.org/design/midg_design_echp.php. Accessed April 21, 2015.
14. http://www.engineersedge.com/heat_transfer/heatlossinsulatedpipe/heat_loss_insulated_pipe_equation_and_calculator_13169.htm. Accessed April 21, 2015.
15. http://www.pipeinsulation.org/. Accessed April 21, 2015.
16. Alison Williams, Camilla Dunham Whitehead, and James Lutz. "*A Systems Framework for Assessing Plumbing Products–Related Water Conservation.*" Lawrence Berkeley National Laboratory. University of California. LBNL 5286E. December 2011.
17. http://www.energy.ca.gov/appliances/2013rulemaking/documents/responses/Water_Appliances_12-AAER-2C/California_IOU_Response_to_CEC_Invitation_to_Participate-Lavatory_Faucets_and_Faucet_Accessories_3013-05-09_TN-70792.pdf. Accessed November 17, 2015.
18. Peter W. Mayer, William B. DeOreo, Eva M. Opitz, et al., "*Residential End Uses of Water.*" AWWA Research Foundation and American Water Works Association, 1999.
19. http://www1.eere.energy.gov/buildings/appliance_standards/product.aspx/productid/64. Accessed December 20, 2014.
20. S. Hills, R. Birks, and B. McKenzie. "The Millennium Dome 'Watercycle' Experiment: To Evaluate Water Efficiency and Customer Perception at a Recycling Scheme for 6 Million Visitors." *Water Science and Technology* 46, no. 6–7 (2002): 233–240. © IWA Publishing.
21. http://www.map-testing.com/assets/files/CA_Statewide%20CodesStandards_2013_CASE_R_Shower_Heads_Sept_2011.pdf. Accessed December 23, 2014.
22. James D. Lutz, "*Feasibility Study and Roadmap to Improve Residential Hot Water Distribution Systems.*" Lawrence Berkeley National Laboratory. University of California. LBNL 54841. 2004.
23. Ronnie Cohen, Kristina Ortez, and Crossley Pinkstaff, "Making Every Drop Work: Increasing Water Efficiency in California's Commercial Industrial Institutional Sector." *Natural Resources Defense Council* Issue Paper, May 2009.
24. "Pre-Rinse Spray Valves Field Study Report." *EPA Water Sense*, March 31, 2011.
25. Environmental Protection Agency, "Water Sense Labeled Pre-Rinse Spray Valves." EPA-832-F-13–001, September 2013.
26. Robert Zogg, William Goetzler, Christopher Ahlfeldt, et al. "*Energy Savings Potential and RD&D Opportunities for Commercial Building Appliances.*" Final Report, submitted to U.S. Department of Energy, Energy Efficiency and Renewable Energy, Building Technologies Program, Navigant Consulting, Inc., December 21, 2009.
27. http://info.nsf.org/certified/food/. Accessed August 14, 2015.
28. http://energy.gov/eere/femp/covered-product-category-commercial-dishwashers. Accessed December 25, 2014.
29. http://www.allianceforwaterefficiency.org/commercial_dishwash_intro.aspx. Accessed December 25, 2014.
30. ENERGY STAR® Program Requirements Product Specification for Commercial Dishwashers, Eligibility Criteria Version 2.0, Environmental Protection Agency, May 2012.
31. Mark W. Hancock, David L. Bohac, and Essam S. Wahbah, "*Gas Dishwasher and Booster Heater Savings Evaluation.*" Center for Energy and Environment, Minneapolis, MN, December 1994.
32. http://www.energystar.gov/products/certified-products/detail/commercial-dishwashers. Accessed December 15, 2014.
33. Avid Boustani, Sahni Sahni, Timothy Gutowski, and Steven Graves, "Appliance Remanufacturing and Energy Savings." Environmentally Benign Manufacturing Laboratory, Sloan School of Management, MIT, January 28, 2010.
34. http://www1.eere.energy.gov/buildings/appliance_standards/product.aspx/productid/67. Accessed December 25, 2014.
35. ENERGY STAR® Program Requirements Product Specification for Residential Dishwashers, Eligibility Criteria, Version 5.0, U.S. EPA, effective January 20, 2012.
36. D. Hoak, D. Parker, and A. Hermelink, "*How Energy Efficient Are Modern Dishwashers?*" *Proceedings of ACEEE 2008 Summer Study on Energy Efficiency in Buildings, American Council for an Energy Efficient Economy*, Washington, DC, August 2008.

37. http://www.appliance411.com/service/date-code.php, or http://www.appliancefactory parts.com/blog/2010/02/how-to-find-the-age-of-your-appliance/. Accessed December 25, 2014.

38. Water Management, Inc., Western Policy Research, and Koeller and Company, "Report on the Monitoring and Assessment of Water Savings from the Coin-Operated Multi-Load Clothes Washers Voucher Initiative Program." Prepared for San Diego County Water Authority, Rose M. Smutko, Water Resources Specialist, October 2006.

39. Department of Energy, 10 CFR Part 431, "Energy Conservation Program: Energy Conservation Standards for Commercial Clothes Washers; Final Rule." *Federal Register* 79 no. 240 (December 15, 2014), Part IV.

40. http://www.regulations.doe.gov/certification-data/CCMS-79222370561.html. Accessed January 23, 2015.

41. http://www.appliance411.com/service/date-code.php, or http://www.appliancefactory parts.com/blog/2010/02/how-to-find-the-age-of-your-appliance/. Both accessed December 25, 2014.

42. http://www.regulations.doe.gov/certification-data/CCMS-79222370561.html. Accessed January 23, 2015.

43. http://www.appliances.energy.ca.gov/QuickSearch1024.aspx. Accessed January 23, 2015.

44. http://www.energy.ca.gov/appliances/database/historical_excel_files/Clothes_Washers/, accessed 1/23/15.

45. http://oee.nrcan.gc.ca/pml-lmp/index.cfm?action=app.search-recherche&appliance= CLOTHESWASHERS. Accessed January 23, 2015.

46. http://www.energystar.gov/products/certified-products/detail/commercial-clothes-washers. Accessed December 26, 2014.

47. Richard Bole, "*Life-Cycle Optimization of Residential Clothes Washer Replacement*." Center for Sustainable Systems, University of Michigan. Report No. CSS06–03, April 21, 2006.

48. David Korn and Scott Dimetrosky, "Do Savings Come Out in the Wash? A Large-Scale Study of Residential Laundry Systems." ACEEE Summer Study on Energy Efficiency in Buildings, p. 9-143, 2010

49. http://www.nrcan.gc.ca/energy/efficiency/housing/home-improvements/12356. Accessed December 24, 2014.

50. http://oee.nrcan.gc.ca/pml-lmp/index.cfm?language_langue=en&action=app%2Esearch-recherche&appliance=DWHR&attr=0. Accessed 12/24/14.

51. R. Bruce Billings and Clive Vaughan Jones, *Forecasting Urban Water Demand*, 2nd ed. American Water Works Association, 2008.

52. Frank Fix, Things to Know about Flushometer Retrofits, *Buildings*, June 1, 2008.

53. http://www.buildings.com/article-details/articleid/6098/title/things-to-know-about-flushometer-retrofits.aspx. Accessed December 24, 2014.

54. Heidi Hauenstein, Tracy Quinn, and Ed Osann, "*Toilets & Urinals Water Efficiency, Codes and Standards Enhancement (CASE) Initiative, For PY 2013: Title 20 Standards Development, Analysis of Standards Proposal for Toilets & Urinals Water Efficiency*," Docket #12-AAER-2C, CASE Report, July 29, 2013. Copyright 2013 Pacific Gas and Electric Company, Southern California Edison, Southern California Gas, San Diego Gas & Electric.

55. Hazen and Sawyer, "Potable Water Conservation Best Management Practices for the Tampa Bay Region." *Tampa Bay Water*, January 2010, pp. 6–15.

56. I. Shapiro, "Water and Energy Use in Steam-Heated Buildings." *ASHRAE Journal*, May 2010, p. 14.

57. Betsy Jenkins, "Leaking Shower Diverters: An Overlooked Energy Waster." *Home Energy Magazine*, March/April 2012, p. 38.

Chapter 13

Electric Loads (Other than Lighting)

Motors and Motor Drives

Motors reportedly use more than half the electricity in buildings. We direct our focus here to motors that we can replace and that are rated independently for energy efficiency, primarily those used for fans and pumps. Other motors, such as those used in air conditioning and refrigeration compressors, or those used in small consumer appliances, are part of assemblies that are rated as a whole for energy efficiency, and are not usually replaced independently.

Motor frames are either *open* or *closed*. Open motors have their windings cooled by air that flows into the motor housing and directly over the windings. Open motors are suitable for cleaner environments, and are at risk if dust or dirt enters them. Closed motors have their windings cooled by air that flows over outside of the housing. The most common type of open motor is called *open drip-proof* (ODP), which are designed and tested to allow liquids to drip on them without doing damage. The most common type of closed motor is called *totally enclosed fan-cooled* (TEFC), which is typically cooled by a small external fan that is mounted on the end of the motor. Other, less common, types of motors include *totally enclosed nonventilated* (TENV), and *explosion-proof motors*, and these are not covered by minimum-efficiency requirements.

To identify whether a motor is open or closed, look at both ends of the motor. If both ends of the motor have openings, it is an open motor. If one end of the motor has an opening, look more closely. It is likely just the opening for the external fan, and the air is then routed over the outside of the motor for cooling. This is a closed motor. External fins are also a sign of a closed motor. The nameplate may also indicate the type of motor: ODP for open, TEFC for closed. Open motors are more affordable than closed motors.

Most motors in buildings are alternating current (AC) motors. AC motors are classified by their speed, the most common nominal speeds being 900 *revolutions per minute* (RPM), 1,200 RPM, 1,800 RPM, and 3,600 RPM. The actual motor speed is slightly slower than these nominal speeds. The nominal speed is also called the *synchronous speed*. The difference between the synchronous speed and the actual speed is called the *slip*. Each of these speeds corresponds to the number of *poles*: 900 RPM is 8-pole, 1,200 RPM is 6-pole, 1,800 RPM is 4-pole, and 3,600 RPM is 2-pole. Motors are also classified by their electrical characteristics: voltage, current (amps), phase (1 or 3), and hertz (50 or 60). Motors are also classified by their frame size.

Motor sizes are designated in *horsepower* (HP). This is a nominal designation that relates to a motor's power consumption (input power), and not to its output power that is delivered at the shaft. To convert from horsepower to kilowatts, multiply the horsepower by 0.746. The output power delivered at the shaft is called *brake horsepower*.

Motors are rated for efficiency by the National Electrical Manufacturer's Association (NEMA). In the early 1990s, the federal government's Energy Policy Act (EPAct) law was passed, requiring most common motors (specifically those from 1 to 200 HP) to meet NEMA's *standard efficiency* requirements, as specified in NEMA's MG 1 standard. Subsequently, NEMA added a second set of efficiencies, denoted *premium efficiency*, which are even higher. Starting December 19, 2010, all common motors must be premium efficiency, with exceptions for less common motors, most of which must meet NEMA's standard efficiency requirements.

Motors smaller than one HP are called *fractional horsepower motors*. The focus of high-efficiency motor efforts, to save energy, has historically been on motors larger than one HP. Only recently were regulations passed to require minimum energy efficiency for fractional horsepower motors. One HP is also roughly the size above which most motors are three-phase. Fractional horsepower AC motors are mostly single-phase.

EXISTING MOTOR EFFICIENCY

Prior to the early 1990s, motor nameplates did not include the rated motor full-load efficiency. Since the early 1990s, the motor efficiency can be found on the nameplate.

Directly measuring a motor's efficiency is possible, but is not frequently done due to the effort required, as the motor needs to be removed from its load and put on a dynamometer, to measure its torque. Indirect motor efficiency measurements are possible, by measuring motor losses, but this method is less accurate.[1, 2]

Manufacturers frequently keep records of motor efficiencies, which they can supply if provided with the model and serial numbers.

For motors prior to the early 1990s, the U.S. Department of Energy (DOE) offers a table of default efficiencies, to be used for estimating power use for existing motors 10 HP and larger.[3] See Appendix P. This table also includes estimated efficiencies for motors between 1 and 7.5 HP from a second source, with extrapolations for part-load efficiencies.[4]

The existing motor efficiencies in Appendix P are on average slightly lower than a prior table, prepared in the late 1990s. They also appear slightly low compared to some tests of actual motors, and compared to example manufacturer data from the 1970s. If existing efficiencies are assumed to be too low, savings will be overestimated. If more conservative estimates of energy savings are sought, we recommend adding 1 to 1.5 percent to these estimates.

In addition to measuring or estimating the efficiency of existing motors, the motor run-time is a critical component of energy calculations. Motor run-time can be measured by data-loggers, using either motor current as an input, or a variety of other inputs (control signal, vibration, etc.). We recommend measuring current, because current can also be used to estimate the motor load.

MOTOR IMPROVEMENTS

High Efficiency Motors

The main motor improvement is to replace a low-efficiency existing motor with a higher efficiency motor. NEMA premium efficiency requirements effective December 19, 2010 are shown in Table 13.1.

The savings to replace a lower efficiency motor with a higher efficiency motor are:

$$Savings = 1 - (Ee/En)$$

Where Ee is the efficiency of the existing motor and En is the efficiency of the new motor.

TABLE 13.1

Motor Efficiency Requirements (%)

	Open			Closed		
	Poles					
HP	2	4	6	2	4	6
1	77.0	85.5	82.5	77.0	85.5	82.5
1.5	84.0	86.5	86.5	84.0	86.5	87.5
2	85.5	86.5	87.5	85.5	86.5	88.5
3	85.5	89.5	88.5	86.5	89.5	89.5
5	86.5	89.5	89.5	88.5	89.5	89.5
7.5	88.5	91.0	90.2	89.5	91.7	91.0
10	89.5	91.7	91.7	90.2	91.7	91.0
15	90.2	93.0	91.7	91.0	92.4	91.7
20	91.0	93.0	92.4	91.0	93.0	91.7
25	91.7	93.6	93.0	91.7	93.6	93.0
30	91.7	94.1	93.6	91.7	93.6	93.0
40	92.4	94.1	94.1	92.4	94.1	94.1
50	93.0	94.5	94.1	93.0	94.5	94.1
60	93.6	95.0	94.5	93.6	95.0	94.5
75	93.6	95.0	94.5	93.6	95.4	94.5
100	93.6	95.4	95.0	94.1	95.4	95.0
125	94.1	95.4	95.0	95.0	95.4	95.0
150	94.1	95.8	95.4	95.0	95.8	95.8
200	95.0	95.8	95.4	95.4	96.2	95.8

For example, if a 75 percent efficiency motor is replaced by a 90 percent efficiency motor, the savings are:

$$\text{Savings} = 1 - (75/90) = 16.7\%$$

Note that the savings are not simply equal to the difference between the two efficiencies.

If this example motor is 15 HP, is operating at full load, and runs for 3,000 hours per year, the savings are:

$$\text{Savings (kWh/year)} = 16.7\% \times 15 \times 0.746 \times 3,000 = 5,606 \text{ kWh/year}$$

When replacing motors, required specifications include the nominal horsepower, electrical characteristics (voltage and phase), the frame size, whether open or closed, the required minimum full-load efficiency, the shaft diameter and length, the direction of the motor rotation (and whether this rotation is viewed from the shaft end or from the lead end), the brake horsepower, and the motor speed.

The DOE has developed software for motor improvement calculations, called Motor Master.

Matching Motors to Loads

In replacing motors, further savings are possible by replacing oversized motors with motors that are better matched to their loads. Older, low-efficiency motors are significantly less efficient at part load than at full load. The loss in efficiency is greatest below 50 percent load. The optimum load is approximately 75 percent. The part-load penalty is larger for smaller motors.[5] Table 13.2 provides estimates for part-load efficiencies for existing motors, over a range of part-load (10 to 40 percent) for which the efficiency degrades significantly, especially for smaller motors.

TABLE 13.2

Estimated Part-Load Motor Efficiencies

	HP		
	1.5–5	10	50
% percent Load	Motor Efficiency		
10%	32%	45%	66%
20%	62%	76%	90%
30%	82%	88%	91%
40%	92%	94%	92%

To assess whether a motor is operating at part-load, its power consumption should be measured, and compared to its full-load power. The load is estimated as the actual power consumption divided by the full-load power consumption.

For example, if a 10 HP, 6-pole enclosed motor is measured to use 1.5 kilowatts, the % Load = $1.5/(10 \times .746) = 20.1$ percent. We round to 20 percent load. From Table 13.2, the motor efficiency for this 10 HP motor is 76 percent. This load is better suited to a motor that is approximately $1.5/0.746 = 2$ HP. We choose to replace the 10 HP motor with a premium efficiency 2 HP motor that is 88.5 percent efficient. The savings are:

$$\text{Savings} = 1 - (76\%/88.5\%) = 14.1\%$$

Note that in this example the savings are 14.1 percent of the existing actual measured power of 1.5 kilowatts, not 14.1 percent of the original nominal motor size of 10 HP. This is another reason to measure the actual power of a motor, to obtain the actual motor power usage, rather than simply assuming that the motor is seeing its full load. Even if the motor is not oversized, for example if it is operating at 75 percent load, and assuming we would not choose to change the motor size but only change its efficiency, our energy savings calculation is more accurate if we know its real power consumption.

It is also possible to estimate the load on a motor using a current measurement, or a rotational speed measurement, also called the slip measurement approach.[5] However, even though both these measurements are slightly simpler than a power measurement, neither is regarded as reliable in the area of greatest interest, below 50 percent load.

Adjustable-Speed Drives

An *adjustable-speed drive (ASD)* is a separate control component that can change the speed of a motor. Adjustable-speed drives are also called *variable-speed drives* (VSDs), and *variable-frequency drives* (VFD). Adjustable-speed drives are primarily applied to existing three-phase motors, over one HP in size. See Figure 13.1.

In recent decades, a motor has been developed, mostly for fractional horsepower use, which can vary in speed, based on a different technology than traditional adjustable-speed drives. Traditional adjustable-speed drives are applied to existing three-phase AC motors. *Brushless permanent magnet motors* (BPMs) incorporate a DC motor, using AC input to the control, which is often integrated into the motor. BPM motors have been widely used for several decades, for fans in smaller equipment, such as residential furnaces, and have started to be used in small pumps as well. BPM motors are also referred to as *electronically commutated motors* (ECM), a trademark of the General Electric Company, which developed the motor. Other terms for BPM motors are *brushless* (BL), *brushless DC* (BLDC), or *integrated control motors* (ICM).

Figure 13.1   Adjustable-speed drive.

TABLE 13.3

Adjustable Speed Drive: Example Savings Calculation

Load	Hours	Power (watts)	Energy (kWh)
100%	100	1000	100
90%	100	729	73
80%	100	512	51
70%	100	343	34
60%	100	216	22
50%	100	125	13
40%	100	64	6
30%	100	27	3
20%	100	8	1
10%	100	1	0
0%	100	0	0
Total:	1,100	3,025	303
If instead full load year-round:	1,100	1,000	1,100
Savings:			798

Common applications of adjustable-speed drives include hydronic pump motors, air handler motors, chilled water pump motors, and condenser water pump motors. Variable-speed drive controls need an input in order to decide on the motor speed. Supply water pressure (for pumps) and supply air pressure (for fans) are the most common input variables on which to base the motor speed. Less frequently, exhaust fan motors are converted to variable-speed control, again, if an input is available on which to base a motor speed.

In calculating energy savings for fan and pump motors, we can assume that the power required varies with the cube of the speed. In performing these calculations, we cannot simply use the average proposed motor speed, because this will overestimate savings. While savings for low speeds are much higher than at the average speed, savings at speeds close to full speed are much smaller, and the smaller near-full-speed savings can outweigh the larger low-speed savings. It is important to at least do a bin calculation, to calculate power consumption and energy usage for various fractions of full-load operation. In the example in Table 13.3, for 1,000 watts at full load, with variable-speed savings varying with the cube of the load, savings are estimated at just over 70 percent. If the average load (in this case 50 percent) were used as the basis of savings, the savings would mistakenly be overestimated at almost 90 percent.

Elevators and Escalators

Elevators are estimated to use between 2 percent and 10 percent of a building's electrical energy consumption (excluding heating and hot water).[6] A study in Europe found that 70 percent of the energy used for elevators in multifamily buildings is used in the standby mode.[7] Typically, low-rise buildings use *hydraulic elevators* due to their low cost, and high-rise buildings use *traction elevators*. Hydraulic elevators have a machine room, typically at the bottom of the building, and a tank with hydraulic fluid, and do not use ropes and pulleys. *Traction drive elevators* typically have an elevator room above the elevator, at the top of the building (except for newer machine-room-less traction elevators), and use a motor with cables and pulleys. Older traction elevators use a direct current (DC) motor, which receives its DC power either from rectifiers or from a separate *motor-generator set* (or *MG set*, or *gen-set*). MG set

Figure 13.2 Elevators can save energy through efficient drives, lighting, ventilation, and controls.

Figure 13.3 Escalators can save energy through efficient drives and controls.

systems driving DC motors are the least efficient form of elevator drive, as the MG set consumes as much as 12 percent of full-load power when it is idling. Timers can be used to limit idle time, as an energy improvement, on MG set systems.

Energy consumption depends on many factors including usage, elevator capacity, and elevator efficiency. Efficient elevators use variable voltage variable frequency (VVVF) drives with AC motors, for their high-energy efficiency and high speed. Energy usage may be reduced by using elevators with high-efficiency drives, and features such as regenerative braking. Elevators can also save energy by using high-efficiency lights, by automatically turning lights off when not in use, by automatically turning ventilation fan motors off when not in use, and by using high-efficiency cooling for elevator machine rooms. See Figure 13.2. Advanced controls can further reduce energy usage by optimizing elevator car travel algorithms, which place cars where they will most likely be needed, and also by turning off power to some cars during periods of low usage, in buildings with multiple elevators. For purposes of estimating energy usage, hydraulic elevators use approximately 0.02 to 0.03 kWh/start, for light-duty low-rise applications, and more for heavier-duty or medium-rise buildings. This usage can be reduced to 0.01 to 0.02 kWh/start with higher-efficiency drives. Elevators with VVVF drives in high-rise buildings use approximately 0.03–0.04 kWh/start, and this usage can be reduced to 0.02–0.03 kWh/start with regenerative braking or DC pulse-width modulated drives.

Electricity consumption of traditional escalators is typically between 4,000 and 18,000 kWh/year per escalator. Variable-speed motors can reduce this energy use, by sensing when no passengers are being carried, and accordingly slowing the motors, or by stopping some escalators during periods of very low use. See Figure 13.3.

Transformers

Transformers (Figure 13.4) change the voltage of an electrical supply. In so doing, there are electrical losses. Most transformers in buildings step a higher voltage down to a lower voltage. Occasionally, a step-up transformer is used, when a specific higher-voltage load is on-site for which the electric voltage is not available.

Transformers have two types of losses: core losses (also referred to as no-load losses), which occur continuously; and winding losses, which occur when there is a load on the transformer. Core losses are higher if a transformer is oversized. Most transformers have been found to be oversized, with average loading at only 16 percent, less than half the 35 percent loading on which transformer efficiency ratings are based.[8] Furthermore, as electricity use in buildings is reduced, transformers are increasingly oversized.

In 1996, the National Electrical Manufacturer's Association (NEMA) developed a standard for calculating energy use of transformers, and specified minimum recommended efficiency levels. This standard was updated in 2002.[9] ENERGY STAR briefly adopted these efficiencies in a program that was suspended in 2007 when federal efficiency requirements went into effect. The new 2007 federal requirements matched what had been the ENERGY STAR requirements, based on the NEMA recommendations of 2002. Standard transformers must meet federal minimum efficiencies ranging from 97 percent for three-phase 15-kVA transformers, to 98.9 percent for 1,000-kVA transformers.

Transformers are also available in premium efficiencies. Premium efficiency transformers are currently defined as 97.9 percent for 15-kVA transformers, rising to 99.23 percent for 1,000-kVA transformers.

Transformers must be well ventilated, and they operate at higher efficiency when located at a lower temperature, so should preferably not be located in hot rooms or in enclosed areas outdoors without adequate air circulation.

Some attention is being directed to liquid-filled transformers for building applications. Liquid-filled transformers were traditionally not considered for building applications, due to concerns over environmental and fire safety hazards of the transformer liquids. Recent advances reportedly include increasing safety for both environmental and fire safety characteristics. Liquid-filled transformers are higher in efficiency than dry-type transformers.[10]

The efficiency of an existing transformer is most accurately assessed by monitoring.[11] Monitoring has the benefit of identifying losses, as well as establishing whether a transformer is oversized. Energy benefits may be obtained by installing a high-efficiency transformer and, on occasion, by reducing the oversizing of the transformer.

In the absence of monitoring, transformers installed prior to NEMA TP 1 requirements (2007 per federal regulations, or approximately 2000 to 2002 for some states that adopted high-efficiency requirements earlier than the federal requirements) may be assumed to have efficiencies shown in Table 13.4.[11]

It should be noted that assuming efficiency for a pre-TP1 transformer is likely to underestimate the savings, as this assumption does not account for oversizing. Actual pre-TP1 efficiencies also were not constant, with reports of liquid-filled transformer efficiencies increasing over time, and dry-type transformer efficiencies actually dropping over time, from 1950 to 1995.[12]

Most buildings do not have transformers on the building side of the electric meter. The transformer is usually the property of the electric company, and there is no incentive to the building owner to replace a utility transformer with a high-efficiency transformer. In general, indoor transformers are the property of the owner and are on the building side of the meter, and outdoor transformers are the property of the utility company. See Figure 13.5.

Commercial Refrigeration

Commercial refrigeration is widely used in food sales and food service in a variety of building types, from supermarkets to restaurants to convenience stores and even to such buildings as churches.

Commercial refrigeration includes primarily medium temperature refrigerators and low temperature freezers. Equipment can be *self-contained* (compressor and condenser included with the refrigerator or freezer, itself) or can have *remote compressors and condensers*. Remote systems can either be *condensing units* (compressor and condenser together), or can be comprised of *compressor racks* and separate *condensers*. The combination of compressor racks and condensers which serve multiple refrigerators and freezers, from open cases to reach-in units to walk-in refrigerators and freezers, are sometimes called *supermarket systems*.

Federal efficiency requirements went into effect in 2010 for self-contained refrigerators and freezers with doors. The maximum daily energy consumption (kWh/day) cannot exceed the following values, where V represents the equipment volume (in

Figure 13.4   75-kVA transformer.

Figure 13.5   Pad-mounted transformer—the electric meter (top right) is an indication that this transformer is on the utility side of the meter.

TABLE 13.4

Presumed Efficiency for Old Transformers

kVA	15	30	45	75	112.5	150	225	300
Efficiency	96.0%	96.5%	97.0%	97.2%	97.5%	98.0%	98.1%	98.2%

cubic feet), and AV represents the adjusted volume of refrigerator/freezers (1.63 times the freezer volume plus the refrigerator volume):

- Refrigerators with solid doors: $0.10\ V + 2.04$
- Refrigerators with transparent doors: $0.12\ V + 3.34$
- Freezers with solid doors: $0.40\ V + 1.38$
- Freezers with transparent doors: $0.75\ V + 4.10$
- Refrigerators/freezers with solid doors: the greater of $0.27\ AV - 0.71$ or 0.70

Additional efficiency requirements for systems without doors, and systems with remote condensers, went into effect in 2012, and will be subject to a further efficiency increase in 2017.

ENERGY STAR ratings are available for some commercial refrigeration equipment types, primarily ones that have doors and are connected to condensing units (either self-contained or remote). These include reach-in, roll-in, or pass-through units; merchandisers; under-counter units; hybrid units; milk coolers; back bar coolers; bottle coolers; glass frosters; deep well units; beer-dispensing or direct draw units; and bunker freezers. Not included in ENERGY STAR's program are consumer or medical refrigeration equipment, drawer cabinets, prep tables, service over counter equipment, horizontal open equipment, vertical open equipment, semi-vertical open equipment, convertible temperature equipment, and ice cream freezers.

ENERGY STAR's initial program for commercial refrigeration was introduced in 2001, some of which requirements match the current federal requirements that went into effect in 2010. ENERGY STAR's second version went into effect in two stages, in 2009 and 2010. ENERGY STAR's third version went into effect in 2014. As with other products, we can no longer assume that an ENERGY STAR label found on an existing piece of equipment assures high energy efficiency. For example, a 2001 product with the ENERGY STAR label is likely similar in efficiency to current minimum-efficiency equipment.

To illustrate the current difference between federal and ENERGY STAR requirements, consider a 40 cubic foot reach-in refrigerator. The federal requirement for refrigerators with transparent doors is $0.12\ V + 3.34$ kwh/day, or 2970 kwh/year, and the ENERGY STAR requirement for the same 40 cubic foot reach-in with transparent doors is $0.15\ V + .32$ kwh/day, or 2307 kwh/year (22% less).

The federal and ENERGY STAR requirements for solid-door equipment are more stringent than for transparent doors. For example, for a 40 cubic foot reach-in refrigerator, the federal requirement for a solid-door unit is $0.10\ V + 2.04$, or 2205 kwh/year, and the ENERGY STAR requirement is $0.01V + 2.95$, or 1223 kwh/year. So, where transparent doors are not required, refrigerators with solid doors generally use less energy, for the same size refrigerator.

For some equipment, such as display cases, the federal energy requirements are given as a function of total display area (TDA). For these products, TDA is defined in ARI Standard 2006. For example, for a horizontal case, the TDA is the projected height of the display opening multiplied by the length of the opening, in units of square feet.

Ratings for current equipment are maintained by the Air-Conditioning, Heating, and Refrigeration Institute (AHRI) in an online database at AHRI's website.

Larger units generally use less energy, per cubic foot (CF), than smaller units. For example, ratings for currently available display merchandisers (fall 2015) show an average 0.38 kwh/day/CF for units smaller than 15 CF, but only an average 0.23 kwh/day/CF for units in the 15-30 CF range, and even lower average 0.20 kwh/day/CF for units over 30 CF.

Energy use of existing equipment is best obtained by measurement under typical use conditions. Measurement has the added advantage of identifying equipment that is malfunctioning. It has been found that typical real-life use conditions are similar to conditions used in refrigeration equipment rating tests.

Rather than measure existing usage, state energy portfolio programs generally recommend using current federal energy use requirements as the baseline for existing energy use. ENERGY STAR's calculator similarly uses the current federal energy requirements as its default baseline. However, such an approach likely underestimates potential savings for energy audits, as older existing equipment typically uses more energy than current federal standards. For example, a transparent-door beverage merchandiser from the early 1990's was found to use over 60% more energy than current federal standards. As another example, a solid-door reach-in refrigerator, also from the early 1990's, was found to use over 70% more energy than current federal requirements. Even current models have widely varying energy use.

The California Energy Commission's historical appliance database contains energy use on several thousand commercial refrigeration units, going back to the early 1990's.

The age of equipment can typically be obtained from the serial number. For example, the True brand has an online database for manufacturing dates. For the Master-Bilt brand, the year of manufacturer is indicated by the second letter of the serial number, with A representing 1984, to Z in 2009, and starting again with AA in 2010. The refrigerant used in equipment, as indicated on nameplates, can provide a clue as to equipment age. Refrigerants 11 and 12 were phased out in the mid-1990's. Newer equipment typically uses refrigerant 134a or 404a.

Walk-in coolers and freezers are typically insulated with polyurethane or polystyrene. Walls are typically 24 or 26 gage steel, and are typically 4″ thick, although newer equipment can have thicker walls, especially for walk-in freezers. Polystyrene is variably stated to be between R5 and R 7 per inch, and polyurethane is stated to be between R6.5 and R8 per inch. Older insulation, prior to the early 1990's, used blowing agents that delivered slightly better R-values, but which were banned because of their harmful effect on the ozone layer. Field inspection of walk-in freezers and coolers should check to see if the walls are less than 4″ thick, if the floor is insulated, if the fans run continuously, if there are strip curtains in place to reduce infiltration when the doors are open, and if the door gaskets are in good condition. To visually distinguish between polystyrene and polyurethane insulation, drill a small hole into the wall – yellow insulation is polyurethane.

In considering high-efficiency replacement options for commercial refrigeration, as with other types of energy equipment, it is interesting to compare efficiencies of commercial refrigeration equipment with similar capacity residential equipment. For example, a 20 cubic foot reach-in solid-door commercial refrigerator has an ENERGY STAR efficiency requirement of 2.35 kWh/day, or 858 kWh/year. By contrast, a 20 cubic foot residential ENERGY STAR refrigerator is required to consume less than 413 kWh/year. So, in locations where a residential refrigerator can be used, it will save far more energy than a comparable high-efficiency commercial refrigerator.

Before examining one-for-one high-efficiency replacement of refrigeration equipment, one should ask whether the most efficient type of equipment is being used for the application. Using the most appropriate type of refrigerator can save more energy than simply replacing a refrigerator one-for-one with a higher-efficiency version of the same type. Closed-door equipment is generally more efficient than open equipment. Solid-door equipment is generally more efficient than transparent-door equipment. Larger equipment is more efficient than smaller equipment, on a per-unit-volume basis. For example, a row of one-door reach-in refrigerators is more efficiently replaced with a larger multi-door unit. At the same time, if an existing piece of equipment is too large, save energy by replacing it with a smaller piece of equipment. And, as mentioned, residential equipment is more efficient than commercial equipment, for applications where a residential refrigerator or freezer will do.

High-efficiency equipment replacement is the next option. Consider ENERGY STAR equipment, or even higher efficiencies where possible.

Other then equipment replacement, other energy improvements include high-efficiency lighting, operation and maintenance improvements (clean coils, straighten bent fins, adjust set-point), heat reclaim from waste heat (for example, DHW heating), avoiding high-temperature locations for condensers (for example, attics, mechanical rooms, etc. – efficiency is estimated to be 2 percent lower for every 1 degree increase in ambient temperature), and strip curtains (both for walk-in when the doors are open, and for open cases—e.g., at night).[13]

Plug-in refrigeration equipment is covered separately, in a discussion of plug loads, below.

Air Compressors

Air compressors (Figure 13.6) are used in older heating and cooling control systems, and in industrial systems. Improvements include finding and eliminating air leaks, and replacing old air compressors with newer, efficient systems. The impact of air leaks can be quantified by shutting compressed air loads and measuring run time or energy consumption of the air compressor.

Ceiling Fans

Ceiling fans (also called *paddle fans*) can reduce air conditioning energy consumption. Ceiling fans have also been used for *destratification*, to counter the effects of varying temperatures in high-ceiling spaces, in an attempt to move warm air from ceiling level down to floor level. Energy savings for destratification applications are discussed in the chapter on heating. Ceiling fans are also used for decorative purposes. This discussion is limited to the electricity use of ceiling fans.

Ceiling fans are generally hard-wired, and have three speeds. The most common size is 52″ diameter, which deliver approximately 6,000 cubic feet per minute (CFM)

Figure 13.6   Air compressor.

on high speed. Older fans use approximately 75 watts on high speed. ENERGY STAR fans must deliver more than 5000 CFM on high speed, and more than 75 CFM per watt on high speed, more than 3,000 CFM and 100 CFM/watt on medium speed, and more than 1,250 CFM and 155 CFM/watt on low speed. The average ENERGY STAR fan delivers 109 CFM/watt, with products available as high as 549 CFM/watt, and several manufacturers offer products over 300 CFM/watt. The program also has efficiency requirements for built-in lights. The ENERGY STAR requirements went into effect in 2002, and the fan efficiency requirements have not changed as of Version 3 in 2012. See Figure 13.7.

Figure 13.7   Efficient ceiling fans can use as much as 80 percent less electricity as older, inefficient fans.

To roughly estimate the existing fan power, multiply the nameplate nominal current (amps) by 120 volts. If a nameplate is not available, a default assumption for an old fan is 75 watts on high speed. Larger-diameter fans tend to be more efficient. Also, fans use less power on low speeds, even though the efficiency is less at these speeds, not only because they are moving less airflow, but because they are more efficient at lower speeds.

Plug Loads

GENERAL

This discussion addresses plug loads, with a few exceptions. For example, clothes washers and dishwashers are covered separately in the chapter on water. Plug loads are a new frontier in building energy conservation. It is unfortunately also a growing load, as larger and more powerful appliances and equipment become popular. It is also highly diverse, and energy use and savings have a strong behavior-related component. Typically regarded as beyond the bounds of commercial energy audits and improvements, plug loads in fact offer good savings potential and often surprising opportunities.

Plug loads are well worth measuring using plug-in watt meters. For example, a museum found that a refrigerator it had received as a donation used almost 2000 kWh/year, five times higher than an equivalent ENERGY STAR refrigerator.[14]

PLUG-IN REFRIGERATION EQUIPMENT

This discussion includes plug-in refrigerators and related equipment, such as vending machines. It does not include hard-wired refrigeration such as supermarket refrigerators or walk-in coolers and freezers.

Refrigeration equipment uses the same principle as air conditioners.

In the case of plug-in refrigerators, all components are located in one assembly (Figure 13.8). The refrigerator removes heat from its interior, and rejects this heat to the room in which it is located.

The two air temperatures, inside the refrigerator and the room in which it is located, are important to the efficiency of the device. If the interior temperature is lower, the efficiency is lower and the device uses more energy. If the temperature of the room in which it is located is higher, the efficiency is lower and the device uses more energy. We want to set the temperature inside the refrigerator as warm as possible, and we want the room to be as cool as possible. For most situations, we have limited control over these two temperatures, but we keep an eye on them for those situations in which we can control the temperatures. We measure the temperature inside the device to see if it is too cool. And we look for locations where the temperature around the device might be too warm (Figure 13.9): a refrigerator located in a hot commercial kitchen, a hotel refrigerator enclosed in a cabinet, a dehumidifier in a closet, an ice maker located outdoors in a hot climate and in the sun, or a spare freezer in a hot attic. Of the two temperatures, the temperature outside the

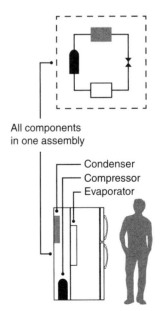

All components in one assembly

— Condenser
— Compressor
— Evaporator

Figure 13.8   Refrigerator.

Figure 13.9   Ice maker with insufficient clearance in a hotel corridor.

Figure 13.10   Hotel refrigerator in enclosed cabinet.

refrigerator has more impact than the temperature inside the refrigerator. It can also vary more widely.[15]

High ambient temperatures increase energy use in two ways: (1) conduction heat gain increases, from the ambient through the insulated refrigerator wall; and (2) the thermodynamic efficiency of the refrigeration cycle is lowered. In an enclosed space, it is a vicious cycle: the ambient temperature is high, so the refrigerator uses more energy, which raises the ambient temperature.

One study of refrigerators found that for ambient temperatures between 60 and 80 F, energy usage increased approximately 3 percent for each increased degree F in ambient temperature. Above 80 F, energy usage increased approximately 6 percent for each increased degree F in ambient temperature.[15] See Figure 13.10.

Similar energy penalties apply to any compressor-bearing device: dehumidifiers, ice makers, and more. Therefore, a first priority with any refrigeration device is to see if it is in a space that is too hot and to evaluate moving it to a cooler space if possible.

Just as high ambient temperatures increase energy use, so do any other negative impacts on condenser heat transfer, including dust, obstructions, and bent fins on fin-tube condensers. Where refrigerator condensers are accessible, cleaning them will reduce energy use. Based on findings with air conditioners, we estimate 4 percent increased energy use for each additional 10 percent loss in heat transfer. Cleaning and repairing condensers is good practice, and should form a standard part of routine maintenance of compressor-bearing equipment.

Refrigerators

Residential refrigerators are widely used in commercial buildings, finding application in office kitchenettes, break rooms, multifamily building apartments, hotel rooms, and more. Commercial plug-in refrigerators and freezers also see widespread use. Both residential and commercial refrigerators are covered by mandated federal efficiency requirements, and by above-mandate efficiency requirements to be accepted in the ENERGY STAR program.

The U.S. Food and Drug Administration recommends that refrigerators be kept at or below 40 F, and freezers at 0 F.[16]

If temperatures inside refrigerators are too low, the set point should be raised to save energy. Many refrigerators do not have their thermostat set points calibrated in degrees, but rather are numbered, such as 1–5, 1–7, and 1–9. Most refrigerators have a control range of 34 to 44 F. Accordingly, the numbered settings correspond to the temperatures shown in Table 13.5.

Freezer set-point ranges are typically –6 to 6 F.

TABLE 13.5

Refrigerator Temperature Set Points

1–5 Range		1–7 Range		1–9 Range	
Setting	Temp (F)	Setting	Temp (F)	Setting	Temp (F)
1	44.0	1	44.0	1	44.0
2	41.5	2	42.3	2	42.8
3	39.0	3	40.7	3	41.5
4	36.5	4	39.0	4	40.3
5	34.0	5	37.4	5	39.0
		6	35.7	6	37.8
		7	34.0	7	36.5
				8	35.3
				9	34.0

The best approach is to measure refrigerator temperatures during energy evaluation work, and to recommend that thermometers be placed inside refrigerators and freezers, to ensure that temperatures stay correctly set.

Refrigerator energy use increases by approximately 2.5 percent for each decreased degree F in interior temperature.[15]

Existing refrigerator energy use is available in a variety of online databases.[17, 18] The Canadian government also maintains old copies of its efficiency rating directories.[19] For very old refrigerators for which ratings are not available, short-term monitoring can be used to measure baseline energy use. Energy use can also be approximated by the age of old refrigerators (Figure 13.11). In 1950, the average refrigerator used 400 kWh/year; in 1960, 1,000 kWh/year; in 1970, 1,600 kWh/year; and in 1980, 1,200 kWh/year. In 1990, the first U.S. efficiency standard took effect (976 kWh/year), and by 2000 the standard had already been lowered twice, to 486 kWh/year.[20]

Unutilized refrigerators should be considered for removal, and food stored in underutilized refrigerators can frequently be combined with other refrigerators. See Figure 13.12.

Vending Machines

Vending machines (Figure 13.13) can be energy-intensive, using energy for refrigeration and lighting. For example, one study found 3,652 kWh/year for a soda machine, and 2,820 kWh/year for another (smaller) machine.[21] ENERGY STAR's program for vending machines indicates an average of 1350 kWh/year. Further savings are possible by putting lights on motion sensors, and even putting entire machines on either motion control or timer control, for machines where the contents are not at risk of perishing if the temperature is temporarily warmer.

Water Coolers and Heaters (Bottle Type)

A field measurement of a standing bottle-type water cooler/heater found usage to be 727 kWh/year.[21] Another field measurement found 1,022 kWh/year for a cooler/heater and 109 kWh/year for a cooler-only.[22] ENERGY STAR requirements are 0.16 kWh/day (cold only), 0.87 kWh/day (hot and cold), and 0.18 kWh/day (hot and cold on demand). Like vending machines, it is possible to put these devices on timers. One study found that a timer used during offices hours reduced cooling-only usage from 109 to 81 kWh/year, and cooling/heating usage from 1022 to 693 kWh/year.[22]

CLOTHES DRYERS

Residential Electric Clothes Dryers

We begin with a discussion of residential electric clothes dryers, because they are subject to federal efficiency standards, and so can form a benchmark for energy evaluation. Residential dryers are also frequently used in commercial buildings.

Most residential electric clothes dryers use electric resistance heat. This is a high-cost form of heat for clothes drying. Eighty percent of residential clothes dryers are electric.[23] A recently introduced form of electric clothes dryer is the heat pump dryer, which does not yet constitute a significant fraction of the dryer market in the U.S.

Federal efficiency requirements for residential clothes dryers were originally established by the DOE in 1988 and updated in 1994. New requirements were put into effect in 2015, requiring a minimum combined energy factor (CEF) of 3.73 lbs/kWh for standard vented electric clothes dryers (over 4.4 cubic feet). The prior efficiency requirement was 3.01 lbs/kWh, using a slightly different rating protocol, which, for example, did not account for standby electricity consumption.

Figure 13.11 GE Monitor Top refrigerator, c. 1937.

Figure 13.12 Unutilized refrigerators should be unplugged or removed.

Figure 13.13   Vending machines can be replaced with high-efficiency machines, use motion sensors for lights, or use motion sensors or timers for refrigeration.

Lower efficiency requirements are in place for compact clothes dryers (smaller than 4.4 cubic feet).

The energy use per year is estimated by dividing the weight of laundry dried per cycle (7 pounds, per DOE test standards) by the rated energy factor (lbs/kWh), and multiplying by the number of cycles per year. For residential applications, the DOE has used 416 cycles per year. More recent research recommends 285 cycles per year, for residential applications.[24] There is no efficiency label requirement for clothes dryers.

Residential Gas Clothes Dryers

Some older gas clothes dryers have standing pilot lights. These have not been used in new clothes dryers since 1988. Pilot lights are usually found in a small access panel, in the front of the clothes dryer. Pilot light energy use can be estimated from Appendix O.

The Federal efficiency requirement for gas clothes dryers effective in 2015 was set at 3.3 lbs/kWh. Between 1988 and 2015, the requirement was 2.67 lbs/kWh.

The gas dryer efficiency rating unfortunately does not separate out the motor usage from the gas usage. In the rating, the gas usage is converted into kWh. We can estimate the fraction of each, roughly. A residential dryer motor uses 200 to 300 watts.[23] The duration of a gas dryer cycle is on average 40 minutes.[25] Assuming 250 watts for 40 minutes, the fan motor energy use is 0.17 kWh/cycle. This can be subtracted from the total kWh/cycle, to calculate the gas usage, in kWh/cycle, and then convert to Btu/cycle. For example, a 1990s vintage gas dryer uses 2.67 lbs/kWh. Assuming 7 lbs per load, the energy usage per cycle is 7/2.67 = 2.62 kWh/load. Subtracting the motor use of approximately 0.17 kWh/load, the gas use in kWh/load is 2.45 kWh/load. In units of Btu/load, the gas use is 2.45 kWh/load x 3412 Btu/kWh = 8,359 Btu/load. The motor use is approximately 8 percent of the total use, in units of kWh.

Gas dryers have far lower carbon emissions than electric dryers, and may even have lower carbon emissions than heat pump dryers. This is not reflected in the efficiency requirement.

Commercial Clothes Dryers

Commercial electric clothes dryers are similar to residential dryers, but are larger. Commercial electric clothes dryers typically do not have automatic termination control. They are also not tested and rated according to any standard. Commercial clothes dryers are classified as multifamily, coin-operated, and on-premises laundries, such as ones used in hospitals and hotels. Multifamily dryers are similar in capacity to residential dryers, with an average of about 8.5 pounds of clothing capacity. Coin-operated and on-premises dryers are typically much larger, with 30 pound loads being the most common, but with sizes going up to over 400 pounds.[26] See Figure 13.14.

Testing of a sample of commercial electric clothes dryers at 8.45 pounds per load found energy usage to be 3.36 kWh/load, on average. This translates into 2.51 lb/kWh. Testing did not account for standby power consumption, as is currently included in the residential CEF, so results compare more directly to the prior residential energy factor. However, we note that the efficiency results for commercial electric dryers are much lower than even the old residential dryer standard of 3.01 lb/kWh. Residential clothes dryers appear to be significantly more efficient than commercial clothes dryers.

Options are limited for commercial clothes dryer efficiency improvements. We presume that over time, features such as automatic termination will become more common, and the dryers will hopefully be subject to minimum efficiency standards. Until then, a reasonable option, certainly for the smaller multifamily commercial

Figure 13.14   Gas-fired clothes dryers.

dryers, is to replace them with high-efficiency residential dryers. For electric dryers, savings of 33 percent can be expected, if replaced with dryers compliant with the 2015 residential efficiency requirements. Savings close to 74 percent can be expected if replaced with Europe's highest efficiency heat pump clothes dryers.

Improvements

ENERGY STAR requirements for residential clothes dryers first went into effect in January 2015. The requirements were set at 3.48 lbs/kWh for gas dryers, and 3.93 lbs/kWh for standard-size electric dryers (over 4.4 cubic feet).

Improvements for clothes dryers include:

- Use dryers in conjunction with high-efficiency clothes washers. The high-efficiency clothes washers wring more water out of the clothes, and so reduce the energy required for clothes drying.

- Replace low-efficiency dryers with high-efficiency dryers (ENERGY STAR), preferably with automatic termination, either temperature sensing or humidity sensing. Temperature sensing reportedly saves approximately 10 percent, humidity sensing saves approximately 15 percent.[23]

- Replace electric dryers with gas dryers. Gas dryers have been found to use less than one half the source energy of electric dryers.[25] Or replace electric dryers with heat pump clothes dryers. Heat pump clothes dryers are more efficient and do not exhaust air outdoors, and therefore eliminate infiltration associated with the dryer exhaust. Heat pump dryers have been estimated to have an efficiency of 4.52 lb/kWh,[24] although even higher-efficiency dryers are available in Europe, in the 5.5 to 9.6 lb/kWh range.[27]

- Drying on low heat settings reportedly saves 13 percent, although drying times are longer.[23]

- If a compact clothes dryer is in use (smaller than 4.4 cubic feet), consider replacing with a high-efficiency larger machine because the efficiency requirements are higher for larger machines.

- Routinely clean the dryer vent.

- For old gas dryers with standing pilots, replace the dryer with one that has an electric ignition.

For electric clothes dryers, to calculate annual energy use, divide the pounds per load by the energy rating, and multiply by loads per year. For example to calculate the savings to replace a standard old clothes dryer (3.01 lb/kWh) with a heat pump clothes dryer rated at 5.5 lb/kWh, using the DOE standard 7 pounds per load, and 285 loads per year:

Old clothes dryer annual energy use = 7/3.01 × 285 = 663 kWh/year

Proposed heat pump clothes dryer annual energy use = 7/5.5 × 285

= 363 kWh/year

Savings = 663 – 363 kWh/year = 300 kWh/year

For a nonresidential setting, such as a multifamily building, where residential clothes washers might still be used, it is important to adjust the number of cycles per year to the actual usage of the machine. Estimating the number of cycles per year can be based on coin receipts, in coin-operated machines (Figure 13.15). Many machines in commercial settings have counters, so usage may be available from laundry service companies, even if machines are not coin-operated. Usage can also be monitored with data-loggers. In the absence of such data, usage can be estimated from DOE

Figure 13.15   Energy use of coin-operated laundry machines can be estimated based on usage from coin receipts.

estimates of averages for multifamily buildings (1,074 cycles per year) and laundromats (1,483 cycles per year).[28] Another source provides slightly higher estimates of 1241 cycles/year for multifamily buildings and 2,190 cycles/year for coin-operated laundries.[22] If the number of people being served by a group of machines is known (e.g., in a multifamily building), the number of cycles per year can be calculated from a research finding of 0.37 cycles per person per day.[23] For example, for a multifamily building housing 100 people, and four commercial clothes washers, each machine serves 25 people on average, and so each machine will run $25 \times 0.37 \times 365 = 3,376$ cycles/year.

OFFICE EQUIPMENT AND COMPUTERS

Computers can unexpectedly consume more power than expected. For example, in a small office seeking to reduce its energy usage to net-zero, short-term monitoring on three desktop computers found them to use over 3,600 kWh/year, amounting to 47 percent of the building's electricity usage, after other energy conservation efforts had already reduced energy use by 80 percent.[29]

It is well worth measuring the power use of office equipment, to find outliers and unusual power consumers. For example, one study found the energy use of a postage meter to be 377 kWh/year.[21]

Since office equipment typically has a short life, one approach to reducing electricity use is to develop purchasing policies within an organization, for example, to standardize on ENERGY STAR rated office equipment.

FOOD SERVICE

Commercial cooking equipment is covered in the next chapter. The discussion here is limited to smaller plug-in food service equipment.

Toasters

Quartz heater elements are reportedly more efficient than sheathed or bare metal elements. A survey of a sample of conveyor toasters found that the average sheathed element uses 30 percent more energy for the same toaster capacity. Whether or not a toaster has a quartz heater element can be identified because a quartz tube has a shiny appearance, typically has a helical element inside (if visible), and has ceramic holders at each end. It is also more typically a straight tube. A nonquartz heater element is either bare metal, like a typical residential toaster heater, or has a metal sheath, much like the element in a residential electric stove and, if sheathed, winds back and forth.

Conveyor toasters (Figure 13.16) typically toast 350 or more slices per hour, whereas popup toasters typically toast 120 slices per hour per two compartments (225 to 250 slices/hour for four compartments). Conveyor toasters use between 1.5 and 3 kilowatts continuously, whereas popup toasters use 1 kW per two compartments, but only use power when toasting. For applications that do not need the high output of a conveyor toaster, or where multiple popup toasters can be used, significant energy savings can be achieved relative to the high continuous power draw of the conveyor toaster. For example, a conveyor toaster has more capacity than necessary for most hotels, for breakfast service. Even if significant capacity is necessary, multiple popup toasters will use less energy. For example, a conveyor toaster operating at 2 kW continuously for four hours per day will use 2,920 kWh/year. A popup toaster operating at 2 kW, but on average only operating two hours per day, will only use 1,460 kWh/year.

Figure 13.16   Conveyor toaster.

Kettles and Coffee Makers

Plug-in electric kettles are more efficient than boiling water on a stove top. More efficient plug-in kettles allow setting the water temperature, rather than just boiling it to 212 F (Figure 13.17). For example, many teas do not require temperature more than 170 to 185 F, and the best temperature to prepare coffee is reported to be 195 to 205 F, and best served at 155 to 175 F.[30, 31]

Coffee warmers use approximately 1 kWh/day. Energy improvements include using timers, or using a thermos. A thermos, prepared twice a day, will save approximately 70 percent, as compared to a coffee warmer kept hot all day.

HAND DRYERS

Electric hand dryers can be either plug-in or hard-wired. They have been shown to have lower carbon emissions and cost-per-use than paper towels. Hand dryers are not subject to energy regulations and are not available as ENERGY STAR products.

Hand dryers can be either operated with a push-button, or activated by a motion sensor. Push-button dryers typically operate for 30 seconds, although reportedly can run for as long as 40 seconds. Dryers activated by motion sensors have been found to be used for 20-25 seconds, for less efficient dryers, and 10-15 seconds for more efficient dryers.

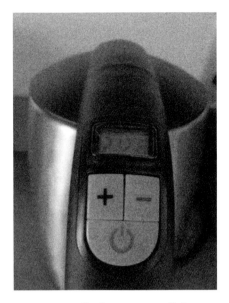

Figure 13.17 Kettles are more efficient than stovetop water boiling, and efficient kettles allow setting the water temperature.

The power consumption of hand dryers can vary from as low as a few hundred watts to more than 2000 watts. If the power consumption is not shown on the nameplate, a reasonable approximation of the power consumption (in watts) is the nameplate voltage multiplied by the nameplate current (amps).

Some newer dryers come without a heater element, and others allow the heater element to be turned off. Limited testing of dryers without a heater element shows longer drying times are required, in the 20-25 second range, despite manufacturer claims of 10-15 second drying times. However the lower energy use, for example, in the 500 watt range, means a lower overall energy use of under 4 watt-hours per use.

High-efficiency dryers tend to either have a blade design, allowing the hands to be slid through two sides of the dryer, or have a high-velocity air stream.

Most motion-activated dryers turn off within 2-3 seconds of when hands are removed. The off-delay is as short as one second for some dryers, and can be as long as over five seconds for others.

For field work:

- Take nameplate data, to obtain both the voltage and the current draw, and the power consumption if available.

- If pushbutton-activated, measure the duration of the dryer cycle.

- If motion-activated, measure the duration to effectively dry hands, and measure the duration of the off-delay, and add both durations to obtain the run-time per use.

Replacement hand dryers should be motion-activated, rather than pushbutton-activated. To estimate energy use for a replacement high-efficiency dryer, use independent test results of typical drying time, and add the off-delay. High-efficiency blade-type hand dryers have been found to typically require 4-5 watt-hours per use. High-efficiency high-velocity hand dryers have been found to typically require 5-7 watt-hours per use. High-efficiency hand dryers without heating elements have been found to use below 4 watt-hours per use, but require a longer drying time.

By contrast, a standard pushbutton-activated hand dryer using 2000 watts and operating for 30 seconds uses approximately 17 watt-hours per use.

OTHER PLUG LOADS

There are almost endless other plug loads. Plug loads are something of a wild card in energy work, as described, a new frontier. However, plug loads are growing, and deserve our attention.

The best way to manage and reduce the energy use of plug loads is to measure their energy use. A best practice is to maintain a log of energy use for all plug loads in a building, rotating a watt-meter to measure each device's energy use in turn, for 24 hours each, or 7 days for equipment that sees different use on weekdays from on weekends.

References

1. S. Corino, E. Romero, and L. F. Mantilla. *"How the Efficiency of Induction Motor Is Measured."* International Conference on Renewable Energies and Power Quality, 2008.
2. A. de Almeida and F. Ferreira, *Efficiency Testing of Electric Induction Motors.* APEC Workshop on Setting up and Running an Energy Performance Testing Laboratory. Manila, Philippines, July 6–8, 1999.
3. Washington State University Energy Program, "Premium Efficiency Motor Selection and Application Guide: A Handbook for Industry." Prepared for the U.S. Department of Energy, Energy Efficiency and Renewable Energy, Advanced Manufacturing Office. DOE/GO-102014–4107, February 2014.
4. Washington State University, "Pre-EPAct Default Motor Efficiency Table for Use When No Nameplate Efficiency is Available." *Prepared for the U.S. Department of Energy, 123_spreadsheet.xls*, downloaded from http://www.motorsmatter.org/index.asp. Accessed July 26, 2015.
5. U.S. Department of Energy, "Determining Electric Motor Load and Efficiency, Motor Challenge Fact Sheet." DOE/GO-10097–517, 1997.
6. Taitem Engineering, "Elevator Energy Use." NYSERDA Multifamily Performance Program Technical Topic, March 2015.
7. A. de Almeida, C. Patrao, J. Fong, R. Araujo, and U. Nunes, "Energy Efficient Elevators and Escalators." E4 Project, European Commission, 2010.
8. The Cadmus Group, Inc., "Low-Voltage Transformer Loads in Commercial, Industrial, and Public Buildings." *Prepared for Northeast Energy Efficiency Partnerships; quoted in Transformer Replacement Program*, 1999.
9. NEMA Standards Publication TP 1–2002, Guide for Determining Energy Efficiency for Distribution Transformers, NEMA, 2002.
10. Cooper Power Systems, "Transformer Technology: Liquid-Filled vs. Dry-Type." Bulletin B210–05059, Revision 1, February 2008.
11. National Grid, Transformer Replacement Program for Low-Voltage Dry-Type Transformers, Implementation Manual, Version 2013.1," April 4, 2013.
12. P. R. Barnes, J. W. Van Dyke, B. W. McConnell, and S. Das. *"Determination Analysis of Energy Conservation Standards for Distribution Transformers."* Oak Ridge National Laboratory ORNL-6847. Developed for the U.S. Department of Energy, July 1996.
13. R. G. Pratt and J. D. Miller, *The New York Power Authority's Energy-Efficient Refrigerator Program, for the New York City Housing Authority—1997 Savings Evaluation.* ENERGY STAR Partnerships Program, September 1998.
14. http://www.gcbl.org/live/home/efficiency/understanding-how-much-energy-we-use. Accessed November 13, 2014.
15. R. Saidur, H. H. Masjuki, T. M. I. Mahlia, and A. S. Nasrudin, "Factors Affecting Refrigerator-Freezers Energy Consumption." *ASEAN Journal for Science and Technology Development (AJSTD)* 19, no. 2 (2002), p. 67.
16. U.S. Food and Drug Administration, "Are You Storing Food Safely?" April 2014.

17. "40,000 Refrigerators from 1979–1992." http://www.waptac.org/Refrigerator-Guide/Energy-Use-Data.aspx. Accessed February 11, 2015.

18. http://www.kouba-cavallo.com/refmods.htm. Accessed July 28, 2015.

19. Natural Resources Canada, Office of Energy Efficiency. "EnerGuide for Equipment." EnerGuide Appliance Directory, 2001.

20. Avid Boustani, Sahni Sahni, Timothy Gutowski, and Steven Graves, "Appliance Remanufacturing and Energy Savings." Environmentally Benign Manufacturing Laboratory, Sloan School of Management, MIT, January 28, 2010.

21. http://www.gcbl.org/live/home/efficiency/understanding-how-much-energy-we-use. Accessed November 13, 2014.

22. http://steplight.com.au/blogs/steplight/14184623-water-cooler-water-boiler-energy-consumption-revealed. Accessed July 27, 2015.

23. U.S. Environmental Protection Agency, "ENERGY STAR Market & Industry Scoping Report Residential Clothes Dryers," November 2011.

24. Steve Meyers, Victor H. Franco, Alex B. Lekov, Lisa Thompson, and Andy Sturges. "Do Heat Pump Clothes Dryers Make Sense for the U.S. Market?" 2010 *ACEEE Summer Study on Energy Efficiency in Buildings*, Ernest Orlando Lawrence Berkeley National Laboratory, August 2010.

25. Natural Resources Defense Council, "A Call to Action for More Efficient Clothes Dryers," June 2014.

26. Yanda Zhang and Julianna Wei, "Commercial Clothes Dryers." Response to California Energy Commission. 2013 Pre-Rulemaking Appliance Efficiency Invitation to Participate, Docket Number: 12-AAER-2D; Commercial Clothes Dryers, May 9, 2013.

27. Rita Werle, Eric Bush, Barbara Josephy, Jürg Nipkow, and Chris Granda. Energy-Efficient Heat Pump Driers—European Experiences and Efforts in the USA and Canada. www.toptenamerica.org. Accessed July 27, 2015.

28. Department of Energy, 10 CFR Part 431, Energy Conservation Program: Energy Conservation Standards for Commercial Clothes Washers; Final Rule, Part IV. *Federal Register* 79, no. 240 (December 15, 2014).

29. Taitem Engineering, "Small Commercial Energy Assessment Program Energy Assessment" for *Ithaca Connected*, 2014.

30. http://www.twoleavestea.com/water-temperature/. Accessed February 14, 2015.

31. http://www.ncausa.org/i4a/pages/index.cfm?pageID=71. Accessed February 14, 2015.

Chapter 14

Gas Loads (Other than Heating and Domestic Hot Water)

Natural gas is mainly associated with space heating and domestic hot water heating. However, a variety of other gas loads are found in buildings, each of which offers opportunities for energy conservation.

Cooking Equipment

Cooking equipment is not subject to federal efficiency requirements, or rated through ENERGY STAR, other than commercial cooking products such as commercial fryers, griddles, hot food holding cabinets, ovens, and steam cookers, which are available as ENERGY STAR but not federally regulated.

A major load within many stoves and ovens are the standing gas pilot flames used to ignite burners and ovens (Figure 14.1). Each pilot can use as much as 2,000 Btu/hr, 24/7, consuming 175 therms per year. Electronic spark ignition uses less energy. See Appendix O for pilot light gas usage.

Where possible, small appliances should be used, instead of larger stoves/ovens. For example, an electric plug-in kettle uses less energy than a stovetop burner.

Energy use in commercial cooking equipment has been well studied by the Food Service Technology Center (FSTC).[1] Over the past few decades, efficiency standards have been developed for different classes of cooking equipment. Testing shows that energy efficiency ranges widely within each class of equipment. For example, standard efficiency gas griddles have been found to be in the 25–35 percent range for efficiency, whereas high-efficiency gas griddles are 40–50 percent efficient, and electric griddles are 65–75 percent efficient.

Energy improvements include replacing low-efficiency with high-efficiency products, and considering products with additional energy features, such as griddles that can heat just a portion of their surface. Cooking at high production rates, where possible, typically occurs at a higher efficiency than at low production rates. Turning off equipment when not in use reduces idling energy use.

.Energy savings can be calculated using online calculators at the FTSC web site[2] or a downloadable spreadsheet from ENERGY STAR.[3] FTSC's calculators include built-in databases of manufacturers/models from which to choose. They also include energy use for default baseline products, based on typical standard efficiency products being sold today, the efficiencies of which have reportedly not changed significantly for decades. And they include efficiencies for default high-efficiency products.

A question frequently posed is, "What is better, gas or electric, for commercial cooking?" The answer depends on exactly what the metric is: Efficiency, carbon emissions, site energy use, or source energy use. The answer also depends on if the electricity is generated by fossil fuels or by renewables or by something else. Electric cooking equipment is typically more efficient than gas, in other words, less electricity is used per unit of cooked food than gas, when measured in equivalent units of site

Figure 14.1 Standing pilot on a commercial stove burner.

energy. This is ostensibly because the electric elements are closer to the food, and there are fewer losses. However, when calculated on a source energy basis, or on a carbon emissions basis, gas is better: Using national-average factors for source energy and carbon emissions, for the same unit of cooked food, 1.5–2.5 units of carbon emissions are generated for electric cooking equipment as are generated for gas cooking. In other words, gas cooking equipment produces about half the carbon emissions, and uses about half the source energy, as electric cooking equipment, on average.

Gas Clothes Dryers

Gas clothes dryers are addressed in the prior discussion of plug loads, with electric clothes dryers.

Gas Leaks

Gas leaks are a type of gas load that are common in buildings. Gas leaks are particularly important to find and eliminate because natural gas has 25 times the carbon emissions impact as carbon dioxide. Gas leaks can be found with leak detectors, or by the telltale odor that is added to natural gas to facilitate detection. Gas leaks can also be detected and measured by natural gas meters. One revolution of the smallest dial on a gas meter usually represents 0.5 cubic feet of natural gas flow. By turning off all gas appliances, for example, for 24 hours, and counting the revolutions and dividing by 24, the leak rate can be measured, in units of cubic feet per hour.

References

1. Don Fisher et al. *"Commercial Cooking Appliance Technology Assessment."* Fisher Nickel Inc. FSTC Report #5011.02.26. 2002.
2. http://www.fishnick.com/saveenergy/tools/calculators/. Accessed 11/22/2015.
3. http://www.energystar.gov/products/certified-products/detail/commercial-food-service-equipment. Accessed 11/22/2015.

Chapter 15

Advanced Energy Improvements

Combined Heat and Power

Combined heat and power (CHP), also called *cogeneration* or *cogen*, refers to the simultaneous generation of electric power and heat. The heat is regarded as free, a by-product of the electricity generation. In a traditional electric power plant, this heat is rejected to the environment, to rivers, lakes, or to the outdoor air. An extension of cogeneration is *trigeneration*, where either the combination of electricity and heat are simultaneously produced, or electricity and cooling are simultaneously produced, typically by using rejected heat to drive an absorption chiller.

CHP has the potential to be a highly efficient use of the fuels of combustion. Instead of reaping only 80 to 90 percent of the content of the fuel for space heating, or the equivalent of 30 percent of the fuel content for generating electricity, CHP can provide both electricity and heat, at the same time. By generating one's own electricity, there is also the potential to reduce the cost of electricity. These are the promises of CHP.

Challenges of CHP include identifying *simultaneous* electric and heat loads. If the heat is to be used to heat spaces, we already start with the loads only being coincident in winter. And in winter, peak need for space heat is typically at night, whereas peak electric loads for commercial buildings are typically during the day. If the heat is to be used for domestic hot water, there are questions of when the domestic hot water (DHW) is used during the day, and whether this is coincident with the electric loads. For example, in the hospitality and multifamily sectors, both of which have high DHW demand, much of the DHW load is in the morning and evening, and this may not coincide with the bulk of the electric load. The underlying advantage of electricity, a source of energy available on demand, intrinsically means that its loads vary widely, as motors start and stop, as people start work and end work, as weather conditions change, and as building occupancy patterns change. Needs for heat also vary widely, depending on weather conditions, depending on when people use hot water, and depending on when people use heat for other processes, for example, cleaning, cooking, or manufacturing.

CHP requires solid analysis during the energy audit phase, analysis that realistically examines the potential simultaneous need for electricity and heat, and that is not wishful in its estimates of combined heat and power loads. There is a risk to overestimate its savings, specifically in overestimating recoverable heat. A tension can also arise between available capacity provided by a CHP system (both electric and heat) and subsequent energy improvement investments: Why invest in energy savings elsewhere that might offset the savings that was to be justified by the investment in the CHP system, or by internal or external sales of CHP-generated energy? Such competition is not good. We want energy savings from our CHP systems, but we do not want a CHP investment to limit subsequent continuous energy improvements.

Good applications for CHP are buildings where there are large and almost continuous electric and heat loads. Hospitals are a prime example, with high levels of continuous lighting, and heat used for hot water, cleaning, and more. Large industrial facilities with electric and process heat needs are another example.

Heat Pumps

As heat pumps have developed in recent decades, we have come to see their potential to recover heat, deliver heat, and remove heat in new and more efficient ways. Advanced heat pump applications continue to identify new heat sources, new heat sinks (where heat can be moved to), and ways in which to reduce the temperature difference between source and sink in order to improve a heat pump's efficiency.

As an example, if a loop of water in a building is used as both a heat source and a heat sink, from which heat pumps can draw heat (for heating) and reject heat (for cooling), multiple possibilities open up. A solar thermal hot water system can be used as heat source for the loop, and will in fact operate at a high efficiency, and can deliver more useful heat, even on cloudy days, because of the relatively low temperature of the water to which it delivers heat. From the heat pump's perspective, this becomes a useful and relatively warm heat source, boosting the heat pump's efficiency. This opens up possibilities for solar-assisted space heating, even on cloudy days in winter.

As another example, radiant heating systems in buildings allow the use of low temperature hot water distribution, raising the efficiency of heat pumps, wherever they are drawing their heat from.

Geothermal heat sources have long been recognized as being efficient for heat pumps, in both heating and cooling, and new approaches further boost this efficiency. Advanced well system design, more affordable well system components, and innovative pumping and piping strategies all offer possibilities to make geothermal systems both more efficient and more affordable.

Heat pump applications for domestic hot water are also developing rapidly. Stand-alone air-source heat pumps for domestic hot water have become common. We anticipate more uses of water as a source of heat for domestic hot water, such as geothermal systems, and water loop systems. Outdoor air-source heat pumps for domestic hot water are another likely development, which avoid heat-source limitations of indoor packaged air-source hot water heaters, such as the inability to put them in closets.

Ventilation is a rich potential heat source and heat sink for heat pumps. Both ventilation exhaust and ventilation intake (outdoor air) can be used in a variety of ways as heat sources and sinks, to recover heat, dispose of heat, and increase the efficiency of heat pumps.

In short, as we have come to recognize the flexible nature of heat pumps, potential advanced applications have only begun. We anticipate advances in energy modeling, to allow such multiple-source and multiple-sink heat pump applications.

Envelope

Stack effect airflow paths, and their impacts, are increasingly recognized. Quantifying the benefits of sealing these pathways, on a production basis as a part of routine energy audits, is a next frontier in energy improvement work. A related improvement is compartmentalization, both to reduce stack effect, and to allow thermal zoning.

Infiltration improvements in commercial buildings also offer untapped opportunities, and are closely linked to bringing stack effect under control.

The blower door test has been successfully demonstrated in large buildings, and is expected to become a production tool for quantifying infiltration, diagnosing it, and guiding appropriate infiltration reduction.

Advances in insulation improvements include new strategies in affordability: pre-manufactured assemblies, cost-effective installation approaches (for example, cutting, taping and fastening rigid insulation), and approaches to reliably control moisture and water intrusion during insulation. New strategies for interior insulation of masonry buildings have opened up the potential to insulate a vast stock of uninsulated old buildings. New approaches to insulate balconies offer the possibility of reducing the balcony thermal bridging that is common.

Heating/Cooling Distribution

Aerosol-sealing of ductwork is a relatively new development that is seeing wider use in commercial buildings, offering both thermal energy savings and the potential for fan motor electricity savings.

Advanced pumping strategies include not only packaged variable-speed pumps, but also a variety of optimization strategies for primary/secondary pumping, and optimized pipe sizing to reduce pumping energy.

A variety of advanced insulation products allow highly-effective insulation of piping components, such as pumps and valves.

Lighting

Solid state (LED) lighting has entirely transformed the lighting industry, in the space of a short few years, and will continue to do so.

The controllability of solid state lighting can be taken advantage of with new controls, such as multifunction controls (integrating motion, photocell control, and/or timer functions in one control), addressable controls, and modulating controls.

Right-lighting is emerging as a new frontier in lighting energy improvements. Instead of relying simply on one-for-one lighting replacements, right-lighting reduces over-lighting that is common in most buildings, and so delivers higher energy savings, reduced installation cost, and reduced maintenance cost.

Controls

Wireless sensors have opened up vast new possibilities in affordable building energy system control. When examining existing control systems, we can now move beyond simply trying to optimize an existing system, and instead we can consider transforming the system with new inputs and, more importantly, new control points. We can add thermal zones, we can control these thermal zones to different temperatures, and to different schedules.

Ventilation controls continue to challenge us. We simply do not know what the ventilation control is doing in most of our buildings, most of the time. Bringing ventilation under control, to deliver the right ventilation at the right time, is a next frontier in energy improvements. Diagnostic systems that reliably and fully identify what current ventilation systems are doing is a step in the right direction, and becomes a launching point for control modifications to bring ventilation under energy-efficient control.

Renewable Energy

Renewable energy improvements include such strategies as solar thermal (see Figure 15.1), photovoltaic, wind, waste heat, and biofuels. They have in common the fact that their net impact on carbon emissions is small or zero. Geothermal hot springs are a form of renewable energy, but geothermal heat pumps, frequently misclassified as a renewable technology, are not.

Figure 15.1 Solar thermal components.

Traditionally, renewable energy improvements have been analyzed separately from energy conservation. Increasingly, they are weighed alongside conservation. Due to reduction in price of renewables, they are now often competitive with conservation. The benefit of evaluating both at the same time is that the cost-optimal combination can be identified. In some ways, renewables compete with conservation improvements, just as conservation improvements compete with each other.

It is still good to recognize differences between renewables and conservation. A big advantage of renewables is the predictability of energy savings. For example, we are now able to accurately predict the annual savings generated by photovoltaic systems, within a few percent. This level of predictability is not the case with many energy conservation improvements. Disadvantages of renewables include higher installed cost, although costs have been dropping; renewables contain embodied energy, which offsets some of the savings; and most renewables require maintenance. Most significantly, renewables have a unique characteristic that places their savings at high risk: Renewables typically have a conventional energy backup, and this presents a problem. When a renewable energy system has a component failure, the backup automatically takes over. Sometimes this failure is not evident. In other cases, the repair might not be immediately afforded, or otherwise deferred. For as long as the renewable system is down, or if it is abandoned, the savings are lost. For renewable electricity, such as wind or photovoltaic systems, the backup is the electric grid. For solar thermal systems, the backup is usually a fossil fuel water heater, or grid-supplied electric water heater. This is a major risk and is not a risk for energy conservation improvements. Controls that automatically alert an owner to problems can reduce this risk, and should be regarded as essential in every renewable energy project. However, in many cases, repairs are simply not made, even with such notification, and savings are lost.

In this way, energy conservation savings can last much longer than renewable energy savings. We call this *persistence of savings*, an important property of energy improvements. In many cases, the savings through conservation can last as long as the life of the building. Consider a heated stairwell, which did not need heat, and where heat was removed. As long as the building lasts, the stairwell will no longer be heated, and savings will accrue, without maintenance or repair, or the savings otherwise being lost. Or consider right-lighting, where we remove lamps or rearrange light fixtures in order to deliver the correct level of light to a space and eliminate overlighting. For as long as the building lasts, even beyond multiple lamp replacements over time, the savings will continue. Or consider daylighting, with judiciously placed skylights. As long as the skylights last, most likely for the life of the building, the building will not need artificial light during the day. Renewable energy improvements require vigilance and maintenance in order to deliver savings that persist over time.

Another characteristic of renewable energy improvements is that they are more visible, and so are favorites for owners who want to show a visible commitment to the environment.

Evaluating Buildings that Already Have Renewables

A separate but related topic is the evaluation of buildings that already have some form of renewable energy or other on-site energy generation. This emerging topic is trickier than it might seem.

For example, consider a building with a photovoltaic system. See Figure 15.2. During sunlight, the solar equipment produces energy that may be metered by a local meter S. This electricity is not metered by the utility. If there are electric loads in the building during sunlight, the solar power is consumed by the loads in the building, in a quantity L that, similarly, is not metered by the utility. If these loads in the building are less than the power produced by the solar panels, the excess $(S - L)$ is sent to the

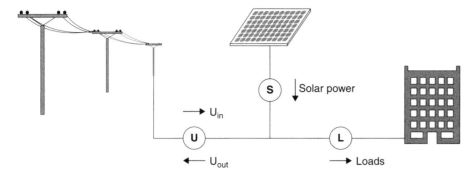

Figure 15.2 Grid-tied photovoltaic system.

electric grid, and this excess is measured by the utility meter U. This quantity $(S - L)$ is equivalent to U_{out}, measured by the utility meter. If the building loads L exceed the solar power S, then utility electricity is drawn into the building and is used, and this is measured by the utility as U_{in}. $U_{in} = L - S$. This occurs, for example, when there is no sunlight, or low sunlight.

It is important to note that the utility only measures U_{in} and U_{out}. Some utilities report both U_{in} and U_{out}, on the utility bill. Some utilities only report the difference between U_{in} and U_{out}, in other words $U_{in} - U_{out}$. As a result, if we only examine the utility bills, we do not know how much electricity was produced by the solar system (S), and, even more importantly, we no longer know how much was consumed by the building (L). Unlike buildings that have no on-site renewables, where L is simply equal to U_{in}, in the case of a building with on-site renewables we no longer have firm ground on which to stand, to analyze the building loads, to help us to true up our energy models, and to begin to plan energy improvements.

The solution is to make sure that the renewables production (S) is monitored, and for the building owner to take readings of this production, as well as to take simultaneous readings of U_{in} and U_{out} at the utility meter. Then:

$$U_{in} + S = U_{out} + L \text{ (the total demand equals the total supply)}$$

Therefore:

$$L = U_{in} + S - U_{out}$$

And we now know how much electricity is being consumed by loads in the building. We can compare this to our energy model, we can disaggregate it using billing analysis techniques, and it forms the foundation of energy planning and energy improvements.

The same issue applies to other on-site generated energy, such as cogeneration, the combustion of biofuels, etc.

Another challenge is the calculation of the electricity rate, from a bill that has net metering, with the sale of site-produced renewable energy back to the utility. If we simply divide the cost of the bill by the electricity bought from the utility (U_{in}), we will underestimate the cost of the electricity. If we divide the cost of the bill by the net electricity bought ($U_{in} - U_{out}$), we will overestimate the cost of the electricity. What we are interested in is the incremental cost of the electricity: If we save one kWh, and that reduces the bought electricity by one kWh, what is the rate per kWh of this one saved kWh? The answer requires examining the purchased rate of the electricity, and ignoring the site-generated electricity and its associated credit.

Improving to Net-Zero or Very Low Energy Use

At the cutting edge of energy work is reducing a building's energy use so low that most or all of its energy needs can be delivered with renewables. Previously regarded

Figure 15.3   Photovoltaic installation.

as unattainable, net-zero performance is routinely delivered in new buildings, and has been achieved in existing buildings as well.

Net zero refers to buildings where the supply of renewable energy, over a whole year, meets or exceeds the building energy consumption. We use the term *net-zero energy*, to describe allowing renewable energy systems to export excess capacity at certain times, and to take credit for this export for those times when the renewable energy system cannot meet the building's energy demands. But over a whole year, the net effect is zero energy consumption. Credit is sometimes allowed for limited purchase of off-site renewables. Net zero can be based on site energy, source energy, carbon emissions, or energy cost. The term *zero energy building (ZEB)* is also gaining use, as a synonym for net-zero energy.

A recognized path for net-zero improvements is to convert a heating system to heat pumps, and to generate the electricity with photovoltaics. (Figure 15.3.)

Challenges and issues with net-zero energy improvements include:

- Modeling net-zero energy. Modeling the interactions between energy improvements is essential.
- The high investment required, and diminishing returns.
- Limited space for renewables. For example, roof space for solar panels on high-rise buildings limit the capacity to reach net zero, if net-zero capacity is defined to require on-site generation.
- Making weather corrections to demonstrate that net-zero was reached. For example, if a building in a northern climate sees a very cold winter, we might wish to correct for the cold weather, during our measurement and verification that the building reached net zero. Weather corrections for conventional buildings that do not have renewable energy are fairly straightforward: We can analyze the utility bills and separate out the heating usage. However, for buildings with renewable energy, we might not know how much of the renewable energy was used for heating, and so this analysis is more difficult.

Despite the challenges, net-zero energy improvements present an exciting cutting-edge opportunity for energy professionals.

A case might even be made that energy audits be redefined to be *a building assessment to evaluate elimination of energy use.* Consider the benefits of energy audits that were sufficiently comprehensive to take buildings to net zero, contrasted with the current approach of perhaps three separate energy audits (ASHRAE Levels 1, 2, and 3), the sum of which do not take a building to net zero.

Consider first the benefits for the building owner. In receiving a comprehensive energy audit that evaluates the improvements necessary to eliminate energy use, the owner now has a long-term planning document. A long-term plan can accordingly be developed, accounting for energy improvements to coincide with long-term major renovations, or equipment end-of-life replacement, minimizing the cost of these improvements. A document is received, along with an energy model, which becomes a part of the building's assets. This can be considered a live document, which can be updated and revised as technology changes, and as the building changes.

For example, take the replacement of refrigerators with an expected life of 15 years, in a multifamily building. The existing refrigerators are medium-efficiency, and are not bad enough to warrant outright replacement immediately. However, if and when the refrigerators are one day to be replaced, the incremental cost of installing high-efficiency refrigerators may well be attractive. The chance of replacement happening within 10 years of an energy audit is 67 percent, in other words, the probability is high. If the audit includes evaluation of refrigerator replacement, there is an extremely good chance that it can guide a high-efficiency replacement at time of replacement. But, conversely, if evaluation is not done in the energy audit (e.g., it is not considered because it is viewed as having too long of a payback), the chance of replacement happening before the next energy audit (and so not being guided by energy considerations) is high. This is where a net-zero energy audit can serve as a long-term planning document.

Furthermore, projected savings for some long-payback improvements might in fact spur capital projects that might otherwise not have been considered as soon into the future. Consider new siding or cladding for a building. If a building owner learns that there are energy and cost savings from adding insulation, even if the savings are modest and may not immediately justify the work, it might spur the re-siding/cladding project. In fact, it might accelerate a project that was possibly going to happen in the future anyway. In this case, the re-siding/cladding project ends up accruing energy savings, due to the insulation, and happens sooner than it might have.

What are cost impacts of a net zero energy audit, relative to current energy audit practice? In other words, how much more does a net-zero energy audit cost? There are, in fact, potentially many cost *savings* with a net-zero energy audit. With a net-zero energy audit standard, there would be no need for negotiating a scope of work for the energy audit. There would be no more need to explain the difference between a walk-through audit and an investment-grade audit, and need to justify the higher cost of an investment-grade audit. A net-zero energy audit might well cost less than the current approach of three levels of energy audits. There would only need to be one site visit, instead of the three that comprise the current level 1-3 approach. There would only need to be one report, instead of three. There would only need to be one energy analysis, instead of three. There would not be a need to reconcile a level 1 or level 3 audit with a prior level 1 or level 2 audit. There would not be a need to present and explain three separate audits to the building owner, but rather just one.

A net-zero energy audit standard would furthermore likely spur advances in the more complex and costly improvements that today do not receive attention. With increased demand for deeper savings, as we indeed are already seeing from building owners, investigating how to make an improvement with a 50-year payback into one with a 10-year payback might, in fact, turn up ways to actually make it happen. We have the experience of photovoltaics, which only a few years ago had 50-year paybacks, and today are strongly competing with other energy improvements.

Furthermore, a net-zero energy audit standard ensures that owners go beyond a low-cost/no-cost level 1 energy audit. In many cases, possibly even in most cases, owners indeed never even proceed beyond a level 1 energy audit. In their minds, they have "received an energy audit", and are in many cases left with the misperception that possible energy savings in their buildings are limited. Rather than opening possibilities, the level 1 energy audit has limited possibilities, it has constrained possible options for energy improvements. It has done the very opposite of what it was intended to do.

Many buildings clearly cannot reach net zero. High-rise buildings with limited roof space for on-site renewable energy cannot reach net zero, under current definitions and with current technology. So a net-zero energy audit standard would need to accommodate such buildings. Perhaps allowance would be made for purchase of more off-site renewable energy in these situations.

Is a net-zero energy audit standard possible? Is it plausible and realistic? We would suggest that if we now know how to reduce lighting energy use by well over 70 percent in most buildings, without sacrificing lighting quality, that a net-zero energy audit standard is possible. We would suggest that heating and cooling savings of well over 50 percent are possible, through a combination of increased plant efficiency, reduced distribution losses, thermal zoning, and other improved controls. We would suggest that ventilation energy improvements of well over 50 percent are possible, through high-efficiency fans, demand-controlled ventilation, duct sealing, and energy recovery. We would suggest that envelope savings of well over 50 percent are possible, through a combination of insulation and air sealing, and improvements such as better windows and better doors, and control of stack effect. We would suggest that we have barely begun to explore the ways in which plug loads, and behavioral components to energy use, can be leveraged, and that a net zero energy audit standard would in fact help us to advance this exploration. We would suggest that, to take a building from 50 to 70 percent savings all the way to net zero, many buildings, if not most, could accommodate some on-site renewable energy, and can certainly consider purchasing some off-site renewable energy. We would suggest that if a small office building in a cold climate can reach over 60 percent energy savings, using a standard set of improvements, and at a modest installed cost, that a net-zero energy audit standard is possible.[1] We would suggest that if the Empire State Building identified 38 percent energy savings, with a fairly modest set of improvements, that a net-zero energy audit standard is possible, with possible accommodations for high-rise buildings, and buildings otherwise without ready access to on-site renewable energy.[2]

Our current approach to energy audits implicitly limits what can be done and conveys the mistaken impression that what can be done is limited. Net-zero energy audits, conversely, convey correctly that much can be done, and we only need to choose and plan just how far we each want to reduce our energy footprint, in order to reach our goals.

Other Advanced Improvements

A wide variety of other advanced and emerging improvements are being developed. Improvements such as dedicated outdoor air systems, various biofuels for heating, and plug load controls are all receiving attention. Advanced ways in which to engage building occupants, in behavioral strategies to save energy, also hold promise. Electronic data interchange (EDI) offers potential to expeditiously obtain energy use data in standardized formats, for energy auditing, analysis, tracking, benchmarking, alarm triggering, and other purposes. Advances in quality control, as applied to building energy improvements, bring opportunities for meeting or exceeding the promises of projected energy savings, rather than finding savings shortfalls at the end of energy improvement projects. Advances in energy modeling allow faster and more accurate energy audits. Mixed mode ventilation promises the reduced use of mechanical ventilation, and greater occupant engagement and satisfaction in control of natural ventilation. Related improvements include night-time precooling of buildings with natural or mechanical ventilation. A wide variety of innovations have been introduced in the area of financing, and in support functions such as standardized building permitting for energy improvements.

Energy auditors and energy managers should stay apprised of such developments. We should also evaluate newly introduced products with professional scrutiny, and avoid the temptation to jump on the latest bandwagon, while abandoning proven energy improvements for the excitement of something new. Our goal is to deliver quality energy savings and to do so in the most responsible and cost-effective manner. The promise of new products is great, but each one needs to prove itself.

References

1. I. Shapiro, "Energy Audits and Improvements in Small Office Buildings." *ASHRAE Journal*, October 2012.
2. T. Malkin, P. Rode, D. R. Schneider, and S. Doig. *"The Empire State Building: Repositioning an Icon as a Model of Energy Efficiency Investment."* Malkin Holdings, Johnson Controls, Rocky Mountain Institute, Clinton Climate Initiative, Jones Lang LaSalle, 2010.

Chapter 16

Estimating Savings

Isolated versus Integrated Estimates

Savings are estimated by either examining a single energy improvement, by itself, or by examining a whole building in more detail. For example, we can estimate savings for a boiler replacement, by itself, making a variety of assumptions, such as the number of hours the boiler runs per year. See Figure 16.1. Or we can model the building, which calculates the building heat loss over a year, and then calculates the savings for the boiler replacement based on the building heat loss, and related factors such as existing and proposed boiler efficiency.

We might call the by-itself estimate an *isolated energy savings estimate*, and the whole-building estimate an *integrated energy savings estimate*. Isolated energy savings estimates are simpler and faster but can be less accurate. Integrated energy savings estimates take more training and time but are more accurate if modeling is done well.

Advances in computers and software have made integrated energy savings estimates more available to a wider group of energy professionals. Experience in recent years has shown that accurate whole-building energy models can readily be prepared by a wide range of individuals, including energy auditors, energy managers, energy technicians, specialty energy contractors such as insulation and weatherization installers, and more.

An example of an isolated savings estimate is a typical photovoltaic assessment (Figure 16.2), perhaps prepared by a solar installation contractor. The contractor will evaluate the available area for solar modules, the orientation of the roof, the selected modules, and shading. But the contractor may not model a building's electric loads to compare the hourly profile of electricity use to what the proposed solar installation might deliver. Rather, the whole-building annual electricity use might be glanced at and compared to what the solar installation can deliver. Energy auditors will likewise typically not do an integrated estimate for photovoltaic savings.

An example of an integrated savings estimate is a whole-building energy model, including R-values for walls and roof, U-factors for windows, internal gains such as lighting and people, and more. The model may be used to estimate multiple improvements, such as attic insulation, lighting improvements, chiller replacement, and changed temperature controls. And the model can account for interactions between these improvements: The reduced cooling load due to lighting improvements is automatically applied to the chiller load, and so the chiller savings do not mistakenly account for cooling savings on the portion of the lighting load that no longer will be in the building.

Figure 16.1 Boiler improvement energy savings can be estimated either in isolation or interactively with other improvements.

Figure 16.2 Photovoltaic energy analysis is frequently performed without hourly electricity use profiles. Models that account for hourly electricity use profiles more accurately predict what will be saved on-site.

Figure 16.3 Writing out units, and making sure units are consistent, reduces the risk of calculation mistakes.

Isolated Savings Estimates

SIMPLE CALCULATION

Isolated savings estimates can be as simple as a lighting calculation:

Existing lighting: 10 fixtures × 100 watts per fixture × 1,000 hours per year
= 1,000,000 watt-hours/year. Divide by 1000 watts per kW,
to obtain 1000 kWh/year

Proposed lighting: 10 fixtures × 15 watts per fixture
× 1,000 hours per year / 1,000 watts/KW = 150 kWh/year

Savings = 1,000 − 150 = 850 kWh/year

When performing hand calculations or spreadsheet calculations, it is critical to check units. A good practice is to write out the units, and to cancel units in numerator and denominator, to make sure the result is in the correct units. In the above example:

$$\text{Fixtures} \times \frac{\text{watts}}{\text{fixture}} \times \frac{\text{hours}}{\text{year}} \times \frac{\text{kW}}{\text{watts}} = \text{kWh/year}$$

Note how "fixtures" and "watts" cancel in the numerator and denominator, and the result is in the sought units, kWh/year. The discipline of writing out the units, and crossing out units that cancel in numerator and denominator, is a helpful way to prevent calculation errors due to mistaken units. (Figure 16.3.)

The calculation is simple and fast, and is likely acceptably accurate, even though interaction with heating and cooling are not accounted for. A fairly straightforward correction can be made to account for heating and cooling interactions, although it will require an assumption about the duration of the heating and cooling seasons, as well as information about heating and cooling system efficiencies.

INVERSION CALCULATION

For many isolated energy savings estimates, we know the current energy use, and need to calculate savings on the basis of replacing a device with one of a higher efficiency, like a motor, chiller, furnace, or other system. We use what may be called the *inversion* technique to calculate the *load* from the current energy *consumption*, and then work from the load back to the new energy consumption.

For example, consider an 80 percent efficient 15–horsepower (HP) motor running for 3,000 hours per year, which we are considering replacing with a 90 percent efficient motor. We first calculate the load on the motor, which is its existing input, 15 HP, multiplied by the 80 percent efficient motor, so the load is 12 HP. Now we know the load, we can work back to calculate the new motor input, which is the load of 12 HP divided by the new motor efficiency of 90 percent, or 12 / 0.9 = 13.3 HP.

Using the inversion technique, the savings can be shown to be:

$$\text{Savings} = 1 - (Ee/En)$$

Here, *Ee* is the efficiency of the existing device, and *En* is the efficiency of the new device. This approach works for any simple isolated energy savings estimate, where the existing energy use is known. The existing energy might be a motor HP, it might be heating energy from oil delivery bills for a boiler replacement improvement, or it might be a chiller system's energy use calculated from electric bills.

SPECIALTY MODELS

Specialty models are available for a wide variety of energy improvements, such as variable-frequency drives (Figure 16.4), photovoltaics, combined heat and power, and the like. Independent sources such as ENERGY STAR, NREL, and CANMET provide a number of such models. Specialty models are also available from product vendors. When using vendor-provided models, check that assumptions are reasonable, and that energy savings are not being boosted with assumptions intended to make a product look good.

BILLING-BASED ISOLATED SAVINGS ESTIMATES

A powerful basis for isolated savings estimates is using billing data. If we are able to *disaggregate* (separate out) one end use from the utility bills, we can use that energy consumption for an inversion calculation of energy savings for an energy improvement to that particular end use. By *end use* we mean the heating system, cooling system, domestic hot water, or the like.

Example disaggregation might include:

Figure 16.4   Variable-frequency drive.

- When a specific form of energy is used for only one end use in a building. For example, fuel oil is frequently used only for space heating. Or often domestic hot water heating is on its own fuel. Occasionally, exterior lighting is on its own electric meter.

- Clearly separate use patterns. For example, wintertime use of natural gas for space-heating typically appears as a clear increase over and above a constant year-round base use for domestic hot water or cooking or laundry. Or summertime use of electricity for cooling appears as a clear increase over and above a year-round base use for lighting and other year-round electrical loads.

For these calculations, once the disaggregated usage is calculated from the utility bills, the savings come from the typical inversion calculation:

$$S = D \times [1 - (Ee/En)]$$

Here, S is the savings, D is the disaggregated end use, Ee is the existing efficiency, and En is the new efficiency.

MONITORING-BASED ISOLATED SAVINGS ESTIMATES

If an end use cannot be disaggregated from the utility bills, it might be possible and worthwhile to monitor an end use, either long-term or even for a short-term period, to refine an estimate of annual usage, as the basis for an isolated savings estimate. Power meters on a specific load, or on a circuit serving one single end use, can be used to monitor electricity consumption. Installing a gas meter on a specific load, such as a gas furnace or boiler, is also not difficult and is frequently a worthwhile investment. On loads with a constant capacity, such as gas furnace or boiler that does not modulate, measuring run-time instead of energy consumption can be used as a surrogate for energy consumption, and multiplied by the rated capacity, or a short-term measured capacity, for a longer-term measurement of energy usage. Such an approach with a heat pump (Figure 16.5) or air conditioner is not as valid, because energy consumption varies with load, such as varying outdoor air temperatures, for these devices. Monitoring-based estimates can be done with appliances, such as refrigerators, vending machines, and laundry machines as long as the monitoring period is long enough to capture a reasonably average usage pattern.

Figure 16.5   Heat pump.

To be clear, what we are describing is monitoring of equipment during a baseline period, in order to establish energy use of existing equipment, for purposes of estimating energy savings, which might be delivered by replacing this equipment. We are not describing monitoring for purposes of measuring *post-retrofit* savings. Post-retrofit measurement and verification (M&V) is discussed in a later chapter.

Once the monitoring is complete, and the current annual energy consumption for the particular end use has been estimated, the inversion equation is applied to calculate annual energy savings.

BIN MODELS

A type of isolated savings estimate that can account for outdoor temperature variation is called *bin modeling*. Separate calculations are performed in a spreadsheet for each of a number of *temperature bins*, or 5-degree temperature ranges, for which we have associated typical annual or monthly hours. Bin data are available from various sources.[1, 2]

The traditional approach to bin modeling is to do annual calculations. The challenge with this approach is that the results are annual, and cannot be compared with monthly billing data. Monthly bin data are now available to allow comparisons to monthly billing data.

DEEMED SAVINGS

Deemed savings are projected savings for an energy improvement that have been pre-determined, based on historical measured experience or based on standardized calculations. Deemed savings are typically applied to individual improvements. In other words, deemed savings are a type of isolated savings estimate. The advantage of deemed savings estimates is their simplicity and standardization of savings estimates. Disadvantages include inaccuracy, if building-specific conditions vary significantly from the values used in the deemed savings estimate.

Deemed savings estimates have been developed by a number of state energy programs for a wide variety of energy improvements.[3,4] Deemed savings calculation guides are a rich source of standardized calculations, as well as assumptions that can be used to support such calculations, such as annual runtimes and baseline efficiencies. They should nonetheless be used with caution, and supported by building-specific measurements wherever possible. If equations from deemed savings manuals are adapted for use in new spreadsheets, they should be carefully checked.

Deemed savings estimates, and the portfolio programs in which they are used, tend to limit energy improvement work to situations where like-for-like replacements are being considered. In so doing, deemed savings approaches can miss larger opportunities for energy improvements. Examples of opportunities that are less commonly evaluated and implemented with deemed savings approaches include right-lighting, heating conversions from low-efficiency systems and fuels (such as steam heat, fuel oil, electric resistance) to high-efficiency systems (such as heat pumps or high-efficiency gas systems), duct leakage reduction, and stack effect reduction.

BEST PRACTICES FOR ISOLATED SAVINGS CALCULATIONS

Isolated savings estimates are typically done in a spreadsheet. Best practices for such spreadsheets include:

- Show equations.
- List assumptions and limitations.

- List sources of equations, assumptions, and inputs.
- Show units.
- Show units being canceled to ensure final units are correct.
- Show constants.
- Avoid hiding constants within equations.
- Link all inputs and constants to calculation equations. Avoid any manual entry other than for inputs.
- Have a colleague review the spreadsheet, without any guidance, for clarity.

The spreadsheet should be subject to careful checking before use, and results of quality control checks should be included at the bottom of the spreadsheet. If custom savings calculation spreadsheets are not checked, mistakes can be repeated and incorrect results disseminated for years.

A helpful format is to group inputs at the top of the spreadsheet, followed by constants and assumptions, followed by equations, followed by results, and with quality control checks at the bottom.

Integrated Savings Estimates

Integrated, or whole-building, savings estimates are performed by computer programs. A widely used program is DOE2, popularized with its eQuest version. A newer Department of Energy (DOE) program is Energy Plus. The Trane Company's TRACE program, and Carrier's HAP, have both been available for decades as well. For research purposes, the University of Wisconsisn's TRNSYS has advanced capabilities, such as custom modeling of advanced control strategies. For multifamily buildings, Performance System Development's TREAT has seen widespread use, using a calculation engine developed by the National Renewable Energy Lab.

Sometimes referred to as *hourly models*, integrated energy simulation programs typically model a building's energy use through each one of the 8,760 hours of a typical year. Energy Plus has the capability to refine the calculations to fractions of an hour. Integrated models use long-term average weather data.

TRUE-UP TO UTILITY BILLS

The results of an integrated model include monthly estimates of energy consumption, for each different form of energy in a building. For energy audit work, these results can be compared to utility bills. By making this comparison, models can be *trued-up* or adjusted to match the utility bills. This provides greater confidence when then modifying the model to estimate savings from energy improvements. A best practice for truing-up is to match the models to within perhaps 5 percent, and we prefer that the total annual energy use predicted by the model not exceed the utility bills. So, we might seek true-up +0 percent/−5 percent. The total bills can be 5 percent less than the model, but cannot be more than the model. This contributes to savings predictions being conservative.

It is ideal if high-quality utility bills are available with which to true up a model. However, frequently utility bills are not available, or are poor quality, for a wide variety of reasons. A building might just have been renovated, for example. Or a new owner might have purchased a building and might not have access to prior bills. Or the occupancy and usage might have changed dramatically, making prior bills unreliable. Or bills are missing. If high-quality bills are not available for true-up, a model can still be built and used. We should not regard lack of bills as a reason not to model.

TABLE 16.1

Default Occupancy Densities

	Gross Square Feet per Occupant	
	Employees	Transients
General office	250	0
Retail, general	550	130
Retail or service (e.g., financial, auto)	600	130
Restaurant	435	95
Grocery store	550	115
Medical office	225	330
R&D or laboratory	400	0
Warehouse, distribution	2,500	0
Warehouse, storage	20,000	0
Hotel	1,500	700
Educational, daycare	630	105
Educational, K–12	1,300	140
Educational, postsecondary	2,100	150

Figure 16.6 Spreadsheets can be used for integrated energy models. However, they can grow complex and require rigorous checking to avoid mistakes.

THE IMPORTANCE OF OCCUPANCY DATA

Occupancy data form a cornerstone of good modeling, allowing us to evaluate improvements such as lighting controls, demand-controlled ventilation, and temperature setbacks. As we have indicated, the best occupancy data are hourly inventories of occupancy in a building. In the absence of good field data, reasonable assumptions should be made about occupancy. Default occupancy densities for different building types are provided in Table 16.1.[5]

SPREADSHEET-BASED INTEGRATED MODEL

Over time, many energy auditors, and some portfolio programs, build their own isolated calculation spreadsheets, to develop a collection of different savings estimates for lighting, heating, cooling, appliances, insulation, domestic hot water, and more—the gamut of common energy improvements. A natural next step is to link these in a spreadsheet-based integrated model. Billing analysis is integrated, to calculate energy rates. These energy rates are linked to each isolated model. Such a model can even provide interactive calculations for reduced loads and increased heating and cooling plant efficiency, to avoid double-counting savings. A bin model in a spreadsheet can do an adequate job of modeling heating loads. Cooling is more of a challenge but is possible. A spreadsheet model could ostensibly even be an hourly model.

This is a valid approach. However, caution must be exercised to prevent mistakes from creeping into the spreadsheet. Maintaining the spreadsheet can also become a major project of its own. These spreadsheets often become large and unwieldy. Sharing and training others on such a spreadsheet is sometimes feasible, but is usually difficult. At the very least, best practices for building, documenting, rigorous checking, and maintaining such spreadsheets should be followed. Experience has shown that, whereas a spreadsheet model can start out by appearing simple to develop and maintain, an integrated spreadsheet-based model eventually can become as complicated as a whole-building hourly model, with less documentation and less quality control. (Figure 16.6.)

TABLE 16.2

Energy Software for Integrated Savings Estimates

Software	Developer	Calculation Engine	Initial Release	Para-metric Analysis	Life Cycle Costing	Hourly	Notes
Automated Energy Audit (AEA)	Retroficiency	Proprietary	2011	Yes	Yes	Yes	
EMAT	Efficiency Mobile Audit Technology	EnergyPlus	2014	No	Yes	No	Bin model
DesignBuilder	DesignBuilder Software Ltd.	EnergyPlus	2005	Yes	Yes	Yes	
Energy Audit	Elite Software	Proprietary - bin analysis	2002 Ver. 7	No	No	No	For light commercial buildings. Bin analysis.
EnergyGauge Summit	Florida Solar Energy Center / UCF	DOE2	2000	No	No	Yes	
EnergyPro	EnergySoft	EnergyPlus for Title 24, DOE2 for other applications	1982	Yes	Yes	Yes	
eQuest	James J. Hirsch & Associates	DOE2	1999	Yes	Yes	Yes	
Green Building Studio	Autodesk	DOE2	2000	Yes	Yes	Yes	
HAP	Carrier	Proprietary	1987	Yes	No (see note)	Yes	Life cycle costing can be performed in separate eDesign Suite.
simuwatt	concept3D	EnergyPlus	2015	Yes (see note)	Yes (see note)	Yes (see note)	Availability 2016. Till then, field data collection solution is available.
TRACE	Trane	Proprietary; EnergyPlus planned for 2016	1972	Yes	Yes	Yes	
TREAT	Performance Systems Development	SUNREL	2000	Yes	Yes	Yes	For multifamily buildings.

SOFTWARE

A variety of software programs are available with which to do integrated energy modeling. A list of some programs available in the United States is shown in Table 16.2.

It is important to note that several programs are primarily used for designing new buildings, and so may have features less suited to energy audits. Energy software is also used for other purposes, such as energy code compliance, tax incentives, equipment sizing, LEED and other standards-compliance, research, and equipment or product sales. Each of these uses gives the different software programs different characteristics, and strengths and weaknesses for modeling existing buildings and calculating energy savings.

The Department of Energy lists software programs that have self-certified to comply with requirements for the U.S. federal 179d tax deduction. ASHRAE Standard 90 also has requirements for energy modeling software.

Energy software can be somewhat limiting in the energy improvements they support. Some software are stronger in envelope improvements, and weaker in heating/cooling or lighting improvements, and vice versa. Few software programs fully analyze specialty systems such as commercial refrigeration or appliances.

Some software will allow improvements such as changing window energy characteristics, but will not allow changing the window area of a building, for example to allow an improvement of removing windows. And so many energy improvements require work-arounds, or require isolated savings estimates. It is not uncommon to combine isolated savings estimates with integrated savings estimates.

CHALLENGES AND RISKS OF INTEGRATED SAVINGS ESTIMATES

Integrated savings models require training in modeling, and experience in building energy systems. They can also be time consuming. The complexity of the programs can lead to incorrect assumptions, resulting in incorrect savings estimates. Calculations and assumptions can be hidden, or can be missed. Quality control of models, such as checking of inputs, is important. Integrated modeling also requires that attention be directed to developing a written scope of work that matches what was modeled. The building owner or energy manager or installation contractor will not have access to the computer model, in order to know what work needs to be done.

A recent development in integrated savings estimates might be called *rapid modeling*. On the basis of limited information, such as square feet of a facility obtained by Internet maps or utility bills, integrated savings estimates are generated. This approach is fraught with risk, because it is replete with assumptions, and cannot identify the different energy anomalies that are found in just about every building. Every building is unique. Remote rapid energy modeling cannot identify actual equipment in buildings, control set points and sequences, equipment deficiencies, occupancy schedules, equipment schedules, equipment efficiencies, human factors that are building-specific, and more. Rapid modeling should be approached with caution, with the risk of producing rapidly incorrect savings estimates.

BEST PRACTICES FOR INTEGRATED SAVINGS ESTIMATES

With experience, integrated savings estimates using whole-building models can be both fast and accurate. These can be achieved using best practices:

- Develop standards, including standard inputs in which you have confidence.
- Carefully check both inputs and outputs.
- Perform reality checks against spreadsheets and other isolated calculations.
- Try to understand all inputs well.
- Try to understand software assumptions and limitations well.

Information Sources for Estimating Energy Savings

As part of estimating savings, we often need to gather information on products, both the existing products that were found in the building, and on potential high-efficiency replacements.

A rich source of product data is equipment catalogs, and installation instructions, today readily available online, which provide ratings such as capacity, power consumption, fuel input, efficiency, default set points such as temperatures, and control settings such as off-delays for motion sensors. Of particular interest is the performance of equipment at what we call *off-design* conditions: The efficiency of a heat pump at different outdoor temperatures, the efficiency of a boiler at different return water temperatures, the actual power draw of a light fixture rather than its rated power draw, the power draw of a dimming ballast at partial light output or zero light output, and more. Manufacturers can also be contacted for product information, either directly or through local manufacturer's representatives.

Established manufacturers frequently produce highly useful technical manuals that go beyond manufacturer-specific information. Examples include an excellent summary of insulation for masonry and steel walls in the Owens Corning *Commercial Complete Wall Systems* (2011) by Herbert Slone, in-depth analysis of boiler efficiency issues in the Cleaver Brooks *Boiler Efficiency Guide* (2010), and others. Product distributors are also good sources of energy information. An excellent hard-copy catalog by W.W. Grainger Inc. has been replaced by an equally useful online catalog.

Online efficiency rating directories are helpful for new equipment energy use, including AHRI, AHAM, ENERGY STAR, the California Energy Commission, and the Consortium for Energy Efficiency (CEE). Some of these directories include all federal-compliant equipment, and others (for example, ENERGY STAR) are limited to high-efficiency products.

Summary

In closing this discussion of modeling, we recognize the power of modeling to give us insight into buildings, and help us to prioritize improvements. We also need to recognize the limitations of modeling. Like benchmarking, modeling cannot tell us about a building, unless we visit the building and inspect it up close. Benchmarking and modeling share a risk, and that is to imply that we can understand a building only from our desks. We need to get up from our desks, and get into buildings, in order to understand them. If we go into buildings, and if we take good measurements, and if we probe and ask the right questions, we can get the information necessary for excellent modeling, and for excellent understanding and prioritizing of improvements. If we do not go into buildings and ask enough questions, our models risk being useless.

On the other hand, a good model is more than just an energy audit tool. It is a capital planning tool. It is a tool that can stay with a building over time, be revised as the building is improved and transformed. This is the reason that we suggest that contracts with energy auditors require that the building model file become the property of the building owner. A good building model can be updated and revised by different consultants. Like a good set of construction drawings, the building energy model is a valuable asset that can be the foundation of a building's transformation.

References

1. https://av8rdas.wordpress.com/2012/03/21/bin-weather-data-for-the-united-states-and-international-locations/. Accessed August 6, 2015.
2. http://web.utk.edu/~archinfo/EcoDesign/escurriculum/weather_data/weather_data_summ.html. Accessed August 6, 2015.
3. Public Utility Commission of Texas, Texas Technical Reference Manual, Version 2.1, Volume 3: Nonresidential Measures, Guide for PY2015 Implementation, January 30, 2015.
4. New York Standard Approach for Estimating Energy Savings from Energy Efficiency Programs – Residential, *Multi-Family, and Commercial/Industrial Measures, Version 3*, Issue Date - June 1, 2015, Effective Date – January 1, 2016, New York State Department of Public Service, Albany, New York.
5. http://www.usgbc.org/sites/default/files/CS20Default20Occupancy20Appendix.pdf. Accessed November 14, 2014.

Chapter 17

Financial Aspects of Energy Improvements

Energy improvements have financial components: They reduce operating costs, require investment, and increase building value.

Energy Cost Savings

Energy cost savings refers to the annual savings from energy improvements. Note the minor difference in phrasing between *energy cost savings*, which refers to dollars per year, and simply *energy savings*, which refers to the annual energy savings, in units of energy. For example, a building might see *energy savings* of 100,000 kWh/year from lighting improvements, with *energy cost savings* of $10,000/year.

Energy cost savings are sometimes referred to as *first-year savings* because they are calculated based on current energy rates, and we expect energy rates to change over time.

Energy cost savings involves multiplying the annual energy savings by the energy rate. In the above example, 100,000 kWh/year was saved at a rate of $0.10/kWh to obtain energy cost savings of $10,000/year.

There can be challenges in determining the rate. A simplified approach is to divide energy use by energy cost, from utility bills. For example, if a tenant in an apartment building paid $100 last month for 1,000 kWh of electricity, the rate might be calculated as $0.10/kWh. However, the monthly cost includes some fixed costs, such as meter charges. Therefore, dividing by the total monthly cost will overestimate the rate. In this example, if the monthly meter charge were $20, the rate at which energy cost savings accrue should really be calculated as ($100 − $20) / 1,000 = $0.08. If we based savings on the simpler calculated rate of $0.10/kWh, we would be overestimating energy cost savings. Other complexities arise if the rate structure is a *declining block rate*, in which the first x units of usage are billed at one rate, the subsequent y units are billed at a lower rate, and so forth. In this case, we are interested in impact of energy savings, which typically affects the lowest block rate. We call the last rate, the one on which energy cost savings are based, the *incremental rate*.

Cost savings must also separate out the effect of *demand cost savings*, which refers to the annual savings from reduced peak energy demand. Peak demand is typically measured in 15-minute increments. Reducing demand is helpful to utilities in order to reduce their peak capacity. Demand cost savings, are an important component of cost savings. Demand reduction generally has little effect on carbon emissions reductions, other than impacts on short-term peak use of dirtier power plants. However, demand reductions are usually accounted for in cost savings calculations, and are an added motivation for investment aspects of energy improvements.

Figure 17.1   To estimate energy cost savings from energy savings, we need to calculate the incremental cost of energy, in other words, the energy rate at which the last (incremental) unit of energy is purchased.

Maintenance Costs

Many energy improvements reduce maintenance costs. LED lamps are estimated to last for 50,000 hours, far longer than the rated life of 10,000 hours for fluorescent lamps, or 1000 hours for incandescent lamps, significantly reducing lamp replacement material and labor costs. Replacing steam heating systems, with hot water systems or heat pumps, eliminates the need to inspect and replace steam traps. Right-lighting a building by removing lamps or light fixtures in overlit spaces will permanently reduce maintenance costs: removed lamps or fixtures will never need lamp replacements again. Automated energy management systems can reduce the time required to manually turn equipment on and off. Automated control systems can also provide alarms in advance of equipment failure, preventing premature failure.

Maintenance cost savings can be treated as an added motivation for energy improvements.

Some energy improvements increase maintenance costs. For example, some sealed-combustion heating equipment has air filters that require cleaning. Renewable energy systems, such as photovoltaic and solar thermal systems, add to a building's maintenance costs. In those situations where energy improvements are likely to add to annual maintenance costs, we must be honest and transparent about it.

Anticipated maintenance cost impacts can be accounted for in financial analysis, where reasonable estimates of the impact are possible, such as with lighting replacements. The American Society of Heating, Refrigerating, and Air-Conditioning Engineers (ASHRAE) has introduced an online maintenance cost database.

Cost Estimating

The *installation cost* is the cost of installing an energy improvement. It is often alternately referred to as the *installed cost*, *implementation cost*, *first cost, capital cost*, or *construction cost*. It is helpful to not refer to installation cost as simply the "cost" because it can easily be confused with the annual energy cost savings of an improvement.

Estimating the installation cost of energy improvements is an important aspect of both energy auditing and energy management. We often give more attention to the projected energy cost savings, but the installation cost has as much influence on metrics such as the payback or life-cycle cost and whether a project happens or not.

While cost estimating is a necessarily approximate endeavor, there are nonetheless ramifications if we estimate significantly too high or too low. If we estimate too low, this can create problems during the project, as bids come in higher than expected, and the building owner is disappointed. It can even cause the termination of a project if it is too far over budget. This can also divert resources from better energy improvement investments. On the other hand, if we estimate too high, we can prematurely eliminate a project from consideration that might well have been a good investment.

A common cost estimating method is illustrated in Table 17.1.

While perfect installation cost estimating is not possible, a variety of best practices can minimize the risk of major overestimation or underestimation. These include:

- Use a structured and consistent approach.
- Document assumptions, including sources of information.
- Adjust labor and material costs for local conditions, including geographical costs, local labor costs, market conditions, etc.
- Account for sales tax on materials if the client is for-profit and if the project is subject to sales tax.

TABLE 17.1

Cost Estimating Example

Heat Pump Water Heater, 80 gallons

	Date:	12/21/14						
	Labor rate:	$60						
	Overhead and profit:	40%						

Item	Description	Unit	Quantity	Material Cost	Labor Hours	Total	Notes
1	Demolition/removals	EA	1	$0	2.0	$168	Rough estimate.
2	Heat pump	EA	1	$1,050	2.0	$1,638	Place water heater.
							Source: Vendor quote.
3	Piping connections	EA	2	$50	0.5	$224	Material—misc piping.
4	Insulation	LF	8	$2	0.2	$157	Source: Means Mechanical 2014.
5	Condensate drain	LF	10	$2	0.1	$112	Source: Grainger online.
6	Electrical hookup	EA	1	$50	3.0	$322	Source: Estimate from local electrician.
	Contingency					20%	
	Total					$3,145	

The Unit is the unit of measure. Examples include EA for "each," LF for "linear feet," and SF for "square feet."

The Quantity is the number of items. In this example, one water heater is being removed, one heat pump water heater is being installed, two piping connections are being made, 8 linear feet of insulation is being installed, 10 linear feet of condensate drain is being installed, and one electrical hookup is required. Material Cost is the cost of the material per Unit.

- Allow a contingency.
- Use recent cost sources.
- Save installed cost estimates for future reuse, but be sure to date them to ensure that they are not out of date if reused.
- Account for prevailing wage requirements, if applicable.
- Account for demolition/removals.
- For large equipment, account for special provisions to bring equipment indoors.
- For equipment on rooftops, account for the cost of crane rental or helicopter placement.
- For large projects, account for general requirements such as trailer rental, staging, material storage and protection, temporary power if required, etc.
- For work on ladders, above ceilings or overhead, in crawlspaces, or in other locations where access is difficult, allow a contingency for access.

A widely used source of information for installed cost estimating is the R.S. Means cost estimating handbooks. There are also useful online sources.[1, 2] Other sources include estimates from vendors, for large equipment, or online distributors.[3–5]

Financial Metrics

PAYBACK

The simplest financial metric for energy improvements is the *payback*. We divide the investment cost by the annual energy cost savings to obtain the payback, in years. The payback is simple to calculate, is simple to explain, and is already understood by most people, and so allows immediate conversations about the attractiveness or unattractiveness of a particular energy improvement. However, there is significant concern in the energy community about the payback not accounting for the expected life of the improvement. Changing a halogen display track light in a retail store to an LED track light, operating 12 hours per day, might have a payback of 1.5 years, but only has an expected life of 11 years. It also brings a risk of subsequent replacement back to a lower-efficiency light, in which case the savings are lost. Air sealing an elevator shaft vent might have the same payback of 1.5 years, but offers an expected life of 50 years or more, likely for the remaining life of the building, with no risk of replacement or loss of savings.[6] Two different improvements with the same payback might offer very different long-term energy savings, expected life, reliability of savings, freedom from maintenance, and other benefits.

RETURN ON INVESTMENT

By inverting the payback, instead dividing the annual savings by the installation cost, we obtain the *return on investment* (ROI). ROI is another simple and widely understood metric. Like payback, the return on investment does not account for the duration, or expected life, of an energy improvement. This is a major limitation.

An alternative to the simple ROI is analysis that accounts for the residual values of investment, the time value of money, taxes, and other factors. Energy improvement investments do not return capital at the end of the investment period, other than added value to a building. This speaks in favor of infrastructure energy improvements, such as insulation, which stay with the building for its life, rather than only investing in consumables, such as lightbulbs.

LIFE-CYCLE COSTING

Life-cycle costing is a form of economic analysis that accounts for the costs of an investment over its life. It can more accurately capture the aggregate costs over time, from the initial investment, to the annual energy cost savings, to maintenance cost impacts, and residual value of the investment in the future, if any. It can also account for the time value of money, and for the impact of inflation on energy rates. Several common life-cycle costing metrics include:

> **Net present value**. The value of the investment when its cost and benefits are translated into a present value, using first principles of economics. The net present value is the difference between net present expenditures and net present savings. If we want a project's net present value to be financially justified, it should be greater than 0.

> **Savings-to-investment ratio (SIR)**. The ratio of net present value of savings to net present value of expenditures. If the only expenditures are the initial investment, as is commonly the case, the SIR is equal to the net present savings divided by the initial investment. If we want a project's SIR to be financially justified, its SIR should be greater than 1.

> **Cost-benefit ratio**. The inverse of SIR: The initial investment divided by the net present value of the savings.

The National Institute of Standards and Technology (NIST) has developed a widely used set of tools for life-cycle costing.[7]

Benefits of life-cycle costing analysis include a more solid financial comparison between energy improvements, accounting for expected lifetimes, and other factors such as inflation. Disadvantages include complexity and a terminology that may not be familiar to building owners. However, experience has shown that owners are quick to learn concepts such as SIR, when presented with the requirement goals, such as SIR needs to be greater than 1.

Life-cycle costing metrics are also highly sensitive to such factors as the anticipated lifetime of energy improvements. And, whereas some energy improvement lifetimes are fairly predictable, such as lightbulbs, others are not. How long will an efficient window last? How long will roof insulation last? In many cases, these infrastructure improvements may well last 100 years into the future. Can we reliably estimate the difference between 20 years and 50 years and 100 years, as the expected life of an improvement, or even as the expected life of a building?

COMPETING INVESTMENT

Implicit in any financial discussions of energy improvements (whether payback, return on investment, or life-cycle costing) is the concept of *competing investment*. People are often interested in what the best investment might be for their financial resources. Presenting an energy project as a financially attractive project seeks to contrast a specific possible energy improvement investment with all possible *competing investments*. This is a limiting way to describe an energy improvement, and may do as much harm as good. If we present an energy project only as a project that is justified by hoping to aspire to be the best possible investment, in a competition with all other possible investments available to an individual or an organization, we are placing much faith in the unlikely possibility that the energy improvement project is indeed the best possible competing investment. Energy improvement investments are and should be justified by all their benefits, and not just as the best competing investment. Energy improvements are an investment in our building stock, the most valuable things we own, and in our health and safety and comfort. Energy improvements are an investment in our environment, in the very quality and viability of our lives on the planet. Energy improvements just happen to also offer tangential financial benefits in the form of energy cost savings. By limiting the conversation about energy investments to their return on investment, we miss their crucial environmental and building-transformation benefits, and so can prematurely quash their consideration.

Figure 17.2 Energy improvements should be justified by all their benefits, and not just as a financial investment. They are investments in our buildings, health, safety, comfort, and the environment.

Expected Useful Life of Improvements

The *expected useful life* (EUL) of improvements, also called the *equipment service life*, is an important metric for several reasons:

- To compare two competing energy improvements
- To assess if an improvement will "pay for itself" in a time period shorter than its expected life
- To support life-cycle costing
- To assess if an existing building energy system is going to need replacement, anyway, for non-energy reasons, in the future. This can support justification of an energy improvement, likely at a lower incremental cost than replacing it immediately.

The expected useful life of a class of equipment is best developed through statistical analysis of large samples. Imagine a sample of over 1,000 heat pumps. Other than some random premature failures, almost all of the units last for 15 years. At 15 years, some of the units start to fail, requiring replacement. After 35 years, all of the units have failed and have been replaced. Between 15 years and 35 years, the units randomly fail one by one. It is found that, at 25 years, half the units have failed. This is called the *median service life*. Statistically, we would then expect half of the heat pumps to fail before 25 years, and half after 25 years. The median service life of a statistically significant sample is an excellent predictor of the expected useful life of this class of heat pumps. This type of analysis is called survivor curve analysis. Survivor curve analysis has been performed for a few classes of energy equipment. More such analysis is needed, for a broader spectrum of energy improvements. Historically, expected useful lives have not been based on statistical analysis. Research has found that widely used expected lives, for several classes of heating and cooling equipment, underestimate actual expected useful lives.

A table of expected useful lives is provided in Appendix Q. ASHRAE has launched a database to track the life of a variety of heating and cooling products.[8] Results already are statistically significant for some equipment classes, and have been included in Appendix Q.

Cost Control during the Energy Audit Phase

A variety of strategies to control cost are available during the energy audit phase. In other words, strategies can be used to reduce the eventual installation cost, and so make improvements more cost-effective, and more likely to meet financial or programmatic criteria for acceptance.

Where redundancy is built into an existing system (e.g., lead/lag pumps, lead/lag boiler, in other words, two identical products serving a single function), we can consider replacing only one of two, making the new high-efficiency component the lead component, and using the redundant component as lag (when needed only), as a backup. Significant cost savings can be achieved, while delivering most or all of the savings if both components had been replaced.

If an energy improvement is too costly to justify, it may be worth justifying as an end-of-life improvement. For example, if it is too costly to justify replacing insulation on a flat roof, but the roof itself is expected to require replacement within a few years, one can evaluate the improvement as an *incremental* improvement, to be installed when the roof is replaced. Here, we spell out clearly in the energy audit that the evaluation was done on an incremental basis, on the basis of the *added* cost *only* of the increased insulation. The same approach can be applied to most improvements, such as boiler replacements, chiller replacements, adding insulation under old siding or cladding of a building, adding insulation to the interior of an exterior wall that is planned to be painted in the future, for appliance replacements, and more. Here, the full power of an energy audit, as a capital planning tool, comes into force. By providing the incremental installed cost, the energy savings, and the expected remaining useful life of a building energy system, the energy auditor has given the building owner a clear path to affordable energy savings, even if it is at some time in the future. Conversely, if this evaluation is not done, a good future opportunity for energy savings is lost.

In some cases, removed equipment has scrap value, in particular heating/cooling/domestic hot water (DHW) equipment and appliances, which have much steel and copper. The scrap value of equipment can be taken as a credit against the installed cost.

In the case of newer buildings that are overlit, the existing light fixtures may be relatively high in efficiency. If right-lighting is planned, with fixture or lamp removals, then the removed equipment can possibly be reused elsewhere in the building, or sold, and its value taken as a credit against the installed cost. See Figure 17.3. Caution should be exercised to avoid promoting re-use of inefficient equipment.

Figure 17.3 Removed light fixtures in buildings where the lighting is already efficient may be re-used elsewhere.

Funding and Financing

Energy improvements can be funded in a variety of ways.

If capital reserves are available, direct owner funding is an option. If planned over a series of years, smaller energy improvements can also be funded directly through operating expenses.

Financing through a loan is also an option. Challenges may arise if the term of the loan is not long enough to generate an immediate positive cash flow. Long-term financing, such as through a mortgage, can provide positive cash flow, although higher long-term financing costs.

A variety of grants and incentives are available, from government agencies, most commonly state and federal, and through electric and gas utilities. These take any number of forms, including rebates, tax incentives, low-interest financing, performance incentives, and more. These incentives are typically not set up to pay for the entire installation, but rather as an outside motivator to reduce energy usage or to reduce peak demand.

Third-party financing is available, especially for larger projects, through performance contracting. In this model, energy service companies (ESCOs) bring financing, along with project management, and share in energy cost savings.

Energy improvements present an unusual opportunity for fund-raising for not-for-profits. The individual donor to a capital campaign for energy improvements will see each dollar matched in multiple ways:

- Federal and state tax deductions for charitable contributions.
- Possible employer matching contributions.
- Future energy cost savings for the not-for-profit. The donor is enabling a long-term reduction in operating costs. In other words, the donor is giving the gift of years of future energy cost savings.
- Government and utility incentives.

For example, a not-for-profit seeks to do $20,000 in energy improvements. The local utility will contribute 25 percent of the project cost, or $5000. The not-for-profit conducts a capital campaign and obtains $10,000 in a charitable contribution from a donor. The not-for-profit contributes $5,000 of its own capital reserves. The energy improvements have a 10-year payback and a 20-year expected useful life. From the donor's perspective, the contribution of $10,000, after federal and state tax deductions at 25 percent, effectively cost $7,500. For their $7,500 effective contribution, the donor saw $12,500 in matched energy improvements completed, plus an additional $40,000 in energy cost savings over 20 years, for a total of $52,500 in matched benefit to their not-for-profit of choice. The $7,500 contribution was leveraged 700 percent.

For donors, energy improvements may represent the single best way to support their not-for-profit of choice. For not-for-profits, energy improvement projects may be the single best way to leverage donations, while reducing their operating costs and serving as stewards of the environment.

Impact of Energy Improvements on Building Value and Rental Premiums

A number of national studies have shown that green buildings see both premium prices, when sold, and rental premiums. Three studies have shown sales premiums of ENERGY STAR commercial buildings in the United States to range from 9 percent to 25 percent. Rental premiums of ENERGY STAR and LEED buildings, in five different studies, were found to range from about 2 percent to 27 percent.[9] Premium appraisal value can be justified through higher net operating income, based both on lower operating costs and on higher rents, and higher occupancy and tenant retention rates.

References

1. http://www.get-a-quote.net/. Accessed December 28, 2014.
2. https://www.swiftestimator.com/. Accessed August 16, 2015.
3. www.grainger.com. Accessed August 16, 2015.
4. www.mcmaster.com. Accessed August 15, 2015.
5. http://www.remichel.com/. Accessed August 16, 2015.
6. Steven Winter Associates and the Urban Green Council, "Spending through the Roof." Report for NYSERDA, 2015.
7. http://energy.gov/eere/femp/building-life-cycle-cost-programs. Accessed August 15, 2015.
8. http://xp20.ashrae.org/publicdatabase/. Accessed August 15, 2015.
9. Institute for Market Transformation, in collaboration with the Appraisal Institute, "Green Building and Property Value: A Primer for Building Owners and Developers," 2013.

Chapter 18

Reporting

Energy Audit Reports

Energy audit reports should include the following sections:

- Summary results, including for each evaluated improvement: annual energy savings, electric demand reduction, annual energy cost savings, installation cost, simple payback, life-cycle metric, and expected life.
- Building and energy system descriptions. These descriptions are important in order to allow quality control review of the energy audit by others. The descriptions do not need to be in written-out text - Outline or checklist descriptions are adequate. At a minimum, building descriptions should include estimated insulation R-values, description of the thermal boundary and deficiencies, description of unconditioned spaces and deficiencies, estimated window U-factors, estimated heating and cooling plant efficiencies, heating and cooling distribution system type and condition (including pipe and duct insulation, and visual observations on duct sealing or air leakage), domestic hot water (including heater, distribution, water temperature, and fixtures), room-by-room lighting survey, other electric loads (e.g., transformers on the building side of the electric meter), plug load survey, and other gas loads (such as clothes dryers and stoves).
- A list of energy improvements already (previously) implemented in a building. This serves to rule out these improvements for new/repeated evaluation, and also gives credit to the building owner for already directing attention to energy issues. It sets a positive tone for the report.
- Improvement workscopes (descriptions). These might include:
  - Boilerplate text that describe each improvement generally, without any text that is specific to the building in general.
  - Building-specific text. Here, the applicability of the improvement to the specific building is described, as well as the building-specific scope of work.
- Methodology for energy savings calculations. Either the equations used (e.g., in spreadsheet form) or a summary of other tools used (such as whole-building energy models), along with printouts of input and output reports. Also, assumptions and limitations of energy savings estimates should be included for each improvement.
- Billing analysis, including benchmarking, calculation of energy rates, and summary of bills.
- Analysis—fraction of energy use, by type of energy, for each major end use. For example, electricity might be divided into the percentage used for lighting, air conditioning, plug loads, large hard-wired loads such as elevators, and other loads. Natural gas might be divided into the percentage used for heating, domestic hot water, cooking, and other loads.

Figure 18.1   A whole-building photograph is a nice touch, on the cover of energy audit reports.

Figure 18.2   Use photographs to document important deficiencies, not only to justify an improvement, but to clarify where the improvement must be applied.

Optionally, reports might include a section on further resources, and program-specific resources, such as incentives, for energy audits done under state or utility programs.

A photograph of the building on the front cover of the report is a nice touch (Figure 18.1). Front covers should be dated, and should include the name of the facility and report author, and the portfolio program name and sponsor, if any.

Photographs of energy aspects of the buildings are also a nice touch but should not detract from the substance of the report. Photographs should focus on conveying important deficiencies, such as infiltration sites, that could help to describe the importance of the recommended improvements, problems that the improvements will solve, and locations that need attention when implementing improvements. See Figure 18.2.

Writing Strong Workscopes

Workscopes are written improvement descriptions, intended to guide subsequent design or installation. Four best practices in writing strong workscopes may be summarized by the acronym "LEFT." These include:

L: location
E: efficiency
F: features
T: testing

The *location* of the improvement is essential for people to understand not only where the improvement is located, but also the quantity. For example, the location for a boiler replacement might be described as "Replace the existing boiler in the first floor mechanical room with two new boilers. Remove the existing boiler and dispose off-site in accordance with local solid waste regulations."

The *efficiency* of the improvement refers to the minimum required rated efficiency, in order to deliver the energy savings predicted in the report. Without specifying the required efficiency, the owner may mistakenly accept a lower-efficiency product, inadvertently chosen by a design professional or substituted by the installing contractor. Examples might include boiler thermal efficiency, chiller integrated part-load value (IPLV), motor National Electrical Manufacturers Association (NEMA) premium efficiency, insulation R-value, or window U-factor.

The *features* include energy-related aspects of a product that may not be captured in the rated efficiency. Examples might include requirements for an outdoor reset control for a boiler replacement (and the proposed set points), seven-day programming capability for a replacement thermostat (and the proposed set points), a variable-speed brushless permanent magnet fan motor on a high-efficiency replacement rooftop unit, a requirement for taping rigid insulation with a specific robust tape, a requirement for a maximum one-minute off-delay for a lighting motion sensor, and the like.

Finally, the *testing* requirements describe the required testing for an energy improvement. Examples might include boiler efficiency testing, airflow measurements for a replacement air handler, and current measurements for a replacement high-efficiency motor. Testing requirements can also encompass documentation and training.

Prioritizing Improvements

Many energy audits prioritize improvements, on the basis of energy savings, annual cost savings, payback, or life-cycle costs. State or utility portfolio programs frequently restrict energy audits to recommending improvements on the basis of a specific criterion, such as a maximum payback. Such restrictions are increasingly seen as an artificial limitation on energy audits, preventing building owners from considering a full range of possible improvements, such as long-payback improvements that may be of interest for non-energy reasons.

Interesting questions arise in prioritizing improvements when *interactions between savings* are accounted for. Interactions between savings occur when two or more improvements have interactive physical effects, such that the sum of the savings is less than the sum of the savings if the improvements were made on their own. For example, installing a high-efficiency heat pump might save 10 percent of the heating energy in a building, and adding roof insulation might save 15 percent of the heating energy. However, the reduced load of the roof insulation means a lower load to which the higher heat pump efficiency would be applied, so the total savings from implementing both improvements will not be 25 percent, but rather will be less than 25 percent. When multiple interacting improvements are considered for a single building, for example, plant efficiency plus control improvements plus envelope improvements, the impact of the interactions can be increasingly significant.

Further, of interest is how to present and prioritize the interacted savings. Consider three interacting improvements. Improvement A, on its own, is projected to deliver 10 percent savings, improvement B will deliver 15 percent savings on its own, and improvement C will deliver 8 percent on its own.

Now, consider if two of the three improvements are implemented. Modeling shows that A + B will deliver 22 percent savings, B + C will deliver 20 percent savings, and A + C will deliver 15 percent savings. Or, if all three are implemented, A + B + C will deliver 26 percent savings

How should the results be presented in the energy audit? Should the results be presented in order of highest-to-lowest individual savings (B, A, C, all separately)? If so, the owner will not be able to see the total (reduced) savings from implementing the full package of savings. Should the results be presented simply as a single package of savings (A + B + C)? If so, the owner will not know the savings by implementing a smaller number of improvements, if the full package cannot be afforded. Should the savings be presented in a series of larger and larger packages, in order of highest-to-lowest package savings (B, A + B, A + B + C)? These decisions will affect how much work goes into the energy audit, as well as the options available from which the owner can choose. They also affect how the energy modeling is done. One approach has been to present the full package, and then to withdraw one improvement from the package, to show the results of the second-most attractive package, withdraw a second improvement to show the third-most improvement package, and so forth.

Figure 18.3   Improvement descriptions in energy audits must contain information to allow effective energy improvements: location and quantity, efficiency, features that impact energy use, and test requirements.

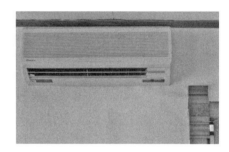

Figure 18.4   Prioritizing energy recommendations is a fundamental purpose of energy audits.

Quality Control

Quality control forms the basis of good quality energy audit reports, which in turn form the basis of good quality energy work. Without quality, an energy audit can deliver low savings, or zero savings.

Mistakes and other problems are common in energy audits. A study of energy audits found the following problems[1]:

- Missing improvements (80 percent of audits).
- Weak improvement scope (77 percent of audits).
- Improvement life too long or not provided (73 percent of audits).
- No life-cycle costing (73 percent of audits).
- Low (or missing) installed costs (60 percent of audits).

- Poor building description (60 percent of audits).
- Inadequate billing analysis (57 percent of audits).
- Overestimated savings, defined as savings more than twice as high as reasonable (53 percent of audits).
- Inadequate review, defined as three or more obvious mistakes (30 percent of audits).

Quality control should happen at several steps. The author of an energy audit report should do a first review. A supervisor should do a second review. External review should be performed by program administrators if the energy audit is part of a portfolio program. Finally, the energy manager should do a quality control review on receipt of the report by the client.

INTERNAL QUALITY CONTROL

Internal quality control is done by the report author, in the process of preparing the report, and preferably also a supervisor, after the draft audit is complete.

Best practices for internal quality control include what is referred to in the field of quality control as *defect prevention*, in addition to *defect detection*. In defect prevention, processes are put in place to prevent mistakes, rather than only seeking to detect mistakes after the work is done.

One strategy for defect prevention is for the energy auditor to prepare the building description, as well as a list of improvements for evaluation, immediately after the site visit is complete. This can form the beginning of the report. The building description and list of improvements are then reviewed internally by a supervisor. This strategy prevents good potential improvements from being missed, the number one problem in energy audits.

Another strategy that falls under defect prevention is to develop internal standards, such as calculation procedures, template workscopes, improvement expected lifetimes, life-cycle costing, and templates for building descriptions. Such standards can prevent many of the common problems in energy audits. However, it is vital that attention is directed to preventing templates from themselves carrying mistakes into an energy audit report. Templates should not be prior versions of documents from other energy audits. The risk is too high of mistaken information being carried into the new energy audit. Templates should be clearly templates, with any building-specific information needing to be entered for each new energy audit.

Another best practice is to calculate the fraction of savings that each improvement represents, and the fraction represented by all the improvements together, relative to the entire energy use of the building. This can also be done for each improvement relative to its specific type of energy (electricity, natural gas, etc.), and the fraction represented by each end use (heating, cooling, lighting, etc.). These percentages can prevent the overestimation of energy savings, another common mistake.

EXTERNAL QUALITY CONTROL

Those individuals providing external quality control, for example, portfolio program administrators with state energy agencies or utility energy programs, should review energy audit reports looking for the same common problems: missed improvements, unclear descriptions of the recommended improvements, mistakes relating to expected improvement lifetime, underestimated installation costs, inadequate building descriptions, inadequate billing analysis, and overestimated savings.

In addition, they should check for compliance with any portfolio program rules. Specific attention should be directed to gaming, the intentional tweaking of an energy audit in order to take advantage of portfolio program rules, typically overpredicting savings in order to maximize program incentives. Gaming can be identified by examining energy audit assumptions, the inputs used in energy models. Common approaches to gaming include underestimating the energy efficiency of existing equipment, overestimating existing energy usage of individual building components (e.g., hours of use), and overestimating the energy efficiency of proposed equipment.

Portfolio program administrators can also contribute to the defect prevention effort. Their contributions to defect prevention can include standardization of important elements of their program:

- Standardization of program rules.
- Standardization of energy audit report format.
- Standardization of calculations.
- Standardization of energy calculation assumptions.

Quality control by an external organization can also serve as a vehicle for training, which is especially useful for newer energy auditors or new energy audit firms. As such, external quality control can further contribute to the defect-prevention effort.

CLIENT QUALITY CONTROL

Even if clients, such as energy managers, are not trained in energy audits, they can and should still perform quality control on energy audit reports. Although the calculations may be somewhat technical, the *measurements* and *assumptions* are generally not technical, and form the underlying basis for the energy savings estimates. The client knows the building better than the energy auditor, and can help to ensure that measurements and assumptions are correct.

Examples of questions to ask the energy auditor include:

- What assumptions were made for each improvement evaluated?
- What occupancy schedules were used for lighting, ventilation, and indoor temperature calculations?
- What energy efficiencies were assumed for existing equipment that is targeted for replacement?

Reference

1. I. Shapiro, "Ten Common Problems in Energy Audits." *ASHRAE Journal*, February 2011, pp. 27–31.

Chapter 19

Sector-Specific Needs and Improvements

Different sectors of commercial buildings have different needs, characteristics, and energy improvements. We examine a few examples of commercial building sectors, recognizing that every building is different and requires site-specific evaluation.

Multifamily

Multifamily buildings find themselves squarely between residential buildings, which they are, and commercial buildings, which they also are. A high-rise multifamily building with a chiller, cooling tower, boiler, four-pipe fan coils, and central ventilation system is certainly a commercial building. And a two-story townhouse duplex appears and behaves, energy-wise, very much like a single-family building.

Low-rise residential buildings find themselves governed primarily by residential standards, contrasted with high-rise residential buildings. For example, low-rise multifamily buildings fall under the American Society of Heating, Refrigerating, and Air-Conditioning Engineers' (ASHRAE's) 62.2 residential ventilation standard, rather than ASHRAE 62.1, which is used for commercial buildings and residential buildings above and including four stories above grade. Likewise, low-rise multifamily buildings fall under ASHRAE's 90.2 energy standard, while ASHRAE 90.1 covers high-rise residential buildings above and including four stories above grade.

There are other important differences between low-rise and high-rise multifamily buildings, and their energy characteristics. Most low-rise buildings have attics, and many associated energy losses, while high-rise buildings with attics do exist, but are fewer. High-rise buildings suffer in a more pronounced way from stack effect, whereas low-rise buildings also see stack effect, just less of it. High-rise buildings typically have more *common areas*, or areas outside of dwelling units, and associated lighting and other common-area energy use: lobby, mail room, stairwells, corridors, laundry rooms, mechanical and electrical rooms, janitor's closets, trash chutes and trash rooms. High-rise buildings are more likely to have elevators. And if low-rise buildings have elevators, they are more likely to be hydraulic elevators, rather than the traction elevators used in high-rise buildings. Many low-rise buildings use boiler systems for heating, but many also use other types of heating, such as forced-air furnaces and electric resistance heat. The majority of high-rise buildings use boiler systems for heating.

In multifamily buildings, we see issues related to affordable housing. Most towns and cities across the country have a public housing authority, usually operating semi-autonomously, but working under rules of the Department of Housing and Urban Development (HUD). These rules include energy-related requirements, such as mechanisms for reimbursement of energy costs and, more recently, ways in which local housing authorities can invest in energy improvements with appropriate adjustments to energy reimbursements, such that they are not penalized for these investments. Not all affordable multifamily housing falls under public housing authorities. Much of it is privately owned, or owned by other not-for-profits, or government entities. Broadly speaking, affordable multifamily housing has been a highly energy-conscious and proactive sector, for energy improvements, in some ways leading the country in multifamily energy innovations, both technically, and in mechanisms for financing energy improvements, and more.

For energy assessments, multifamily buildings have the convenience of repeating elements of the buildings: Multiple apartments, each with a kitchen, each with bathrooms, each with living rooms and bedrooms, and each with an entry area. These repeating elements can make energy work easier. However, we must be sure to account for variations in these repeating elements. We cannot get by with just a single-apartment inspection in evaluating energy use in a multifamily building.

Multifamily buildings, like all landlord/tenant buildings, are faced with challenges of what has come to be called the *split incentive*. In buildings with a landlord and tenant, either one or the other has an incentive to save energy, based on who pays the energy bills. But this means that one of the two parties does *not* have an incentive to save energy, and this can cause problems.

Further landlord/tenant energy issues arise when energy improvements shift energy costs from one party to the other. For example, converting a central gas-fired domestic hot water heating system, on the landlord's gas meter, to individual electric heat pump water heaters, on the tenants' electric meters, represents such a shift.

The complexity of metering issues plays itself out in other ways in multifamily buildings. There is much individual electric metering, and a historic trend toward individual metering and submetering. Individual gas metering is also not uncommon. In some buildings, gas metering is for the gas stoves in apartments only, with the tenant paying more for the monthly meter charge than for the gas. And some apartment complexes have installed individual gas heating systems, to allow adding a gas meter to each apartment, and so shift the cost of energy totally to the tenant. This trend is increasingly being questioned, for several reasons. Individual metering of tenants means that the landlord loses incentive to improve landlord-owned energy infrastructure, from refrigerators to light fixtures to air conditioning and more. In electrically heated buildings (electric resistance or heat pumps), or individually gas-heated and gas-metered apartments, this further means that the landlord has no incentive to improve or fix inadequate insulation, or insulate the roof, or replace windows. The elasticity of savings achievable by the tenant is less than the elasticity of savings that the landlord can achieve through infrastructure improvements. And there are ways in which tenants can be held responsible for energy use even if they do not pay the energy bills. Evidence is mounting that, in individually metered buildings, landlords do not improve infrastructure as much as they do in master-metered buildings. One study found that refrigerators in individually metered buildings are on average four years older, and use 20 percent more energy, than refrigerators in master-metered buildings.[1] Central renewable systems, such as photovoltaic systems, are not as feasible with individual metering systems, and instances have been reported of buildings removing individual electric meters, to go back to a single master meter, for this reason. As energy use continues to drop, the magnitude of individual meter charges, that is, the monthly utility charge just for the meter itself, begins to loom large, relative to the energy costs. For all these reasons, the trend toward individual metering is being questioned.

Multifamily buildings see some mixed use, where commercial spaces, typically on the first floor, are part of the same building. Again, energy metering issues become complex, as does decision making about energy improvements, especially for systems shared by both parts of the building.

Refrigerators are an important energy load in multifamily buildings, as every apartment has one. A best practice is a 100 percent inventory of refrigerators, when doing energy work, to evaluate which ones need to be replaced, and to avoid simply postponing the inventory to the implementation phase.

Laundry machines are common in multifamily buildings. The laundry is typically shared, which provides a good opportunity for replacement with high-efficiency machines, because clothes washers and dryers see much higher use in multifamily buildings, and so energy improvements pay for themselves faster. There is often the complexity of an outside laundry service company that owns the machines, so energy improvements can involve negotiations with the laundry vendor. The energy manager should make the case that the owner is paying for the energy costs, so the laundry vendor should be sympathetic to requests to reduce energy costs.

Water use in multifamily buildings is typically higher than in many other types of commercial buildings. This works in favor of water conservation in general, as well as hot water energy conservation in particular. It also more cost-effectively supports advanced improvements, such as solar hot water, and hot water energy recovery.

Multifamily buildings are inordinately high users of steam for heating, especially in the northeast, and in large, older industrial cities, with over 50 percent of buildings being steam heated.[2] And steam is an inordinately high user of both energy and water, with multiple studies over several decades showing that buildings with steam heating systems often use twice as much energy for heating as do buildings with non–steam heating systems.

Figure 19.1   24/7 lighting of common areas in multifamily buildings presents an opportunity for significant energy savings.

Lighting in multifamily buildings is somewhat unusual, in that common-area lighting (such as in the lobby, stairwell, corridors, and laundry rooms) is often on 24/7, whereas in-apartment lighting sees very low hours of use (estimated at two to three hours per lamp, or less). The long-duration common-area lighting makes this lighting attractive for energy improvements, but the low-duration in-apartment lighting is more of a challenge to justify improving. See Figure 19.1.

Ventilation strategies are also somewhat inconsistent in multifamily buildings. Low-rise buildings typically just have bathroom and kitchen exhaust fans, on tenant-controlled switches, which are used very little. High-rise buildings typically have central exhaust for bathroom and kitchens, typically controlled by the landlord, and run 24/7. Such central ventilation can consume 20 percent or more of the energy in a high-rise multifamily building.[3] Many high-rise buildings also have central makeup air, supplied to the corridor, which is ostensibly intended for the apartments. Many central makeup air systems have been found to have failed, and therefore are not operational. If operational, much of the outside air supplied to the corridor does not end up in the apartments, but rather finds its way out of the building, through the elevator shaft, stairwells, trash chutes, corridor windows, and other paths. Duct leakage and lack of balance are also known problems in high-rise multifamily ventilation systems, with emerging solutions, such as duct sealing, balancing dampers, and fan speed reduction.

Multifamily buildings do not see as much cooling as in other types of commercial buildings. However, cooling is frequently provided by room air conditioners, which are inefficient, and which have been found to be major sites of infiltration.

Multifamily buildings have some unique ownership structures, such as condominiums, with implications for energy improvements (Figure 19.2). Condominiums present challenges, as individual preferences for appliances and lighting give energy improvements less access to economies of scale, with more individual negotiation with owners necessary too, all of which can make in-apartment energy improvements more difficult.

Figure 19.2   Condominium ownership presents challenges to energy improvements, as individual owner preferences provide less economy of scale.

Envelope improvements are already common in low-rise buildings, and are a new frontier for high-rise buildings, with the role and mechanisms of stack effect increasingly recognized, and the many points and paths of unwanted airflow knowable and

Figure 19.3   Common-area lighting presents energy conservation opportunities in hotels.

Figure 19.4   Thermal zone control in the hospitality sector allows energy savings when rooms are not occupied.

stoppable. Those wishing to push the bounds of energy improvement can look at balcony losses on high-rise apartment buildings, a known weakness and point of energy loss, and itself a frontier of energy improvements in this building stock.

Hospitality

The hospitality sector is represented by hotels, motels, inns, resorts, and the like. Unlike multifamily buildings, billing is much simpler, with master metering typical, and the owner carrying the full responsibility for paying the bills, and so seeing more incentive to make energy improvements.

Like multifamily buildings, the repeating elements of hotel rooms and suites make economies of scale good, for both evaluation and for installation of energy improvements. And, also like multifamily buildings, hospitality buildings have high water use, both cold and hot water. So water fixtures and water heating systems deserve attention, and advanced improvements, such as solar hot water and hot water heat recovery, may be cost-effective. Specialty improvements, like the repair of the widespread leaks in bath diverter valves, can be attractive in hotels, because most hotels have bathtubs, and showering is common. And, also like multifamily buildings (especially high-rise), common-area lighting is important in hospitality buildings: lobby, stairwell, corridor, laundry, as well as areas such as the fitness room, business center, and pool (Figure 19.3). Exterior lighting is also important. And, like multifamily buildings, ventilation energy use and improvement opportunities requires attention, although more so in buildings with central continuous ventilation.

The two most common forms of heating and cooling in the hospitality sector are packaged terminal air conditioners (PTACs) and fan coils. The heating component of PTACs in each building is important to identify. The heat is frequently electric resistance, a costly and inefficient form of heat. Conversion to packaged terminal heat pumps (PTHPs) is an option. However, PTHPs rely on backup electric resistance heat at cold temperatures, so this option is only partially effective. Fan coils are typically located above the ceiling as one enters the hotel room, and are more characteristic of luxury hotels, and older hotels. They are served by central boilers and chillers, which call for attention of their own, using standard chiller and boiler energy improvements, including to the cooling tower and distribution systems. Whether heated and cooled with PTACs or fan coils, the intermittent occupancy of buildings in the hospitality sector calls for thermal zone control, and specifically unoccupied room control, to shut off heating and cooling when not in use (Figure 19.4). Remote override control, from the front desk or by the facility manager, is best, to supplement local control by the guest. An innovative approach to controlling heating and cooling, and all energy use, in guest rooms is to only enable power to the rooms when the guest card is inserted in a card-reading switch as the guest steps into the room, a strategy that has been used in European hotels.

Converting either PTAC systems or central chiller/boiler systems to high efficiency air source heat pumps might be an option, eliminating the inefficiencies of PTACs and their infiltration, and the problems with large central systems, while retaining good zone control.

Less frequently, water loop heat pumps are used in guest rooms, either located above the ceiling on entry, or in vertical cabinets/chases. The latter are called vertical stack water source heat pumps, and have been found to have a host of energy problems, most prominently air bypass. Guidance has been developed to identify and mitigate such problems.[4]

Hotels often have in-room refrigerators, which are typically concealed in cabinets, where their operation is highly inefficient, as refrigerators need open air to operate efficiently. It is preferable to not have the refrigerators in cabinets at all.

Specialty spaces, such as restaurants, conference centers, and pools, require special energy attention. Their energy use and systems are entirely different than hospitality room energy use.

Offices

Office buildings have more floor area (12.2 billion square feet) than any other building type in the United States and have the highest total energy consumption (1.1 quadrillion Btu) of any building type. Furthermore, the largest buildings have a higher energy use intensity (energy consumption per square foot) than any other size of building.[5]

The challenges of energy evaluations in large office buildings are many. Large heating, ventilating, and air conditioning (HVAC) plants and controls can be complex for new energy auditors, and even for experienced engineers. High-rise buildings have unpredictable and uncontrolled airflows, driven by interactions among stack effect, exhaust fans, wind pressures, outside air ventilation, and air distribution systems.

Although large office buildings present a broad set of challenges, they also bring unusual opportunities. The size of the buildings allows for economies of scale in energy audits and implementation, and energy savings can be large. A single owner, frequently a private entity or individual, can allow for easier decision-making. Repeating space types, from area to area, and floor to floor, and building to building, can simplify the energy audit: closed offices, open offices, corridors, stairwells, kitchenettes, toilets, first floor/lobby, and conference rooms. A few large loads can offer large energy savings opportunities: Ventilation, HVAC plant, HVAC distribution components such as large motors for air handlers and pumps, and adjustments to incorrectly operating HVAC systems. Repeating (often identical) loads also make things easier: computers and peripherals, kitchenette appliances such as refrigerators, lighting, and windows. Lighting, in particular, sees long hours of usage, unlike in many other building types for which occupancy is more sporadic, and so offers greater opportunity for energy savings.[6]

A large office can be a poster child for right-lighting. Lamps and fixtures can be removed from old overlit old buildings, but also from new overlit buildings, paving the way for decades of energy savings and maintenance savings. Nighttime lighting deserves scrutiny, as does decorative lighting.

Envelope improvements should not over overlooked. All the strategies of stack effect reduction and envelope evaluation deserve attention. We can take a page from the Empire State Building, which rebuilt its old windows with an added internal glazing and gas fill, as part of a project to reduce its energy use by 38 percent (Figure 19.5).

Air conditioning is important in office buildings, and the combination of lighting, high occupancy, and office equipment all extend the cooling season forward from summer and back past the end of summer, well into the shoulder seasons. All-interior office space can even be in cooling year-round. Therefore, cooling improvements take on greater importance, such as high-efficiency cooling plant, distribution improvements, and free (economizer) cooling in the shoulder months. We can even think out-of-the-box and consider nighttime cooling.

Ventilation is also important, and the combination of ventilation and cooling-intensive operation both point to the possibility of dedicated outdoor air systems. Energy recovery ventilation may also make sense, due to the combination of high ventilation loads and high occupancy. Demand-controlled ventilation is another option, though which may compete with energy recovery ventilation, depending on the degree to which occupancy is intermittent. Conference rooms need special attention for ventilation, with their intermittent high occupancy.

Small office buildings bring their own characteristics, challenges, and opportunities. Housed in low-rise buildings, even in old houses, attention must also be directed

Figure 19.5   Envelope energy improvements represent a new frontier for office buildings.

to envelope improvements, such as attics and basements, while recognizing the lighting, ventilation, and cooling-intensive characteristics of offices.

One load that is unusually low in offices is hot water use. Point-of-use water heaters may make sense, because use is so low, allowing elimination of standby losses and distribution losses.

Office kitchenettes offer energy opportunities, especially in tea/coffee brewing and warming, also in water coolers and heaters, and refrigerators. Creative solutions are available, such as combining the use of nearby underutilized refrigerators, replacing old and oversized refrigerators with smaller efficient ones, and encouraging the use of reusable bottles for water in refrigerators, rather than using refrigerated water coolers.

Plug loads are important, for office equipment. Because of the relatively rapid turnover of office equipment, three to five years at most, what is important is a purchasing policy to promote high-efficiency equipment, such as ENERGY STAR. Plug-load monitoring is worthwhile. Frequently, we see a single old computer that is underused, but drawing a large amount of power. Emerging technologies allow remote monitoring and control of plug loads.

Server room cooling should take advantage of free (economizer) cooling, for the long hours where cooling is needed but outdoor temperatures are cool. Larger server rooms should be treated with the special attention, and special energy improvements, required for data centers, one of the most energy intensive and rapidly increasing type of spaces. With data centers, internal machinery gains are everything, and a variety of approaches have been developed to control energy use, while providing the uniform and controlled temperatures needed to protect the equipment.

Split incentive metering issues can arise in office buildings, where tenants pay part or all of their energy costs, and this requires creativity and commitment, to reduce energy use, with all parties adequately compensated for contributions to energy improvements.

Food Service and Sales

Food service goes beyond restaurants, and occurs also in churches, fraternities/sororities, dormitories, nursing homes, hotels, and more.

In food preparation areas, we start with the kitchen hoods. Emerging approaches to demand-controlled ventilation allow control of kitchen hood airflow with sensors that measure temperature, contaminants, or both. Control of makeup air, such as delivery of unconditioned makeup air directly behind the hood, reduces energy use and improves comfort in the building, by eliminating uncontrolled infiltration as makeup air.

Cooking and warming offer opportunities for energy improvements. Old gas stoves can have multiple large standing pilot lights, consuming energy 24 hours a day.

Refrigeration is common to all food service facilities, and can range from smaller plug-in refrigerators and freezers to large commercial refrigeration equipment. We want to move refrigerators out of hot kitchens where possible, where they operate inefficiently. High-efficiency refrigeration, and refrigeration improvements such as correct temperature settings, can be attractive.

Cooling energy use tends to be higher in food services facilities. Smaller food service facilities are often served with low-efficiency rooftop units, out of sight and out of maintenance. Look at outdoor air dampers that might be failed in the open position, or economizers that are not working. High-efficiency replacements for the cooling equipment are an option.

Lighting levels can be high, and runtime may be both long and intermittent, justifying the full suite of lighting improvements: right-lighting, replacement, and controls.

Health Care

Health care can be divided into hospitals, on the one hand, and smaller clinics and doctor's offices, on the other. Their energy needs are different.

Hospitals are one of the most energy-intensive types of buildings. Contributors to high energy use include day-and-night occupancy and activity, high lighting levels, high use of machinery and equipment, the need for comfort in all areas at all times, the need for cleanliness, intensive laundry use, high ventilation rates, high occupancy rates, high foot traffic rates, high use of hot water, and the need for cooling to counter all the internal gains.

Accordingly, everything is on the table: high-efficiency cooling, high-efficiency motors, all lighting improvements, laundry improvements, domestic hot water improvements, and more. Hospitals are one of the best applications for cogeneration (combined heat and power), due to the large and coincident demand for electricity and heat.

As with industrial buildings, steam systems are common, and should be scrutinized closely. Where steam for space heating can be separated out from process steam, conversion to non-steam space heating should be considered. Steam use in absorption chillers is a candidate for replacement with high-efficiency electric chillers, delivering considerable savings.

The important ventilation needs of hospitals should not mean that we do not look at ventilation. Ventilation rates are high, and so opportunities for saving are high, for example, if some areas are being overventilated. Duct sealing can also contribute to both indoor air quality and energy savings.

Smaller clinics and doctor's offices are not nearly as energy-intense as hospitals. The range of energy improvements are perhaps similar to those of offices, just with higher cold and hot water energy use, and possibly on-site laundry that deserves attention.

Education

All classrooms tend to be intermittently occupied, and when occupied they can have a high occupant density. Ventilation is a must, and the intermittent occupancy points to demand-controlled ventilation as an important improvement. As in offices, energy recovery ventilation is an option that sometimes competes with demand-controlled ventilation, although they can be done together as well.

Lighting in classrooms can be a big energy user. Look for opportunities for intermittent control, when classrooms are not occupied, possibly with bilevel control in corridors and stairwells, for low level operation during classes and during the period between the end of school and when cleaning staff come through in the late afternoon and evening.

High-bay gym lighting is often metal halide, with the warmup and restrike times that discourage turning them off. Replacement with T5 fluorescent lighting has been a successful way to increase efficiency and allow intermittent use, and light-emitting diode (LED) lighting is now seeing increasing use in the same way. Gym locker rooms see high cold and hot water use, as do restrooms in schools and universities.

University labs can be energy intense, and deserve special attention of their own, with a focus directed to fume hoods, associated exhaust fans, and leaky ductwork. Variable-speed fan control, coupled with duct sealing, can deliver measurable energy savings.

College dorms have much in common, energy-wise, with both hotels and multifamily buildings, such as lighting-intensive common areas, common laundry, high cold and hot water use, lower air conditioning use, higher heating use, and the emerging envelope improvements for medium- and high-rise buildings.

Schools and universities are large users of steam, and so steam energy replacements can be put to good use.

Many schools and university buildings have attics, and these deserve attention. Uncapped chases and other large air bypasses up into the attic are common, as is inadequate attic insulation.

Religious Buildings, Performing Arts Centers

We examine religious buildings and performing arts centers together, because they have so much in common: largely unoccupied for many hours of the week, typically housing a small office that might be occupied for much of the week, a large space for gathering, high occupancy for short periods, and decorative elements including lighting.

Therefore, improvements such as setback temperature control are important and, to allow this, thermal zoning of heating and cooling to allow small occupied areas such as offices to be comfortable during setback of the other larger spaces (sanctuary, auditorium, etc.).

An unusual opportunity presents itself for heat pumps, due to the deep indoor temperature setback for much of the week. Heat pumps operate at high efficiency when indoor temperatures are low in heating and high in cooling. Consider an example heat pump at 47 F outdoor temperature: At 70 F indoor temperature, its efficiency is rated at 3.94 coefficient of performance (COP), but at 55 F indoor temperature, its efficiency is a full 20 percent higher, at 4.74 COP.

Air conditioning use in these types of buildings is generally low, and so justifying air conditioning energy improvements can be difficult, unless air conditioning is being replaced anyway for comfort reasons.

The need for ventilation is high when the chapel or theater is occupied, but only for short periods. Again, this makes energy improvements such as heat recovery ventilation harder to justify. At the least, ventilation should only happen when it is needed—timer control or demand control of ventilation may make sense.

Theater lighting energy use can be high, if only for the hours of performances or rehearsals. But a well-used venue, seeing one to two performances or rehearsals a day, will accumulate enough hours per year to consider lighting replacement. LED lighting is seeing widespread adoption, for its controllability, while delivering energy savings and the comfort benefits of lower temperatures. See Figure 19.6.

Decorative lighting is common in religious buildings and performing arts venues. From outdoor uplighting to nighttime lighting of church towers to indoor chandeliers and artwork display lighting, energy improvements to decorative lighting are a next frontier for these buildings. See Figure 19.7.

Water use can be high, even if intermittent, due to the high occupancy. Water fixtures, distribution, and water heating are worth evaluating.

Religious buildings often see multipurpose use as schools and daycare centers. Those areas require energy evaluation through the lens of educational buildings.

Religious buildings, as well as all not-for-profits, often find themselves on the receiving end of donations of used equipment, and often this equipment is energy-consuming: old refrigerators, old stoves, old vending machines, and even occasionally old lighting and heating/cooling equipment. This old equipment is unfortunately often inefficient. Energy auditors and managers should be vigilant in looking for inefficient equipment in these buildings. Their donation may be well-intentioned, but results in high energy costs over subsequent years, offsetting

Figure 19.6 Replacing theater lights reduces energy, improves comfort, and can increase the ability to control the lights.

Figure 19.7 Decorative lighting presents energy opportunities for religious buildings and performing arts centers.

the value of the initial donation. Not-for-profits are best served by donations of high-efficiency equipment, which ends up being a double donation, both for the initial equipment and for the energy cost savings over time.

There are usually attics up above the vaulted ceilings in church sanctuaries (Figure 19.8) or old theaters. Access requires climbing up ladders and passing pigeon nests, but opportunities abound for air sealing and insulation up in these attics.

Retail

General and display lighting are important energy loads in retail buildings. Big-box retail is single-story, and therefore is a good candidate for skylights for daylighting, as has already been demonstrated successfully by large chains. Other lighting improvements include high-efficiency replacements, right-lighting, and controls. Reflective surfaces (walls, floors, ceiling) make sense, to allow reduction in artificial lighting while maintaining or even improving illuminance. Display lighting can be replaced with high-efficiency LED lighting, available now in a wide variety of display configurations. Exterior lighting, such as parking and evening access, should be examined.

Figure 19.8   Attic above arched church ceiling.

Air conditioning is also important in retail buildings. High-efficiency cooling, free cooling, and cooling controls are all worth examining.

Demand-controlled ventilation makes sense to address the variable occupancy of stores.

Front entrance envelope measures are worth examining, due to high foot traffic and associated infiltration, for both energy and comfort reasons. Unconditioned vestibules or revolving doors will save energy and improve comfort.

Hot water loads can be low, such as in department stores, so point-of-use hot water may deliver energy savings. In some retail buildings, such as supermarkets, hot water loads are high, but can be served with heat recovery from refrigeration condensers.

Specialty areas, such as food courts and in-store restaurants, deserve separate attention, for a focus on commercial kitchen hoods and cooking/warming equipment. Likewise, refrigeration in food sales stores requires separate attention, whether smaller plug-in refrigerators, or central commercial refrigeration such as walk-in coolers and freezers, large display refrigeration, and the like. High-efficiency refrigeration, refrigeration heat recovery, and smaller improvements such as gasket repair and temperature adjustments should be considered.

Industrial and Process Efficiency

Industrial and process efficiency offer a diversity of energy improvements for consideration.

Compressed air systems are common. The energy penalty of air leaks can be measured by measuring compressor run time, and/or energy use, over a period when there are no compressed air loads running. Leaks should be eliminated. High-efficiency air compressors are another option.

High-bay lighting is common, in manufacturing areas and in warehouse areas. High-efficiency replacements include T5 fluorescent and, increasingly, LED lighting. As always, look at right-lighting and at lighting controls as well, to double or triple energy savings, rather than just settling for high-efficiency replacements. Looking at lighting improvements also makes sense in office and lab/test areas, and exterior lighting. Lighting improvements can deliver even more significant energy savings if run time extends beyond a single shift, to two or three shifts.

Steam is common in many, if not most, industrial facilities. If the steam is being used for space heating, conversion to hot water or other heating systems is warranted. If the steam is needed for process applications, steam improvements, such as pipe insulation, steam trap evaluation and repair, flue gas heat recovery, and steam leak identification and elimination, can be considered.

Infrared heating can save energy in areas such as workshops, and other intermittent occupancy areas, by delivering comfort while keeping air temperatures low.

Envelope improvements include infiltration control at overhead doors: Insulated doors with good air seals, air curtains (preferably unheated) or physical curtains, possibly unconditioned areas for unloading.

Ventilation needs can be high, and should be examined for energy efficiency: high-efficiency motors and demand-controlled ventilation if possible.

Cogeneration (combined heat and power) might make sense for facilities with large and long-duration heating and electric loads that happen at the same time.

Many industrial processes require heating a fluid or material, and at some subsequent point in the process the fluid or material is cooled. This heating and cooling lends itself to heat recovery, in which heat is extracted from the cooling process to preheat the fluid or material for its heating.

A big success that we seek for industrial facilities is the energy improvement that also serves as a process improvement. Variable-speed motor drives will save energy and improve process control. Eliminating compressed air leaks and steam leaks will save energy and ensure that steam and air are available for process needs. Many more options are available, and this should be a prime lens through which industrial facilities are examined.

References

1. Taitem Engineering, "Refrigerator Age and Efficiency in Multifamily Buildings." Unpublished/in progress, 2015.
2. I. Shapiro, "Blowing Off Steam: The Case for Variable Refrigerant Flow (VRF) Heat Pumps to Replace Steam." Taitem Engineering. Report for the NY Indoor Environmental Quality Center, Syracuse Center of Excellence, Innovation Fund, July 23, 2015.
3. I. Shapiro, "Evaluation of Ventilation in Multifamily Buildings." Final Report to the New York State Energy Research and Development Authority. Report 93–5, June 1993.
4. K. Varshney, I. Shapiro, Y. Bronsnick, and J. Holahan. "*Air Bypass in Vertical Stack Water Source Heat Pumps*." HVAC&R Research, October 2011.
5. Energy Information Administration, Office of Energy Markets and End Use, Forms EIA-871A, C, and E of the 2003 Commercial Buildings Energy Consumption Survey.
6. I. Shapiro, "Energy Audits in Large Commercial Office Buildings." *ASHRAE Journal*, January 2009, p. 20.

Chapter 20

Project Management

We begin the discussion of project management by presuming that an organization will start with an energy audit, then choose improvements, and then oversee the implementation (installation) of the improvements. Another form of procuring energy improvements is through *performance contracting*, which means engaging a third party to both finance and install energy improvements, and this will be addressed separately.

Project management is about project control. With project control, projects go well and energy savings are achieved. Without project control, schedules can be missed, the budget can be exceeded, work can be of poor quality, disputes can arise, and energy savings are not delivered.

The five elements of project management are quality control, schedule control, cost control, scope control, and risk control. Phases of an energy project include developing the scope, bidding, installation/construction, and project completion. Developing the scope usually starts with an energy audit.

In this chapter, we refer to a *project manager*. The project manager represents the owner's interests, and supervises an energy project, typically from start to finish. The project manager may well be the energy manager, or may be another individual within an organization, or may be contracted to manage the project.

Procuring Energy Audits

An example *request for proposal (RFP)* for energy audits is provided in Appendix R.

In writing an RFP for energy audits, a good first step is to set goals, and to put the goals in writing as part of the RFP. What are your goals? Do any goals have priority? How deep are the reductions you seek? What are your organization's goals? What are the goals of key individuals in your organization? Are everyone's goals the same? Are the goals to reduce carbon emissions, or to reduce operating costs, or both, or something else? Reconciling and harmonizing goals is important, so that the energy auditor is given clear goals, and can deliver an energy audit that meets these clear goals.

The quality of an energy audit depends on the active involvement of the energy manager. If the energy manager writes a clear energy audit RFP, the energy audit will be better. If the energy manager provides solid information about the building, including not only energy bills but also occupancy schedules, the energy audit will be better. If the energy manager supervises the energy auditor, and asks relevant, probing questions, the energy audit will be better. If the energy manager reviews and critiques the energy audit, the energy audit will be better. And if the energy manager correctly interprets and acts on the recommendations of the energy audit, the delivered energy savings will be better, and the project goals will be reached.

While being actively involved in the energy audit, there are ways in which the energy manager should also stay out of the way. The energy manager should avoid biasing the energy audit with personal preferences, or with directives such as "We will

Figure 20.1 Project management is about project control. With project control, energy savings are delivered. Without project control, energy savings can fall short.

Figure 20.2   Energy managers have significant control over the quality of energy audits they procure.

only be interested in improvements that pay back in under two years." Such directives will only result in a less useful energy audit, one in which less information is available upon which to base future decisions.

A major role of the energy manager is to review the energy audit, and to perform quality control on it. A checklist for quality control review of energy audits is provided in Appendix S.

The next step is to interpret the energy audit. The energy manager will frequently need to explain the energy audit in language understandable by the building owner or other management. Make time for the interpretation. Avoid overly simplistic summaries. The energy manager should convey the pros and cons of each improvement, in explaining the energy audit to the building management. The energy manager should explain not only the installed cost and payback of each improvement but also the non-cost issues, the risks, and the potential benefits in comfort or health and safety. If management understands the full scope of an energy improvement, and its risks, then its implementation will be more realistic. For example, if management understands that a chiller replacement must happen during the winter, with design and bidding starting the prior summer to allow time to order the new chiller, then the risks are minimized for the chiller replacement not happening in time for the summer cooling season.

Project Delivery Options

Energy improvement projects can be delivered in many different ways.

Smaller projects, for example lamp or appliance replacements, might be done in-house, by the owner's own staff. Others, such as purchasing ENERGY STAR computers and office equipment, might form part of an office purchasing policy, and so be implemented over time. Others, such as periodically cleaning air conditioner coils and adjusting temperature set points and schedules to meet changing occupant needs, will become part of operation and maintenance procedures, that are the responsibility of facilities staff.

The majority of improvements that transform energy use in a building are construction projects. Widely used construction project delivery options include[1]:

Design/bid/build. The traditional way in which buildings are built and major renovations are done. An owner contracts with an architect or engineer to design the work, the work is bid out, and the contractor with the lowest bid, qualified as being responsible, is separately contracted by the owner to do the work.

Construction Management at Risk (CMAR), also called CM/GC. An approach in which a construction manager serves as a consultant to the owner during the project development, but then takes the role of a general contractor during construction.

Design/build. The owner hires a single contractor, and the contractor does both design and installation. Two common forms of design/build include:

Bridging. With bridging, the owner hires a design professional for the early schematic design, and the contractor then takes over to complete the design and construction.

Public private partnership (P3). With a public private partnership, a private sector firm brings financing to deliver a completed a project, and then shares in the revenue, in this case represented by the energy savings. The Construction Management Association of America's definition of P3 relates to public sector clients, but this approach could well also be used with private clients. This approach is what is now known in the energy field as *performance contracting*.

Each approach offers pros and cons, and different levels of control and risk (Figure 20.3). Generally, more control brings more risk, and less control brings less risk, both for the owner and, conversely, for the contractor. For example, consider a large energy project, to replace a chiller, domestic hot water (DHW) system, lighting, and a roof with new insulation. With design/bid/build, the owner can have an architect and engineer develop a set of drawings that detail all aspects of the project. The owner potentially has good control of the scope of work, but has elements of higher risk, as the owner needs to coordinate between multiple contractors (heating, ventilating, and air conditioning [HVAC] for the chiller, plumbing contractor for the DHW, electrical for the lighting, and general contractor for the roof), and resolve disputes between the contractors and the design professionals. With design/build, the owner has less control over the project, but also lower risk: A single contract for design and construction means that the owner (or the owner's construction manager) only has to manage one contract, and the general contractor has to resolve all design and construction issues. For the contractor, control is higher, but risks are also higher.

Figure 20.3 Different approaches to project delivery carry different levels of control and risk, for the owner and for contractors.

Construction Management

Transformational energy improvement projects are almost always construction projects. Construction projects take time, require investment, can disrupt a facility's operations, make noise, involve construction hazards, require contracts, require insurance, and more. Treating energy improvement projects as construction projects gives them the attention and priority they deserve. Conversely, treating energy improvement projects as simple commodity purchases invites significant risks. These risks mostly involve underestimating the effort required, and in doing so missing schedules, going over budget, risking conflict between contractors and design professionals and management and building occupants, and finally not achieving the goals of energy savings.

All energy improvement projects need to be managed. Whether or not a formal construction manager is hired, the management and coordination work needs to happen. On some projects, a construction manager is called a *clerk of the works*, a term dating back hundreds of years. The well-known English poet Geoffrey Chaucer was the Clerk of the King's Works, managing royal construction projects in the late 1300s. Managing the implementation of energy improvements is one of the most important, and unrecognized, activities in reducing energy use in buildings. If time and effort is allocated to this work, the energy projects are more likely to go well, and if not they will more likely not go well.

We will suggest a construction management process, with the steps outlined below. All projects do not need to fully follow this process, but all projects end up needing to follow most, if not all, the steps below. All projects have a schedule and need schedule control. All projects have a budget and need budget control. All projects have a scope and need scope control. All projects need quality control. Even the smallest project, the proverbial "changing one lightbulb," needs attention to detail in order for it to happen on budget, on schedule, and with sufficient quality such that energy savings happen.

Scope Control

Scope control means defining project goals, adhering to the goals, and achieving the goals.

Options for energy conservation are developed with the energy audit. The owner and energy manager then need to choose improvements. This happens through a process of clarification. What did the energy audit mean, for each recommended improvement? Are these recommendations acceptable to the owner? Are clarifications necessary? Does the owner wish to proceed with implementing all of the energy audit recommendations, or just some of them? Is there sufficient budget to implement the chosen recommendations? What are non-cost obstacles, such as disruption of operations, and how can these obstacles be overcome?

The process of choosing the energy improvements takes time, thought, and skill. It is a process of reviewing and harmonizing goals. And it is a process of obtaining the approvals necessary to proceed.

Once the improvements are selected, they then need to be specified in sufficient detail to be implemented in a way that meets the intent of the energy audit—to deliver energy savings. In some cases, this specification may be as easy as changing a light-bulb: in these cases, all the information necessary is right in the energy audit. In other cases, the energy manager might need to translate the energy audit into a set of performance specifications, a written document that sets out performance goals, but is not a set of full engineering plans. And, finally, in some cases, full architectural or engineering design may be required. For example, a large project requiring the replacement of chillers, cooling towers, air handlers, and a control system, might best be handled by full professional design.

In all cases, whether it is as simple as changing a lightbulb or as complex as a project requiring a full set of architectural/engineering drawings, we follow our practice that was begun with energy audits, using the acronym "LEFT" to ensure that the specification of the work is complete: L—location, E—efficiency, F—features, T—testing. The location and quantity of energy improvements need to be spelled out, the minimum efficiency of the new systems needs to be established, features that impact energy use need to be defined, and required testing set forth.

Specifying improvements is not only a description of the improvements themselves, but also of other requirements, such as insurance, schedule, payment terms, where and how materials are stored, what facilities the contractor can use during installation, what form of contract is to be used, who obtains the building permit, and more. This is where an energy improvement project shows that it is really a construction project. Standard construction forms and contracts can be useful. For example, standard *general conditions*, such as those provided by the American Institute of Architects (AIA) or the Department of Housing and Urban Development (HUD), can be used, qualified as necessary with project-specific *supplemental conditions*.

Working with architects and engineers can help to define the scope of work for complex energy improvement projects. Architects and engineers have unique skills in describing project scopes, and in bringing a critical professional eye to the work. Whereas professional design offers an opportunity to add necessary detail to the energy improvement scope, it is also an opportunity for changes to be made that can negatively impact energy savings. In working with architects and engineers, the project's primary energy goals should be repeatedly reiterated, including the assumptions made in the energy audit regarding the proposed locations, efficiency, features, and testing required for the energy improvements.

Scope control continues, through the process of bidding work, and through construction. Changes to the scope, for whatever reason, should be discussed, and documented in writing. The entire quality control effort, outlined below, is intended to ensure that the originally sought scope is delivered.

Cost Control

Cost challenges of energy projects include the hidden conditions that are found once work begins in existing buildings. For example, hazardous materials such as asbestos

Figure 20.4 Choosing energy improvements requires asking questions and, then, decisiveness.

or lead may be uncovered. Another challenge is that project budgets are established based on rough estimates that are produced with energy audits. If the energy audit estimate is too low, the budget may not be sufficient to do the work.

Three fundamentals of cost control are:

- Estimate early and realistically.
- Include a contingency to cover hidden conditions and other unexpected issues.
- Control the scope, to minimize change orders.

Review the energy audit estimate to verify its detail. Did it include a contingency? Did it include soft costs such as the building permit, design (if required), and commissioning? Did it include removal of existing equipment? Did it include all components of material and labor required for the project? If the project is primarily mechanical, such as a boiler replacement, did it include provision for electrical hookup? Did it include overhead and profit for the installing contractor? A detailed update to the cost estimate may be required, and can be done by the design professional, by the energy manager, or by the construction manager.

The following contingencies have been provided as typical for construction projects[2]:

- Program phase (equivalent to the energy audit): 10 to 15 percent.
- Schematic design phase: 7.5 to 12.5 percent.
- Construction documents phase: 2 to 5 percent.

In other words, as a project progresses from energy audit towards the bid phase, the contingency can be reduced. However, because energy projects involve hidden conditions, with work in existing buildings, higher contingencies are recommended, perhaps 15–20 percent during the energy audit phase, 10–15 percent during schematic design, and 5–10 percent heading into construction. Contingencies can also vary depending on the project. Projects with many hidden conditions might warrant higher contingencies, whereas simpler projects can use lower contingencies.

Cost control also means understanding sources and uses of funding for the project. *Sources* are where the funding comes from, and can involve a combination of loans, grants, tax incentives, operating cash, and capital reserves. *Uses* are where the funding goes. A complex energy project may have multiple simultaneous contracts (uses), and multiple sources. Sources need to be available in a timely manner. A tax incentive, or grant, may only flow at some time after a project starts, or even after it finishes. In these cases, bridge financing is required. Timing the availability of sources to meet the uses is an important part of cost control.

Another best practice for cost control is to remind all parties involved, repeatedly, of the project's budget, and cost priorities.

The project manager is typically also responsible for reviewing and approving *applications for payment* (invoices). Each payment should preferably be associated with a deliverable, so an inspection can verify that the associated work has been completed.

The contract should stipulate that a percentage of payments will be retained until work is complete. While 10 percent retainage is common, it may not be enough to get a contractor to complete work. 15 to 20 percent retainage is recommended.

BID PHASE

Soliciting multiple bids is a cornerstone of cost control. This can be done through advertising, through bidding firms, or through word of mouth. A best practice is to seek recommendations for contractors from others who have had similar work done.

In many ways, an owner or energy manager needs to market their project to potential contractors. Making a project attractive ensures more bidders, and attracts more responsible bidders. The ways in which to make an energy improvement project attractive include presenting it in a professional way, ensuring that the scope is clear and well defined, assuring potential bidders that the project is well controlled and has an adequate budget, and creating a collaborative atmosphere, even while arm's-length relationships are maintained. It might make sense to even share the project budget openly, even if competitive bids are sought. Scheduling a walk-through for bidding contractors allows greater familiarity with the project. Allowing sufficient time for bids gives a sign that detail should be given to bids. Treating contractors professionally includes not requesting estimates for purposes of cost estimating, and not requesting detailed design information for purposes of getting free design services.

A public bid opening is required for many public-sector bids, and also gives a sign that the process is transparent and fair, and that no bidder is being preferred.

Once the low bidder is identified, the project manager should qualify the low bidder as being responsible. This requires checking multiple references, possibly even inspecting completed work of a prior project by the contractor, may include an interview, and may include a credit check. Requesting and reviewing a *schedule of values*, representing the bidders cost breakdown, allows an opportunity to ensure that the bidder did not miss something in the bid, or did not otherwise underbid. Underbidding introduces great risk to a project, as a contractor that loses money is at risk of not completing a project, or of completing it inadequately, or of going out of business.

Quality Control

Quality control was revolutionized in the latter half of the twentieth century, becoming its own science, particularly in the manufacturing industry. Quality control can be equally well applied to energy improvements, and we have seen that this process has begun, in the energy field.

Fundamental aspects of quality include defining requirements, adhering to requirements, measuring if requirements are met, and continually improving the process based on the deviation of the measured results from the defined requirements. For energy work, the requirements are initially defined in the energy audit. The requirements might be further defined in a set of drawings, or performance specifications. The possible measurement of results can be well defined by the main goal of the project: To save energy. We can measure energy savings, and so we can measure the quality of a project. Therefore, we can know whether a project succeeded or missed its mark.

Energy improvement projects also present challenges for quality. Unlike manufacturing, where products may be manufactured in quantities of thousands, on a daily basis, and so measurement and improvement can be done with statistical accuracy, energy improvement projects happen in the ones and tens and perhaps hundreds per year, for a given owner or energy professional, but not in the thousands per day. There are a small number of projects from which to draw data, to improve the process. Projects also take time: From energy audit, through owner decisions about which improvements to make, through bidding and construction, through measurement of energy savings, the process definitely takes months, and usually takes more than a year. This long time frame also makes energy improvement projects less able to take advantage of feedback, and makes quality and quality control harder. Finally, energy improvements involve many parties: Energy auditor, energy manager, owner, engineer, contractor, building inspector, building occupants, facility manager, and more. Any weak link in the process can erode energy savings. If an energy auditor misses good energy improvements, potential energy savings are lost. If an owner chooses not to implement good energy recommendations, potential energy savings are lost.

If a contractor substitutes low-efficiency products, potential energy savings are lost. If control set points, such as temperatures, are not set correctly, potential energy savings are lost. If equipment is not maintained over time, potential energy savings are lost.

The science of quality control helpfully distinguishes *defect detection* from *defect prevention*, and encourages both, to reach quality and achieve a project's goals. *Defect detection* for an energy improvement project typically means review and inspection (Figure 20.5). We review energy audits and look for mistakes. We inspect completed energy work and look for incomplete or defective work.

Defect prevention is in some ways a revolutionary, and more robust, path to quality and delivering energy savings. Instead of looking for mistakes and correcting them, we seek to not make mistakes in the first place. Using a common phrase, defect prevention means doing it right the first time. Or, as is said in quality circles, defect prevention is doing the right things right the first time. Defect prevention is choosing good energy improvements. Defect prevention is realistic budgeting, with realistic contingencies, to prevent going over budget. Defect prevention is realistic scheduling to prevent missing deadlines. Defect prevention is overcommunication, so that building occupants and neighbors are not surprised when energy improvement work starts or disrupts their operations. Defect prevention is checking that the correct high-efficiency chiller was recommended in an energy audit, that the correct chiller was specified on the drawings, that the correct chiller was bid, that the correct chiller was ordered, and that the correct chiller was delivered, before it is installed. Defect prevention is training building facilities staff and occupants, to make sure that energy–saving equipment is used and maintained in a way to deliver its full potential energy savings. Defect prevention is reviewing project scopes with contractors, over and over, at the prebid walk-through, at the construction kickoff meeting, and at team meetings throughout construction.

Another best practice for defect prevention is to hold a construction kickoff meeting. This sets a professional tone for the project. An agenda should be issued before the meeting, including: Project goals, project requirements including design documents or specifications, project schedule, health and safety on the construction site, submittal procedures, inspection procedures, payment procedures, change order procedures, commissioning requirements, closeout requirements, and project priorities (schedule, budget, quality). The project manager should review the scope before the meeting, and know it well: Re-read the energy audit, re-read the specifications, re-read the contract, and re-read correspondence.

Another best practice for defect prevention is submittal review. Submittals, sometimes called "cut sheets" or "product cuts," are descriptions of a product that a contractor proposes to install. Submittals may be identical to what was specified, or may be a proposed alternate. Submittals allow the owner's representative to review and approve a product before it is ordered or installed. Submittals should be reviewed for substitutions, energy characteristics of the product (such as the energy efficiency ratio of an air conditioner), energy-related features (such as a requirement for a programmable temperature control), and any features that may be of importance to the owner (such as product color or finishes). Contracts should require that submittals be provided for all products, for written approval by the owner and project manager, prior to ordering.

Despite our best efforts with defect prevention, we still need to inspect completed work to identify defects and have them corrected. Best practices for inspections include:

- Request a statement from the installer that the installation is ready for inspection.
- Bring documentation to the inspection: energy audit, contract, drawings and specifications if any, approved submittals.

Figure 20.5 Quality control comprises both defect detection, such as inspection, as well as defect prevention.

- Prepare an inspection checklist beforehand.
- Request that the owner or installer arrange for access to all parts of the installation, for example that they have given advance notice to tenants.
- Request that the owner or installer have someone knowledgeable about the installation present for the inspection.
- Pack installation tools: camera, flashlight, tape measure, other tools as required for specific improvements.
- Allow enough time for inspections.
- Schedule routine inspections. Daily inspections are best, weekly inspections at a minimum.
- Take notes and photos of inspected work.
- Was the correct equipment installed?
- Were required energy features installed?
- Issue inspection notes in a timely fashion, within a day after inspection.
- Follow up on unresolved defects uncovered during inspection, until they are corrected.
- Reject bad work. This is key to quality control.

Commissioning is the emerging discipline of inspecting and testing completed construction work, specifically in the area of energy systems. Commissioning is quality control, brought to energy improvements.

Schedule Control

Energy improvement projects take time. Scheduling and completing an energy audit can easily take two to three months. Choosing among energy audit recommendations, obtaining management approval, lining up funding, and specifying the improvements can easily take another two to three months, although six months or more is not unusual. This phase is one of the most unpredictable and time-consuming phases of a project. Bidding and construction can take two to three months for small projects, and over six months for larger projects that have large energy systems, such as chillers, with long lead times to order and have them shipped to the site. Commissioning can take another few weeks. And, most significantly, measurement and verification of savings will take months, and preferably will take over a year, in order to get at least one full cooling season and one full heating season into the energy savings measurements. At a minimum, the process will take 18 months, and only if highly accelerated. Real-life examples have repeatedly shown projects to take two to three years or more. Energy improvement projects have many steps, and involve many people. Accordingly, best practices for schedule control include:

- Develop a realistic schedule. Provide time for management to deliberate and decide over recommended energy improvements. Provide time for design. Provide time for bidding and contracting. Provide time for ordering and having long-lead-time items delivered. Provide time for construction, including staging and adjustments to existing occupant routines, if any. Provide time for commissioning. And provide time for measurement and verification.
- Follow up at every step of the way. Follow up with contractors after bid requests are issued, to make sure they were received. Follow up after contracts are signed, to make sure they are received, signed, and returned. Follow up to make sure equipment is ordered after contracts are signed. Follow up to make sure equipment is received. Follow up to make sure equipment is

on-site. Follow up to make sure occupants and neighbors are notified of the start of work. Follow up to make sure action items on meeting minutes are complete. All through to project closeout, the project manager is following up. Follow-up is a best practice for schedule control.

Risk Control

INSURANCE

Types of insurance common to construction projects include *general liability* and *worker's compensation*. If the project involves design professionals, *professional liability insurance*, also called *errors and omissions insurance*, should be required.

Insurance definitions relative to construction projects include:

- Insurer: The insurance company.
- Insured: The contractor required to carry insurance.
- Producer: The insurance agent or broker who places the coverage for the insured.
- Holder: Any party that is looking for proof of coverage from the contractor (owner, general contractor, etc.).

Required insurance should be spelled out in the contract documents. Evidence of insurance should be obtained before contract signing.

Worker's compensation is not required by law for one-person firms. Many firms claim to be one-person, but if they subcontract without an arm's-length relationship, then the subcontractors are technically employees. Also, as a best practice, the worker's compensation coverage should not have exclusions for the owners, partners, or officers.

Other best practices with insurance:

- Have the certificate sent from the producer or insurer directly; ensure it is authentic.
- Check signatures.
- Look up insurers at www.ambest.com to check their rating.

COMMUNICATION

Energy improvements in occupied buildings can inconvenience occupants, at best, and at worst can be a construction hazard. Good communication is a form of risk control. A best practice is to overcommunicate, with building occupants, with building management, with contractors, with neighbors, and with members of the community. At a minimum, communication should include advance notification of work that will affect occupants and neighbors.

On the positive side of communications, energy improvements are good for the environment, and organizations should take credit for investing in energy conservation. Announcements of planned projects, projects under way, and completed projects are good news, and can be shared in newsletters, social media, and press releases.

HEALTH AND SAFETY

Health and safety include both construction-phase risks and risks to ongoing building operations.

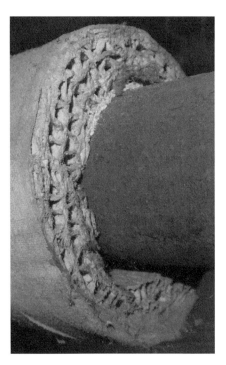

Figure 20.6　Asbestos pipe insulation

TABLE 20.1

Common Health and Safety Risks, and Strategies to Overcome Them

Health and Safety Risk	Improvement Strategy
Inadequate lighting	Measure light levels in each space and compare to current standards; improve when upgrading lighting
Poor indoor air quality	Identify ventilation deficiencies; identify and control contaminant sources
Asbestos (Figure 20.6)	Identify possible sites; remediate or encapsulate
Inadequate air for combustion equipment	Identify possible inadequate combustion air situations; improve when upgrading equipment
Humidity and water intrusion issues	Identify possible sites, where observed; recommend remediating, as part of insulation renovations
Lead paint	Identify possible sites; follow lead-safe practices
Hot water temperature too high	Recommend reducing temperature, both to save energy and to reduce scald risk
Inadequate heating—spaces too hot or cold	Recommend rebalance heating system, both to save energy and to resolve comfort issues; add thermal zones if necessary

Construction Phase Risks

Construction-phase risks include construction hazards (falls, electrical hazards, etc.) and impacts on indoor air quality, such as dust or the disturbance of hazardous materials, including lead or asbestos. Since most energy improvements happen in occupied buildings, exposure risk is not only for contractor personnel, but also for building occupants and neighbors.

Best practices include requiring that someone responsible for health and safety be appointed by the contractor, and that the contractor submit a company health and safety policy, and project-specific plan. Isolating work areas from occupied areas, with sheeting, for example, reduces indoor air quality risks.

Ongoing Health and Safety Risks

Energy improvements can improve health and safety in buildings or can hurt it. An important principle of energy improvements is to not only reduce energy use, but to also improve health and safety.

A variety of common health and safety risks are shown in Table 20.1, along with strategies to reduce these risks while making energy improvements. Many resources are available to address health and safety while doing energy work.[3–7]

Working with Regulatory Agencies

Managing an energy improvement project can mean working with several regulatory agencies.

Construction work typically requires a building permit. It makes sense for the building permit responsibility to be assigned, in the contract documents, to the installing contractor.

Progress and final inspections by the local building inspector should be expected. Code compliance typically involves not only the building code but also the energy conservation code, electrical code, mechanical code, plumbing code, and, where applicable, the life safety code.

Public agencies frequently provide financial assistance, in the form of grants, rebates, low-interest financing, and other incentives. This assistance can come with a variety of compliance requirements, from applications, to reporting, to inspections, to measurement and verification, and other forms of quality assurance.

Interaction with the local gas or electric utility can also be required. At the very least, utility bills may be needed. Beyond billing data, a new gas line may be needed, or an interconnect agreement for a new renewable system such as photovoltaic, or a modified electric service for such energy improvements as converting an absorption steam chiller to a new electric chiller.

Performance Contracting

Performance contracting generally refers to an outside firm implementing energy improvements in a building, largely or entirely with their own financing, and sharing in the energy cost savings. In other words, over time, the owner pays the performance contractor through savings in the energy costs. Performance contracting has also been referred to as *shared savings contracts*. Various forms of contract details are used, with varying levels of risk, control, investment, and recoupment, for both the owner and the performance contractor.

Performance contractors are also known as *energy service companies* (ESCOs). Many of the large ESCOs in the United States are also manufacturers and installers of controls systems, primarily targeting heating and cooling, but also able to control lighting and other building systems. Some of the large ESCOs are manufacturers of heating and cooling equipment.

Advantages of performance contracting include the outsourcing of financing, and the turnkey aspect of an outside firm providing the full range of services, from energy audit to design to installation and operation/maintenance. Additional advantages include upgrading of building systems, saving energy, and benefiting from the energy savings through the shared savings mechanism.

Disadvantages of performance contracting include a possible loss of control by the owner, and a loss of some of the savings, since savings are shared with the ESCO. Performance contracts are also at risk of favoring the products of the ESCO: ESCOs that are control manufacturers might seek to favor controls improvements, ESCOs that are lighting manufacturers might seek to favor lighting improvements, and ESCOs that are heating and cooling manufacturers might seek to favor heating and cooling improvements.

Another aspect is that some ESCOs use the performance contract as a vehicle to promote a long-term service contracts. This can be a positive or negative. Building owners should decide what level of control they wish to keep over their energy systems.

Best practices in procuring and managing performance contracts include:

■ Competitive procurement. Develop and issue an RFP to multiple ESCOs.

■ Use an independent third-party consultant to review the RFP, proposals, contracts, energy audits, design, and installations.

Other than the contracting mechanism, financing, and shared savings, all other aspects of performance contracts are identical to other energy improvement projects. We seek good energy savings, we seek realistic energy savings, we want to avoid health and safety issues and, preferably, improve health and safety in our buildings. There is an energy audit, there is selection of improvements, there is detailed scope development, there is installation, there is evaluation, and there is ongoing operation and maintenance. Energy managers and building owners are advised to treat performance contracts in the same way as any other energy improvements. Performance contracts should be subject to the same scrutiny: the same scope control, quality control, cost control, risk control, and schedule control.

Social Aspects and Energy Myths

We end our discussion of project management with a brief overview of what we might call social aspects of energy work, and some energy myths.

Energy auditors and energy managers, and the people in the buildings they serve, are all human beings. We all have traits that are all too human: Pride, shame, competitiveness, hope, fear—the list goes on and on. Many of these traits can be put to good use in the stewardship and improvement of our buildings. Some of these traits can be obstacles to good energy work. And so we often need people skills, in navigating though energy improvements in buildings.

Facility managers and building owners are often proud. They do not always like to learn about building deficiencies from an outside energy audit firm, or from an energy manager. Energy auditors and managers should affirm energy-efficient aspects of buildings in their reports. Facility managers and building owners should also try to open to outside input. Pride can alternatively be turned into a positive, as a motivator for people to improve their buildings.

There are frequently complex dynamics between facility managers and building owners. Owners might be reluctant to invest in energy improvements, where facility managers feel that their building systems need to be maintained and upgraded.

There are also frequently complex dynamics between tenants and landlords, with what can be significant mutual suspicion: Tenants might believe that landlords do not invest in the property, or care about comfort, whereas landlords might believe that tenants do not care about energy conservation. This is made complicated because it is possible to shift expenses from landlord to tenant, or vice versa, when energy improvements are made or if metering is changed.

There is a tendency for energy auditors to evaluate 5 to 10 improvements in an energy audit. A survey of 20 large commercial energy audits, all by different firms, found the vast majority (15 out of 20) recommended between 5 and 10 improvements, only two recommended between 10 and 20, three recommended fewer than 5, and no audits recommended more than 20 improvements. It seems that 5 to 10 improvements is a generally accepted number of improvements for an audit to look complete. Perhaps fewer than 5 "looks inadequate," and more than 10 "feels like too many." However, there is nothing to say that audits, and specifically the buildings they describe, need to have 5 to 10 measures. Maybe a building needs only three improvements. Maybe it needs 50. It is important for energy auditors to avoid including a particular number of measures because "it looks right."

Among energy auditors and energy managers, there is sometimes a tendency towards wishfulness, putting too much faith in particular improvements. This wishfulness leads to a growing expectation of savings that may well be more than is realistic for a particular improvement or package of improvements. This, further, leads to overestimating savings. We cannot wish savings. Savings only result from real physical changes to a building. This typically involves difficult and costly work. We must use real and realistic predictions of savings, based on sound building science and real physics, rather than hope or expectations or wishfulness.

As we do our energy work in buildings, we are faced with some interesting questions. For example, if an energy system is highly inefficient, but was recently installed, there is typically a reluctance to consider replacing it. It looks bad to remove something that was recently installed. There may also be valid questions about the embodied energy of the system, the energy that was used to make it. However, if the system is not evaluated for replacement or improvement, this may mean that the building is now condemned to use this inefficient system for years and decades into the future. The question is raised: Is newness of equipment reason enough not to replace or improve it? There may not be a single right answer, but we suggest that just because something is new and it might seem embarrassing to replace it is not a sufficient reason to reject evaluating its improvement.

Greenwashing is a relatively new term, given to an artificial claim to being green. We want to avoid greenwashing in our projects. We want to avoid artificially claiming that we have greened a building, when the energy improvements are token and barely measurable.

We also need to resist the temptation to focus on improvements that are "aesthetically satisfying." For example, weather-stripping a single front door that has a visible crack is a good thing to do but is rarely the only air sealing that a building needs, and in fact is typically a very small part of needed air sealing. We need to identify large opportunities, even if hidden, and resist the temptation to focus on small ones, just because they are visible.

There is a tendency in energy work to focus on preferred improvements. Faith in a specific improvement can develop, based on one good experience, or recommendations by trusted sources, or as a result of effective salesmanship by product vendors. Associated with this focus is the wishfulness we have described—we tend to ascribe more possible savings than are realistic. We *wish* the improvement to succeed. However, this single-improvement focus and wishfulness can become a major obstacle to measurable energy savings. One single improvement can rarely provide deep energy savings. If we are interested in deep energy savings, we need to install multiple energy improvements. Our focus must be broad. We must avoid the temptation to be sold on a single improvement. We must be whole-building focused.

Finally, a few myths prevail in the building energy field, and are worth demystifying:

- *Setback (unoccupied) temperature control does not save energy. The energy required to bring a space back to occupied temperature offsets any savings.* Setback temperature control always saves energy. The more the setback, the more energy is saved. Savings come in at least two different ways: (1) reduced envelope losses, including not only reduced conduction, but reduced infiltration losses, reduced ventilation loads, and reduced stack effect; and (2) greater thermodynamic plant efficiency. Plant efficiency gains can depend on other control settings, but envelope savings accrue whenever temperatures are set back. On rare occasions, such as a heat pump that uses electric resistance heat as a backup, the recovery will use added energy that will offset savings, unless controls prevent such incorrect operation.

- *Heat pumps do not work in cold climates.* Heat pumps work in all climates. They work more efficiently in warmer climates, but they still work in all climates and can save energy in all climates relative to many other heating sources. Attention must be directed to proper sizing and control.

- *If basement walls are insulated, the basement is within the thermal boundary, and heating/cooling losses in the basement are negligible.* Even if basement walls are insulated, if the basement is an unconditioned space, heating/cooling losses such as duct leakage or pipe losses are substantial, and reducing or eliminating such losses will deliver substantial savings.

- *Steam boiler systems can be significantly and reliably improved.* The depth and breadth of problems with steam boiler systems are so many that attempts to improve them inevitably deliver small savings, or deliver savings that do not persist over time. Steam systems are best removed and replaced with systems that do not operate at high temperature, do not suffer fugitive steam losses, do not lack zone controls, do not operate at an intrinsically low thermodynamic efficiency, and do not rely on a nineteenth-century technology.

■ *Building owners will only make energy investments that pay back in less than 3 to 5 years.* Building owners frequently invest in energy projects that take 10 years or more to pay back. We cannot presume what building owners will or will not do.

■ *LED lighting is so efficient, that it does not make sense to also examine right-lighting (reduced overlighting) or lighting controls.* This is a newer myth. But in many cases, right-lighting and controls still make sense to evaluate and implement, even with high-efficiency LED lighting. For example, in areas such as stairwells or exterior lighting, motion sensor control can reduce runtime from hours per day to minutes per day, and in fact works particularly well with LED lighting because of its excellent controllability. In the case of right-lighting, it can both save additional energy and reduce the cost of installation, as fewer new LED lights are needed in situations where the existing space is over-lit.

References

1. Construction Management Association of America (CMAA), "An Owner's Guide to Project Delivery Methods," 2012.
2. American Institute of Architects (AIA), "Construction Costs: Emerging Professionals Companion," 2013.
3. U.S. Environmental Protection Agency, "Moisture Control Guidance for Building Design, Construction and Maintenance." EPA 402-F-13053, December 2013.
4. Weatherization Assistance Program Technical Assistance Center, Lead Safe Weatherization Curriculum, 2012.
5. Building Performance Institute, "Technical Standards for the Building Analyst Professional," 2005.
6. U.S. Consumer Product Safety Commission, "CPSC Safety Alert: Avoiding Tap Water Scalds," 2012.
7. Illuminating Engineering Society (IES), New York, NY, *The Lighting Handbook*, 10th ed., 2011.

Chapter 21

Operation, Maintenance, and Energy Management

Preventive Maintenance

Preventive maintenance (PM) is a cornerstone of ongoing energy conservation efforts, as we seek to make energy savings persist after implementation is complete.

We usually think of preventive maintenance as such tasks as changing or cleaning air filters. Interestingly, changing or cleaning air filters does not necessarily save energy. In fact, it increases fan energy power consumption. However, it can reduce energy of direct expansion vapor compression systems (air conditioners, heat pumps, refrigeration equipment). Either way, cleaning air filters should not be the sole focus of preventive maintenance programs.

A strong preventive maintenance program looks for a broad set of energy deficiencies, from dirty coils to incorrect temperature set points to incorrect operation of lighting. It goes beyond heating, cooling and lighting to include preventive maintenance of the building envelope. Domestic hot water is also included in preventive maintenance, as are plug loads.

An energy preventive maintenance schedule is provided in Appendix T.

Vigilance for Anomalies and Catastrophes

As important as routine preventive maintenance—if not more important—is to maintain vigilance for failures in energy systems, what we have previously called anomalies and catastrophes. Whereas minor equipment degradation will cause minor increases in energy use, anomalies and catastrophes will cause major increases in energy use. These anomalies and catastrophes are so frequent that they can be included in our preventive maintenance schedule.

We are particularly interested in such things as obstruction to heat transfer in air conditioning and refrigeration condensers, which we know significantly hurts energy efficiency: blocked flow, failed fans, dirty heat exchangers, failed condenser water pumps, and the like. We recall that, for heat pumps, the condenser is the indoor heat exchanger, so we direct attention to both heat exchangers.

We look for signs of incorrect refrigerant charge in air conditioning and heat pump systems, such as low-capacity, high-supply air temperature in cooling, and low-supply air temperature in heating. We look for changes in these conditions over time, as a sign of refrigerant leakage. If incorrect refrigerant charge is suspected, it should be professionally tested.

We check combustion systems, testing for efficiency, and examining the products of combustion. Unburned natural gas in the vent is not unusual in combustion systems, causing both inefficiency and high carbon emissions. We check the flame on combustion systems, and we check that burners are clean. And we look for natural

Figure 21.1   Neglect is a sign of possible energy inefficiencies.

gas leaks, not only by odor, but looking at the gas meter leak dial for any consumption when no gas loads are running.

We inspect for water leaks, with an emphasis on hot water, but not neglecting cold water leaks.

The energy manager should be the protector of the thermal boundary. During renovations or during the installation of new wiring or piping, look for removed or disturbed insulation in locations such as attics and basements. Take periodic infrared scans, to look for settled insulation in walls, or detached insulation in crawlspaces, basements, or knee walls.

And, when initially taking over responsibility for a building, we look for signs that equipment might not have been maintained (Figure 21.1). Deferred maintenance is a polite term for neglect, and neglect is a sure sign of possible energy inefficiencies, and the need for testing equipment to make sure it is operating efficiently.

Metering

Metering is used to measure energy delivered to a building, for purposes of billing, primarily for electricity and natural gas. Water is another metered quantity that is of interest in energy work.

Master metering refers to a single meter, even if a building has multiple tenants. *Individual metering* or *direct metering* means one meter for each tenant in a building. *Submetering* means when a master-metered building has separate meters installed by the building owner, for billing or monitoring purposes, but the submeters are not read or owned by the utility company. *End-use metering* refers to metering individual loads within a building, usually done by the building owner, not the utility, and done for monitoring purposes. *Net metering* refers to metering of both the electricity supplied by a utility to a building *and* the electricity supplied conversely by the building to the utility, for example, through on-site photovoltaic electric power generation. See Figures 21.2 and 21.3.

Bulk measurement refers to the measurement of bulk fuels delivered to a building, such as fuel oil, coal, propane, or wood pellets. Bulk measurement is also

Figure 21.2   Although energy metering has been mainly used for billing, it is increasingly used for energy diagnostics, to understand energy use in buildings, and to plan and evaluate energy improvements.

referred to as *batch measurement*. A key difference between continuously metered energy, such as electricity and natural gas, and bulk measurement is that continuously metered fuels are measured as they are being consumed, whereas bulk fuels are measured when they are delivered, before they are consumed. A utility bill for continuously metered energy shows the consumption of energy quantities after they have been consumed, with payment after consumption, whereas bulk measurements and their associated bills are for energy that has not yet been consumed, with payment before consumption. Bulk deliveries are also more sporadic than the typical monthly gas or electric bills. These different characteristics of bulk fuels can make analysis of their consumption a little more difficult than the analysis of continuously monitored energy.

Historically, we have relied on meter readings by utilities, and bulk fuel delivery receipts, for energy analysis for buildings. Increasingly, there is a need for building owners to take their own meter and bulk fuel consumption measurements, for two reasons:

- Utility meter readings are frequently bimonthly, and sometimes meter readings are missed and happen even less frequently. Bulk fuel deliveries are typically even more infrequent. These erratic energy consumption readings make energy analysis less accurate.

- In buildings with on-site renewable energy, such as photovoltaics, there is a need for readings that can be correlated with readings of energy produced by the renewable source. If simultaneous readings of all three primarily electricity flows are not made (the renewable source, the electricity purchased from the utility, and the electricity sent to the grid), we no longer can tell what the electricity consumption is, as we are able to for buildings without renewables, where consumption is simply equal to what was purchased.

Fuel oil can be measured with float gauges on the fuel oil tank. Propane use can be measured indirectly by using the gauge on the tank (see Table 21.1).

Analog natural gas meters are read from right to left. If the hand is between two numbers, record the lower of the two numbers. If the hand is between 9 and 0, record it as 9. If the hand appears right on a number, look at the dial to the right: If the dial to the right has passed 0, then use the number that the hand is on. If the dial has not yet passed 0, then use the number lower than the number that the hand is on. In the example below, the reading is 0151. Ignore the two test dials; the test dials have no

Figure 21.3 Energy managers should take their own meter readings, to increase the frequency of readings, for purposes of energy diagnostics.

TABLE 21.1

Propane Tank Level Measurements

Propane Tank Gauge	60-Gal Tank	125-Gal Tank	320-Gal Tank	500-Gal Tank	1,000-Gal Tank
	Gallons Remaining				
80%	48	100	256	400	800
70%	42	87.5	224	350	700
60%	36	75	192	300	600
50%	30	62.5	160	250	500
40%	24	50	160	200	400
30%	18	37.5	96	150	300
25%	15	31	80	125	250
20%	12	25	64	100	200
10%	6	12.5	32	50	100

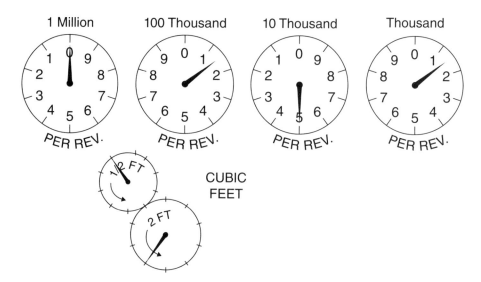

Figure 21.4   Gas meter reading.

numbers, and are only used to test the consumption rate of gas loads and for inspecting for gas leaks (covered in Chapter 14). See Figure 21.4.

Analog dial-type electric meters are read the same way. Newer digital meters can simply be read directly.

While tracking energy use, it is also useful to measure water use, both for water conservation and in support of hot water use analysis and energy improvements. The units of measurement of water meters are stamped on each meter. The most common unit of measurement is the cubic foot (CF), even though billing is often in other units, such as hundreds of cubic feet (HCF). In larger commercial facilities, compound meters are used, in which two meters work together to measure water use. With compound meters, one meter measures high flowrates, and the other measures lower flow rates. A shunt directs water flow to the appropriate meter. The total water usage is equal to the sum of the readings through both the two meters. Multiply cubic feet by 7.48 to obtain gallons.

Benchmarking

Benchmarking is the comparison of a building's energy use to the energy use of similar buildings and, as importantly, to a building's own prior use over time. Benchmarking can highlight if a building is a high energy user, relative to other buildings of the same type, or relative to its own historic use, and so can be used as an indirect indicator of potential for energy savings. Conversely, benchmarking can show if a building is an average energy user, or is a low energy user. And so benchmarking offers the promise of prioritizing which buildings deserve first scrutiny, among a portfolio of buildings.

Goals of benchmarking include:

- Comparing a building's energy use to that of other buildings of the same type.
- Comparing a building's energy use to its own prior use, over time.
- Establishing energy efficiency goals, and measuring progress toward these goals.
- Labeling, such as to obtain an ENERGY STAR label.
- Complying with municipal regulations, such as recent regulations in cities such as New York.
- Obtaining incentives for specific portfolio programs.

- Recognizing buildings that achieve good benchmarking scores and, conversely, for buildings with poor benchmarking scores, prompting building owners to direct attention to energy deficiencies, by publishing benchmarking results.

- Initiating a process of diagnostics, for buildings that receive poor benchmarking scores.

- Collecting data for portfolio program management, research, and policy development, at the portfolio level, but also at the city, state, and federal government levels.

- Protecting consumers, by allowing energy scores to be used as a form of building label, much as a car receives a fuel efficiency score and label.

We anticipate the possibility of benchmarking being used in the future for carbon tax determination.

Benchmarking is performed using a variety of software. Most popular is the Environmental Protection Agency's (EPA's) Portfolio Manager. A variety of other public and private software is available. For example, EnergyIQ is benchmarking software developed by Lawrence Berkeley National Laboratory that has added capabilities in the area of benchmarking individual building systems, on the basis of sub-metered data. Most benchmarking software programs today have a link to Portfolio Manager. Most also have the capability to report such metrics as carbon emissions. Other features that are provided in different software include forecasting, budgeting, tenant billing, tracking savings, diagnostics, and identification of billing spikes.

Benchmarking can and does serve a useful purpose. Benchmarking also faces a variety of serious challenges, mostly in the area of data quality. These include missed meters, missed meter readings or batch fuel deliveries, mistaken meter readings, estimated meter readings, mistaken data entry when moving data from billing records into benchmarking software, batch fuel deliveries that are far apart in time, rounding in batch fuel delivery data, mistaken floor area measurements, inability to access billing data (such as for a prior owner or tenants), and more.

Further challenges arise due to changes in building use, such as occupancy changes, new additions to a building, empty occupancies, and the like. Filtering comparable buildings according to additional properties, such as geographic location or number of occupants, can sometimes reduce the comparable buildings to a number too few to be statistically meaningful. And additional challenges include impacts of the weather. Portfolio Manager and other benchmarking software attempt to correct for some of these issues.

Benchmarking can be provided on the basis of energy use normalized to building area, for example, as kBtu/SF/year (called the *Energy Utilization Index*), or as a score, for example between 1 and 100. Benchmarking results as scores, for example, on a scale from 1 to 100, suffer from the inability to distinguish between buildings at the extreme ends of the scoring scale. For example, a building that uses much energy might score a 1 on a scale of 1 to 100. However, another building that uses twice as much energy as the first building will also score a 1.

With all the information that benchmarking can provide, it is also important to recognize what benchmarking cannot provide. Benchmarking of whole-building energy use cannot provide a breakdown of end uses, to identify an unusually high specific energy use within a building, which might be offset by unusually low energy use by another component, and so not show up in benchmarking data. Benchmarking cannot identify building failures, such as the sites of infiltration, or a heat pump that is low on refrigerant charge, or the impact of leakage of ductwork in a ventilated attic. Benchmarking cannot perform infrared scans to identify missing insulation or measure infiltration with a blower door. In short, benchmarking cannot do all the things that people can do who visit and inspect and come to understand a building.

Benchmarking cannot substitute for seeing a building up close. Benchmarking can and does serve an important role in energy work, but its limitations are important to recognize.

Billing Analysis

Billing analysis can be used to tell us more about energy use in a building, in addition to and beyond benchmarking. Goals of billing analysis include obtaining billing rates for use in energy audits, identifying billing errors, measuring energy savings, truing-up energy models, and disaggregating (separating) some energy uses, where possible. It is also possible to use energy bills for some specialty uses, such as estimating the required capacity of a heating system, estimating the added appraised value of a building, predicting future energy bills, and calculating utility reimbursements, such as those used by the Department of Housing and Urban Development. Billing analysis is also called *preliminary energy-use analysis (PEA)*.

Disaggregation of some energy uses is an important part of billing analysis. For example, billing analysis can be used to disaggregate heating energy use from cooling energy use, and both from nonheating and noncooling use. By examining different meters, we can typically differentiate energy use in common areas from energy use in tenant areas. Sometimes, because of the use of a single fuel for a single end use (e.g., domestic hot water or cooking), or through monitoring of specific loads, further disaggregation is possible.

The classic approach to disaggregating heating energy use is to plot monthly data, identify summer baseload energy use (ostensibly without heating energy use), and subtract the baseload energy use from the total annual energy use, to obtain heating-only energy use. This approach does not account for the possibility that winter baseload energy use is higher than summer baseload use. Outliers can also make the analysis difficult. However, this approach is still widely used and can provide a broad-brush estimate of heating energy use. If summer electricity use is higher than winter use, this approach can also be used to disaggregate cooling energy use.

A more accurate approach to heating disaggregation involves plotting daily-average energy use (on the y axis) for each billing period against either average outdoor temperature (on the x axis), or against heating degree days (also on the x axis) for the period. Billing periods without heating use should fit a horizontal line, and billing periods with heating use should fit a sloped line. The intersection of these two lines represents the inflection point, at which heating ends. Baseload (nonheating energy use) can then be estimated, per day, extrapolated to a whole year, and subtracted from total energy use to obtain heating energy use. An interesting side note is that this analysis can identify an undersized heating system, if a second horizontal line appears for midwinter billing periods.

Billing analysis can be helpful in advance of an energy audit, specifically before fieldwork is done, in order to allow checking that all meters have been identified, and in order to provide hints of possible problem areas in energy use.

Whatever the purpose of billing analysis, it can be made more accurate by reading one's own energy meters, and tracking both metered energy use and batch energy use such as fuel oil. Taking one's own readings eliminates many data quality issues, such as missed meters, missed meter readings or batch deliveries, mistaken meter readings, estimated meter readings, batch fuel deliveries that are far apart in time, rounding in batch fuel delivery data, and inability to access billing data.

Billing analysis sees many of the same challenges as seen by benchmarking, including a broad number of data quality issues, building use changes, and weather impacts. A new challenge has arisen in the introduction of on-site renewable energy, such as photovoltaics, making billing analysis more complex.

Advanced billing analysis seeks to account for demand charges, time-of-use rates, and other class-specific aspects of energy costs. Most of these do not have a significant impact on energy use and savings but can be useful for those interested in cost savings.

For those interested in carbon emissions reductions, we note that energy improvements can often reduce energy cost without reducing carbon emissions. In some cases, carbon emissions might even be increased while energy costs are reduced. Examples might include demand-reduction improvements, where daytime demand is reduced, but nighttime demand and usage might be increased. So, for those interested in reductions in carbon emissions, it might be worthwhile calculating and reporting these carbon emission impacts, in addition to energy and demand cost reductions.

As part of billing analysis, we consider several possible metering improvements. For example, consolidating meters can reduce metering charges, if there are more meters than necessary. It is also possible to eliminate meter charges if a specific fuel is eliminated, for example, a gas-heated building that is converted to heat pumps. Occasionally, incorrect utility classifications can lead to cost savings, when we identify and correct other billing errors.

Monitoring

Billing analysis can be significantly augmented by on-site energy monitoring of specific loads, in developing an understanding of a building's energy use. Short-term and long-term monitoring can also be used to prevent unwanted increases to energy use. For example, by installing a water meter on the makeup water line to steam and hot water systems, we can be vigilant for unusual increases in hot water use and associated energy use.

Plug loads can be monitored over short-term periods of one day to a week, using 120-volt plug-in watt meters for loads such as computers, other office equipment, refrigerators, and power strips. For loads that vary between weekdays and weekends, such as office equipment, seven-day monitoring is important. Useful readings are the accumulated electricity consumption (kWh) and the elapsed time over which the energy use accumulated. This electricity use can then be prorated by the elapsed time to estimate annual electricity use:

$$\text{Estimated annual use (kWh/year)} = \frac{\text{Accumulated energy use (kWh)} \times 8,760}{\text{Elapsed time (hours)}}$$

It is important to differentiate electricity consumption (kWh) from demand (kW), and not confuse the two. For 240-volt loads such as stoves and clothes dryers, and for hard-wired loads, data-loggers with true power input sensors can be used.

Appropriate times to do such monitoring are when an electricity-consuming device is first purchased, and then periodically thereafter, perhaps every three years. We look for changes in a device's power usage that indicate efficiency deterioration or change in use.

Permanent submeters can be installed to monitor an entire subpanel, for example serving one area of a building, or a tenant in a building. Submeters can even be applied to a single circuit of a panel, in order to disaggregate usage more finely, by load.

Trending can be performed using either data-loggers or central energy management systems. Of particular interest are temperatures that will indicate deterioration in the efficiency of energy systems: return and supply air temperatures on direct expansion systems and on furnaces, and return and supply water temperatures on boilers and chillers. Measuring outdoor air temperature, return air temperature, and mixed air temperature can be examined in order to assess air leakage at ventilation dampers, air leakage at room air conditioners, and the like. (See Appendix C.)

Retrocommissioning

Commissioning can be defined as ensuring that building energy systems work correctly. It is a term usually applied to new buildings. *Retrocommissioning* is defined as commissioning an existing building.

Retrocommissioning involves methodically testing controls to make sure control actuators (valves, dampers) work as intended, as well as heating and cooling plant (chillers, boilers, furnaces, direct expansion equipment) and support equipment (pumps and fans). At its simplest, we call for heating and cooling for each thermal zone and see how each controlled piece of equipment responds. We turn fans on and off. We check actual occupancy schedules against programmed schedules, such as for lighting, ventilation, and setback temperature controls. Retrocommissioning means taking the time to check things we wish we had time to check in an energy audit, or in the everyday work of an energy manager.

Retrocommissioning has been found to deliver typical energy savings of "about 15%."[1] Caution is advised against simply assuming savings of 15%, for an example in an energy audit, in advance of retrocommissioning. We cannot definitively predict how much energy retrocommissioning will save before the work is done because we do not know what deficiencies will be found and corrected. Savings estimates are more prudently done after deficiencies are identified.

Some say that retrocommissioning should happen before an energy audit, so that the energy audit can use the findings and changes performed in retrocommissioning to estimate its savings, and then proceed with evaluating structural changes to the building's energy infrastructure. Others say that retrocommissioning should happen after the energy audit, and indeed after structural changes have been made to a building's energy systems, in order to avoid retrocommissioning equipment that may subsequently soon be replaced.

An independent, accredited *commissioning provider* typically provides retrocommissioning. Executing a retrocommissioning project is done in three steps:

1. Establish the current owner's and occupants' current building use requirements. These requirements have typically changed from when the building was originally designed. This process is similar to the development of *Owner's Project Requirements* for a new building design project and its commissioning, and we might call the retrocommissioning version of this document the *Building's Current Requirements*. A review is also done of original design documents, original test and balancing reports, and changes to the building.

2. Test the function of the building's energy control systems, heating and cooling plant, lighting, and peripheral/support equipment such as pumps and fans. This is called *functional testing*. The goal is to identify deficiencies.

3. Implement changes in order to correct deficiencies. This should include revisions to building documentation, and operator training.

Short-term monitoring may form a part of retrocommissioning. Short-term monitoring can substantially strengthen the diagnostics of retrocommissioning. Trending capabilities of an existing energy management system may be used for this monitoring and/or short-term monitoring equipment purposefully installed to obtain such data.

The Building's Current Requirements sets forth the owner's and occupants' goals, including the main purpose of the building, relevant history, future needs, expected remaining building life, and intended use for all spaces. Environmental goals are also set forth, including voluntary or mandatory certifications (e.g., Architecture 2030, ENERGY STAR, LEED, and Passivhaus), energy efficiency (e.g., a specific percentage below code, or net-zero site energy), carbon emissions, thermal comfort, specialty lighting, and owner priorities for assessing green options (such as lowest emissions or lowest life-cycle costs).

The owner should preferably identify, on a space-by-space basis, the occupancy (number of people) and type of activity for each hour of a typical weekday and for weekends. This information is used for retrocommissioning of the ventilation system, lighting controls, and heating and cooling controls. The more detail, the better. If occupancy information is guessed and is too conservative (too many people assumed), the ventilation airflow will be too high. Accurate occupancy information also facilitates quality control through subsequent retrocommissioning tests.

Target light levels should also be established in the Building's Current Requirements document. Target light levels should be documented, with the default being the low end of Illuminating Engineering Society (IES) recommendations. Lighting controls should be selected and documented on a space-by-space basis, for example "manual on, occupancy off, one-minute off-delay" or "manual control, multilevel switching to allow 1/3 and 2/3 the maximum light level that shall be reconfigured to the low end of IES recommendations." Outdoor lighting needs should also be identified, with a discussion about which lights are required for security, which lights are required for access, which lights are required for outdoor evening recreation, and which lights are required for decorative or signage purposes. Security needs for outdoor lighting should be further explored: Can use of motion sensors deliver greater security and lower energy usage? If motion sensors are not used, are all outdoors lights required for security all night long, or can some lights be turned off at the end of the evening?

Current target indoor temperature and humidity should be identified, also on a space-by-space basis, for summer and winter, and for both occupied and unoccupied modes. As part of this process, the owner should clearly identify which spaces currently need heat, which spaces currently need cooling, and which spaces currently need neither heating nor cooling. Further, temperature and humidity control capability should be identified, on a space-by-space basis. In other words, which spaces need control capability? The retrocommissioning professional should spell out the tradeoffs for each type of control so the owner can make informed decisions, as these decisions will strongly impact both energy savings and comfort.

Best practices in the development of Building's Current Requirements include:

- Organizing workshops where key stakeholders can participate in a discussion of the building's current requirements, and where the importance of the requirements can be reviewed. A two-phase workshop works well, with distribution of draft Building's Current Requirements after the first workshop, and distribution of the final document after the second workshop.

- Avoiding generalities in the Building's Current Requirements. For example, "*to operate at a high level of efficiency to minimize utility consumption,*" does not say as much as "*building improvements to meet an ENERGY STAR score of 85*" or "*to meet a maximum energy utilization index of 30 kBtu/SF/year.*"

- Focusing on the owner's needs, and not on topics that should be covered by professionals, such as outdoor design temperature conditions. Every entry in the Building's Current Requirements should be understood by the owner. This promotes owner involvement in developing the requirements and reduces the risk of the document simply being completed by the retrocommissioning provider.

With the Building's Current Requirements, the more detail, the better. It is a rare opportunity for the owner to learn, understand, and choose among important building operation options. It is an invaluable vehicle for the retrocommissioning provider to communicate these options to the owner. The Building's Current Requirements document forms the foundation for the best possible retrocommissioning, and for the energy audit and improvements as well.

Periodic Energy Audits

Periodic energy audits should be scheduled, preferably every five years. Between energy audits, the energy manager should keep an eye out for potential energy improvements. The energy manager should keep up to date with new energy technologies, by reading trade publications, attending conferences and expositions, and talking to equipment vendors.

Reference

1. EPA ENERGY STAR, *Building Upgrade Manual.* Chapter 5: Retrocommissioning. Revised October 2007.

Chapter 22

Portfolio Programs

Portfolio programs are initiatives to save energy across multiple buildings. Private-sector corporations, utilities, public housing authorities, universities, the federal government, state governments, cities, and increasingly even counties and towns run portfolio programs. An example might be initiatives by a national retail chain, seeking to meet sustainability goals across all its buildings, and committing to disclose results of its initiatives. For such firms, energy efficiency is an attractive way in which to meet sustainability goals because of the associated energy cost savings.

The role of portfolio programs in energy conservation cannot be overestimated. Portfolio programs largely drive energy improvements in buildings. Portfolio programs also drive innovation, by proving concepts, testing them, demonstrating them, and building a body of evidence that supports acceleration of new technologies and new programmatic approaches. The technical advances of portfolio programs further allow energy code requirements to rise, as new approaches have been proven and accepted.

Public and utility portfolio programs typically offer incentives to encourage energy conservation. Incentives might include cash rebates, tax incentives, low-interest financing, or nonmonetary incentives such as a high-performance building certification, group purchasing, coordinated outreach/marketing, or other support. Portfolio programs are one of the largest drivers of energy conservation in the United States.

Elements of effective portfolio programs include the establishment of clear goals; standards; quality control; incentives; effective outreach/marketing; requirements for certification of participating energy professionals; cost control and cost reduction; recognition of participants; identifying non-cost benefits; identifying challenges/barriers and associated solutions; evaluation; continuous improvement; research and development, such as acquiring data from participating buildings, to assess what works and what does not work; the development of best practices; shared experience through training, conferences, white papers, demonstrations/pilots, and so on; and ultimately market transformation, as successful technologies and practices are adopted more widely.

Goals of Portfolio Programs

A primary goal of portfolio programs is energy savings. They are successors to earlier initiatives, which began to address the energy crises of the 1970s. Increasingly, the primary goal has become carbon emission reductions, through energy savings.

A variety of secondary goals underlie these primary goals and are implicit in various aspects of portfolio program design:

- Market stimulation.
- Cost control, through competition, bulk purchasing, economies of scale, innovation, and the control of soft costs, such as permitting, design, etc.

Figure 22.1 Portfolio programs are one of the largest drivers of energy conservation in buildings.

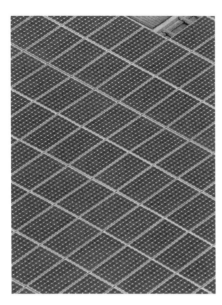

Figure 22.2 The photovoltaics industry was successfully transformed through judicious use of portfolio programs.

- Data acquisition, resulting from the portfolio programs, which allows increasingly targeted and cost-effective efforts.

- Standardization and quality control. These elements, administered by the portfolio programs, are intended to maintain quality, serving as a form of consumer protection for building owners entering the energy conservation field, as the field itself emerges and matures.

- Acceleration of new technologies, through demonstration and adoption by portfolio programs.

- Economic development.

Many of these secondary goals have been successfully demonstrated. For example, federal and state portfolio programs supporting renewable energy have substantially contributed to the transformation of the photovoltaic industry, with significant market growth, reductions in material costs, and advances that have reduced installation costs and soft costs as well. See Figure 22.2. All six of the secondary goals have at least been partially achieved: The market was stimulated, costs have come down, data have enabled wider and more cost-effective efforts, standardization has increased and quality has generally been good, new technologies have been accelerated, and jobs have been created. The federal tax credit has certainly helped. Significantly, success has been stronger in states that have had their own portfolio programs to supplement the federal tax credit. In states with weaker or no solar programs, all of which have still benefited from the substantial federal tax credit, success has been weaker.[1]

We would suggest that portfolio programs have another important goal, which strongly supports transformation: training. As portfolio programs stimulate energy improvement efforts, everyone learns from the process. Building owners and managers, tenants, facility personnel, energy auditors, government/utility program managers, and the general public all learn about energy conservation. This only serves to further the primary goal of energy savings, themselves.

In discussing goals of portfolio programs, we recognize the power of goal-setting, in achieving goals. It has been found that if concrete energy-savings goals are set, that goals are more likely to be achieved. Conversely, if no goals are set, energy savings achieved are typically low. For example, a random sample of commercial energy audits identified only 9 percent savings, in a portfolio program without any specific target for energy savings. In contrast, energy savings averaged 28 percent in a sample of energy audits in a portfolio program with a goal of 20 percent savings goal. And energy savings averaged 34 percent in a sample of energy audits with in a portfolio program a goal of 30 percent savings. How far can we go? In a residential program with a goal of over 75 percent savings, actual savings averaged 83 percent. It appears that we save as much as we want to save. More importantly, if we set a goal that is low, we likely will only save a low amount of energy, and if we set no goal at all, we will save very little energy. Goal-setting is of paramount importance in energy portfolio programs.

Definitions

In discussing portfolio programs, common definitions include:

Naturally occurring conservation. These are energy improvements that might happen even in the absence of the portfolio program.

Collateral effects. These effects are consequences of portfolio programs other than the sought energy conservation in the target buildings. Collateral effects can be positive or negative.

Take-back (also called snap-back, or rebound). Take-back is a collateral effect in which there is *increased* energy use as a result of increased equipment use, allowed by lower energy costs resulting from higher efficiency. In other words, because equipment is higher in efficiency, and so energy costs are lower, users are at risk of using the equipment more, and so offsetting some of the savings that would have been delivered. Take-back is undesirable.

Free drivership. Free drivership is a collateral effect in which equipment dealers or distributors become "free drivers": As a result of the demand of a portfolio program, free drivers introduce energy-efficient equipment, and then make this equipment available to customers beyond the portfolio program. Free drivership is desirable.

Free ridership. Free ridership is a collateral effect in which participants in a portfolio program benefit from incentives, but who would have engaged in naturally occurring conservation even had the portfolio program not occurred. Free ridership is regarded as undesirable, a use of incentive resources for people who would have pursued energy conservation anyway, even without a portfolio program.

Spillover. Spillover is a collateral effect in which energy-efficient equipment and approaches are adopted beyond a portfolio program, as a result of awareness that was generated by the program. For example, it has been observed that some owners who install photovoltaic systems then seek to reduce energy use in order to maximize what the photovoltaic system can deliver. Spillover is desirable.

Gross savings. Gross savings are the total savings of a portfolio program, without measuring negative collateral effects (such as takeback or free ridership) or positive collateral effects (such as spillover and free drivership).

Net savings. Net savings are the useful savings of a portfolio program, after subtracting out collateral effects.

Net-to-gross ratio. Net-to-gross ratio is the ratio of net savings to gross savings. A high net-to-gross ratio is desirable.

Low-hanging fruit. Low-hanging fruit refers to energy improvements that are easy to implement.

Cream-skimming. Cream-skimming is a term for pursuing only low-hanging fruit, and therefore sacrificing more comprehensive improvements and energy savings. Cream-skimming is not desirable.

Low-cost/no-cost measures. Low-cost/no-cost measures are energy improvements that, as described, are affordable. This term is often synonymous with low-hanging fruit.

Realization rate. Realization rate is the ratio of delivered energy savings to the projected energy savings, in other words, the ratio of net savings to projected savings in a building-specific energy audit or across a portfolio program.

Debates exist whether collateral effects can be measured accurately or whether common approaches to measure them are adequately accurate.

Portfolio Program Types

Portfolio programs might be divided into two broad types: Limited and comprehensive.

Limited portfolio programs typically target a single technology. Examples include utility rebates for high-efficiency lighting or cooling equipment, or the federal tax credit for high-efficiency renewable technology, such as photovoltaic systems.

Figure 22.3 Limited portfolio programs target a single technology. Comprehensive portfolio programs encourage multiple energy improvements, building-wide, to deliver measurable energy savings.

Comprehensive portfolio programs encourage multiple technologies, building-wide. For example, the federal 179d tax deduction rewarded whole-building energy improvements in envelope, heating/cooling, and lighting. Leadership in Energy and Environmental Design (LEED) for Existing Buildings (EB) and ENERGY STAR are comprehensive portfolio programs, awarding certification to buildings that achieve a minimum whole-building energy reduction or efficiency, through either modeling (LEED) or utility bill benchmarking (ENERGY STAR).

Limited portfolio programs are characterized by simplicity, limited savings, and difficulty in verification, as savings rarely reach a whole-building level, and therefore cannot be verified with billing analysis. An exception to the rule on verification is distributed generation systems, such as photovoltaic systems, which now frequently come with reliable verification systems that measure delivered energy. Limited portfolio programs are sometimes characterized as cream-skimming because a single, ostensibly high-return, technology is deployed. Effectiveness in delivering energy savings is often also called into question, as a technology might be eligible for and receive a rebate or other incentive, despite ineffectiveness. Consider a low-efficiency lightbulb in a closet that is only used for 10 minutes a day, replaced by a high-efficiency lightbulb. The owner receives a rebate, but the energy savings will take well over 100 years to pay for the cost of the lightbulb.

Comprehensive portfolio programs are characterized by greater complexity, deeper savings, and more reliable verification that is frequently based on actual reduction in energy consumption, as shown on utility bills. Well-designed comprehensive portfolio programs, with well-executed quality control, have been shown to reliably deliver realization rates over 90 percent, with comprehensive building-wide energy reduction over 20 percent.

Target Building Portfolios

Target building portfolios might include all the buildings owned by a developer or by an entity such as a college or university; all commercial/industrial buildings within a utility territory; large commercial buildings in a city; a class of building such as schools or commercial office buildings; and so forth. The federal tax credit for renewables, for example, is a portfolio program that ostensibly targets all buildings, although practically the target portfolio is restricted to entities that pay income tax.

Portfolio programs can either be voluntary or mandatory. Traditionally, most programs have been voluntary, such as utility rebates for high-efficiency equipment, federal and state incentives for energy-efficient equipment and renovations, and certification programs for high-performance buildings, including, for example, LEED and ENERGY STAR. As the urgency has increased for a coordinated societal response to climate change, we are beginning to see more mandatory programs, such as requirements for benchmarking and energy audits in large buildings in cities such as New York, Philadelphia, Chicago, and others; and quasi-mandatory goals such as 20 percent energy reduction by 2020 in state-owned buildings in New York, and the like. The portion of the energy conservation code that addresses existing buildings might be considered to be a mandatory portfolio program, targeting those buildings that undergo renovation.

As we examine target-building portfolios, we are interested in issues of *scale* and issues of *underserved populations*. Scale refers to the ability of a portfolio program to ramp up to allow us, as a society, to achieve measurable energy savings. This is desirable, as we go about combating the emissions that contribute to climate change and other forms of pollution. As we pursue scale, we also are interested to ensure that the needs of underserved populations are met. Underserved populations might include those who cannot afford the capital investment of energy conservation improvements, such as people who live in low-income housing. Underserved populations might also include those who cannot qualify for portfolio programs. For example, not-for-profits cannot take advantage of tax credits and tax deductions. Small commercial buildings are frequently underserved, as they do not provide the scale that is desirable in large commercial/industrial portfolio programs, and they are not big enough to qualify for performance contracting initiatives. Buildings in territories served by municipal electric utilities can be underserved, as they are not eligible for state-mandated portfolio programs run by investor-owned utilities.

Incentives

Incentives can be financial or can be recognition based, such as a certification for high-performance buildings.

Financial incentives can be rebates, state or federal tax credits or deductions (including accelerated depreciation, or excluding the value of energy improvements from real estate tax assessments), low-interest financing, and the like. The federal government has had a 30 percent tax credit for renewable energy installations. The federal government has also had a tax deduction for comprehensive energy reductions in commercial buildings, termed the 179d program, requiring whole-building savings of 50 percent, or smaller savings for envelope-only, heating/cooling-only, or lighting-only improvements. Many utilities across the country offer rebates for high-efficiency lighting and heating/cooling equipment. Some utilities offer performance-based incentives, rewarding conservation on a case-by-case basis for estimated savings for larger energy conservation projects. Low-interest financing has been provided by states, as another form of financial incentive. Access to financing can itself be considered a form of incentive, regardless of the interest rate. An emerging financing mechanism is Property Assessed Clean Energy (PACE), in which energy projects are financed through property taxes.

The basis of financial incentives, in other words, how the incentive is calculated, varies depending on the portfolio program. The federal 30 percent tax credit is based on the installation cost of the renewable energy installation. Most utility rebates are based on the quantity of installed components, such as lightbulbs or high-efficiency heating/cooling products, on a per-unit basis. Performance-based portfolio programs vary the incentive depending on the delivered energy savings. Sometimes, financial incentives are applied to soft costs such as energy audits, or engineering design.

Another form of incentive might be called "recognition based." In recognition-based incentives, a building receives a form of recognition, or certification, by achiev-

Figure 22.4 Tax incentives are powerful but may not provide any quality control. Tax incentives combined with state or local portfolio program incentives can powerfully deliver both scale and quality in energy savings.

ing a defined level of energy efficiency. Examples include EPA's ENERGY STAR certification for buildings, the U.S. Green Building Council's LEED program for existing buildings (LEED EB), and others.

Tax incentives are powerful, and clearly have contributed significantly to the success of such initiatives as renewable energy, through the 30 percent federal tax credit. However, tax incentives typically do not have any associated quality control or other aspects of successful portfolio programs that engage the building owner in training or that disseminate program results. A powerful approach is to combine tax incentives, such as federal tax credits, with local portfolio programs, such as state programs. In this way, the power of the federal tax incentives is complemented and augmented by the quality control and other components of a local state program. The federal tax incentive serves as a carrot, and the state program serves as a stick, in order to maintain quality and ensure that savings are delivered.

Do incentives work? Are incentives always needed? Do incentives need to remain long-term? These questions are central to the entire effort to save energy, and reduce carbon emissions, in our society. The case is made that incentives represent the cost to society of the pollution of the commons caused by energy use. Without incentives to reduce energy use, the consumer of energy is only paying for the extraction and delivery of fuels and energy, and is not paying for its cleanup, health impacts, or environmental impacts.

Are incentives needed long-term? Can the entire energy conservation effort not simply be turned over to the markets, as markets are transformed? The initial lens of energy conservation efforts has been very much focused on market drivers. We have focused on energy cost savings, on payback, on life cycle costs, on market transformation, on market stimulation, and on a vision that incentives can be reduced and then withdrawn, with the markets picking up where incentives leave off. However, funny things have happened along the way. Instead of energy prices going up, in many areas, and for many energy sources, prices have come down. The discovery of new fossil fuel sources, and new techniques for extraction, have driven down the cost of many fuels. Indeed, we should fully expect that if our overall energy conservation efforts are successful, demand should drop, and so prices should drop. Market-driven energy conservation is entirely dependent on fuel prices. As prices have dropped and continue to drop, the prime driver for market-driven energy conservation falls away. We need to be aware of these market dynamics. We need to know that it is very possible that as we wait for markets to take over, the markets may simply not show up. Careful calibration and adjustment of incentives may be the only way to continue energy conservation efforts, as we observe the degree to which markets do or do not participate in the process. Incentives should be seen not only as a tool for market stimulation, but in fact as a responsibility of society to reduce the devastating consequences of unconstrained burning of fossil fuels without concern for their environmental impacts. Over-reliance on markets, and premature termination of incentives, may well come to be viewed as irresponsible.

Equipment Disposal

An important but little recognized component of portfolio programs is the disposal of removed equipment, such as lamps, ballasts, low-efficiency heating and cooling equipment, and low-efficiency appliances. Options include:

- Require removal and decommission, or
- Allow reuse, ostensibly under the justification that there is embodied energy in the device. However, the reuse of low-efficiency equipment reduces demand and acquisition of new, and energy code–compliant, higher-efficiency equipment. Reuse typically also violates laws regarding the resale and distribution

of energy equipment that does not comply with federal efficiency standards. The embodied energy of removed low-efficiency equipment is typically far outweighed by energy savings. Unless defensible calculations definitively show otherwise, all removed low-efficiency equipment should be required to be decommissioned and recycled or disposed of in compliance with state and local solid waste regulations.

The question of possible reuse of removed low-efficiency equipment speaks to the power of installations that go beyond one-for-one equipment replacement (limited portfolio programs) and instead promote deeper energy conservation (comprehensive portfolio programs). Installations that both replace equipment and improve installation aspects, such as controls or building-specific application, have a greater likelihood to actually deliver persistent savings. As an example, one-for-one lamp replacements are subject to take-back collateral effects, and also risk the reuse of the old, inefficient lamp in another application or building. If, instead, lamp replacement is combined with right-lighting (reduced overlighting) and lighting controls (e.g., vacancy sensors), then savings are obtained even if take-back occurs or even if the old, inefficient lamp is reused elsewhere.

Quality and Effectiveness

We have already touched on how the science and study of quality are slowly but steadily arriving in the field of energy improvements. First principles of quality include defining requirements, adhering to requirements, and measuring results in order to continuously improve adherence to requirements. In the energy field, and specifically in portfolio programs, this means defining the program's goals, setting forth clear program requirements, checking for adherence to requirements, and measuring results. In seeking to deliver results that adhere to requirements, that meet our energy improvement goals, we want to structure portfolio programs that not only provide *defect detection*, for example, inspections of completed work and measurement of energy savings, but that provide *defect prevention*. Defect prevention means defining strong program requirements, providing training, requiring professional certification, providing clear standards, providing defensible energy savings calculations guidance and methods, and encouraging robust and persistent energy improvements, rather than energy improvements that are token or limited.

For portfolio programs, two distinct approaches to quality are defined: *quality control* and *quality assurance*. With *quality control*, we seek to provide a structure and methods for checking quality for every project, through energy audit reviews and through inspections of completed work. With *quality assurance*, we seek to provide a higher level of evaluation and feedback, in order to improve the program as a whole. We might think of quality control as being directed to defect detection, and quality assurance as being directed to defect prevention.

Quality control in portfolio programs should include reviews of energy audits. The review should include both a technical review of the energy audit (see Appendix S), and a review for compliance with rules of the specific program.

Best practices for on-site inspections are the same as those described for the project manager:

- Allow enough time for inspections.
- Schedule routine inspections. Daily inspections are best, weekly inspections at a minimum.
- Take notes and photos of inspected work.
- Issue inspection notes in a timely fashion, within a day after inspection.

Figure 22.5 Energy savings are most effectively ensured through quality control, and quality control is most effectively delivered through a combination of defect detection (inspection) and defect prevention (designed-in quality).

■ Follow up on unresolved defects uncovered ruing inspection, until they are corrected.

■ Reject bad work. This is a key to quality control. We must have sufficient character to reject defects.

The final test of whether projects deliver energy savings is in *measurement and verification* (M&V). Options for M&V have been well defined in the International Performance Measurement and Verification Protocol (IPMVP).[2] Options include:

Option A—Retrofit Isolation: Key Parameter Measurement. Here, we measure some parameters in the field, but not all parameters, and use estimates for the parameters we do not measure. For example, for a lighting improvement project, the lighting power draw is measured, but the lighting operating hours are estimated.

Option B—Retrofit Isolation: All Parameter Measurement. Here, all parameters are measured. In the preceding example, both lighting power draw and duration would be measured.

Option C—Whole-Facility Measurement. Here, the utility bills for the whole building are used, for a 12-month period prior to the energy improvements, and again for a 12-month period following the improvements. Adjustments are made as necessary, for example, if weather patterns for the baseline period and post-retrofit period are different.

Option D—Calibrated Simulation. This approach might suit a building in which no baseline measurements are available, for example, a building that has just been built or that was unoccupied prior to an energy retrofit. Measurements that are available, for example, for the post-retrofit period, are used to calibrate a computer simulation, and the computer simulation is then used to estimate energy for a hypothetical baseline period, in order to calculate savings.

Each option has advantages and disadvantages. Options A and B, for example, are difficult to apply to envelope improvements, such as insulation and air sealing, when heating and cooling changes have also been made.

For the case of transformational energy improvements, where substantial energy reduction is the goal, as is increasingly the case, Option C makes the most sense. Most existing buildings do in fact have valid baseline energy usage data. And Option C is based on real measurements, rather than subject to a variety of vagaries and gaming of simulation.

Evaluation

Many portfolio programs integrate some form of evaluation. The purpose of evaluation is to ensure that programs are well-designed and investments are well-spent. The focus is both on the *process* of the portfolio program's design and execution, and on the *impact* of actual energy savings, as compared to estimated savings, while seeking to quantify collateral effects. Evaluation can include billing analysis, on-site measurement and verification, documentation review, interviews/surveys, and more.

Evaluation of the energy impacts of isolated savings improvements, such as lighting replacements or any limited one-for-one replacement, is always difficult. The evaluator faces many of the same difficulties as are faced by the energy auditor: not knowing exact equipment runtimes, not knowing the exact efficiency of baseline equipment (which is typically no longer in place, making the baseline energy use even more difficult to know), not knowing the operational efficiency of new equipment, not knowing exact usage patterns. The accuracy of evaluation of isolated savings improvements is typically no better than the projection made in the energy audit, and in many cases is worse, because the baseline cannot be evaluated.

By contrast, the evaluation of integrated savings for buildings that have sought to measurably reduce energy use on a whole-building basis is very simple. If the improvements have made any dent at all in the utility bills, these impacts are readily available to the evaluator, both for the pre-improvement period and for the post-improvement period. Corrections might need to be made for differences in weather conditions, between the pre-improvement and post-improvement periods, but protocols for these corrections are fairly standard and straightforward. The confidence in whole-buildings savings impact evaluation speaks in favor of transformational whole-building energy improvements, where energy use reduction is measurable and clearly falls outside the noise of utility bills. We know when we have succeeded, and we know when we have failed. This is in stark contrast to limited one-for-one energy improvements where, other than perhaps knowing that the new equipment was in fact installed, we never know much about the impact of the installation. We never know if we have really saved energy. The savings are so small that they are lost in the noise of the utility bills.

Challenges

Challenges with portfolio programs include serving difficult-to-reach buildings: Low-income populations, small buildings, not-for-profits, and other underserved building portfolios.

Another challenge arises due to the *split incentive*. In buildings with a landlord and tenant, either one or the other has an incentive to save energy, based on who pays the energy bills. If the building is master metered, with the landlord paying the energy bills, the tenant has no incentive to save energy. In this case, the tenant might leave lights on or not adjust heating and cooling temperatures. If the building is individually metered, with the tenant paying utility bills, then the landlord has no incentive to save energy. In this case, the landlord might not make structural improvements, such as adding insulation to the building or changing the heating and cooling system. Energy conservation requires incentives for both landlord and tenant to save energy. The tenant needs an incentive primarily for *behavioral* energy conservation. The landlord needs an incentive for *structural* energy conservation, in other words, to make changes to the building infrastructure. Finding a way to incentivize both is the challenge of the split incentive.

Another challenge is to maintain stability in portfolio programs. A typical pattern is for staffing changes at a state or utility agency to bring about an interest in changing programs, for new program staff to have an opportunity to exercise creativity. However, this can cause instability in programs, creating confusion among building owners, property managers, consultants, and vendors. While it is important for programs to adapt to changing conditions, instability itself is not good. Uncertainty in the market can drive participants out, and training and experience are lost. Consistency and continuous improvement are hallmarks of stable and successful portfolio programs.

Portfolio programs also run into challenges when they run out of budget. Rather than manage programs to a fixed budget, program success in delivering portfolio-wide energy savings should be encouraged by expanded budgets.

A variety of common mistakes appear in portfolio programs:

Territoriality. This is where programs compete, for example, state agencies and utility programs seeking the same customers, causing confusion among customers.

Instability. Instability arises when programs continually change or are terminated prematurely. There can be a tendency for programs to be subject to what might be called the reinvent treadmill, with revolving management at government agencies and utilities seeking to put their own stamp on programs. Instability causes confusion among customers and energy firms.

Incentives pendulum. Incentives in portfolio programs frequently swing wildly. At the beginning of a program, incentives are set too high, as program managers are concerned about insufficient participation. Over time, incentives are over-reduced, or eliminated, as funds dry up or as program managers seek to show that they can reduce program costs. As participation dries up and energy savings goals are not met, incentives are re-instituted, often at a level that is again too high.

Gaming. Gaming is where incentives are sought to be maximized through bending program rules, such as by overestimating energy savings, through overestimating baseline energy usage or underestimating proposed energy usage, by program participants. Gaming is often done even when incentives are not involved. A consultant is frequently tempted to overestimate savings in order for an energy audit to look promising. Gaming is a negative aspect of portfolio programs. Gaming hurts the credibility of the energy conservation industry, while also hurting specific programs, consultants, and, of course, the building owner.

Micromanagement. This is where too many constraints are placed, through program rules, and arbitrary oversight is applied to individual projects.

Lack of scale-up. This is where too much focus is directed to demonstration projects and pilots.

Research and Data Needs

Portfolio programs have a variety of research and data needs, from which they benefit in order to deliver programs that deliver energy savings reliably and cost-effectively. Moving forward, we expect that increasing effort will be directed to provide portfolio programs with access to a variety of research and data:

Equipment efficiency ratings data. Efficiency data on old equipment (boilers, air conditioners, furnaces, chillers, cooling towers, appliances, refrigeration equipment, and more) is difficult to find, is not stored for many types of equipment and, where stored, is made available only at great cost to portfolio programs and their participants. Portfolio programs need more ready and affordable access to these data. Without it, we struggle to calculate energy savings accurately.

Utility data. Portfolio programs need utility data in order to model energy savings. These data can be exceedingly difficult to access. Convenient electronic-form data is required not only from investor-owned utilities but from suppliers of bulk fuels, like oil and propane.

Best practices. Best practices are needed for a wide variety of energy improvements. These include indoor temperature set points for different space types (both setup and setback temperatures), best-practice lighting power densities for different space types, efficient lighting patterns for different space types, safe but efficient illuminance levels for different space types (industry-standard recommendations tend to be fairly broad, for purposes of energy audit work), and more statistically valid expected equipment lifetimes.

Conclusion

Portfolio programs form the basis for large-scale energy improvements across many buildings. Their importance cannot be overstated. Great strides have been made in recognizing the components of successful portfolio programs: clear goals; standards; quality control; incentives; effective outreach/marketing; certification of energy professionals; cost control; recognition of participants; identifying noncost benefits; identifying challenges/barriers and associated solutions; evaluation; continuous improvement; research and development; dissemination of best practices; training; demonstrations/pilots; and ultimately market transformation, as successful technologies and practices are adopted more widely.

With the emergence of hundreds, possibly even thousands, of portfolio programs at every level of government, in utilities, in the private sector, at universities, and more, there is a strong need for harmonization of these programs. There is a need for rigorous and transparent evaluation, with peer-reviewed research-level assessments of what works and what does not work, in order for robust best practices to be developed, disseminated, and harmonized. With more consistency between programs, energy auditors and energy managers can meet the growing need for energy improvements in compliance with the programs. In an ideal world, effective portfolio programs would be portable and replicable.

With only a few decades of collective experience, the energy improvement field is young. We are all still learning. There is much important and good work to do. As we share experiences, share best practices, learn what works, and measurably reduce energy usage, our work just becomes more fulfilling. The promise of transforming our existing buildings, into making them buildings that work for us, and in the process cutting carbon emissions and dependence on fossil fuels, is a promise worth pursuing.

References

1. http://cleantechnica.com/2013/06/25/solar-power-by-state-solar-rankings-by-state/. Accessed August 16, 2015.
2. Efficiency Valuation Organization, "International Performance Measurement and Verification Protocol (IPMVP): Concepts and Options for Determining Energy and Water Savings, Volume 1." EVO 10000 – 1:2012, January 2012.

Chapter 23

Resources

The following resources are unusually rich in the breadth and depth of materials that each includes. It is not intended as a comprehensive list, but rather a valuable set of starting points for energy improvements in existing commercial buildings.

Air Conditioning, Heating, and Refrigeration Institute (AHRI). Maintains online searchable directories with energy ratings of new equipment. Data are exportable, but unfortunately only 250 records at a time. The directories include a wide variety of heating and cooling equipment and also include water heaters. Packaged terminal air conditioners (PTACs) are covered in the AHRI directories, but not room air conditioners, which are covered by the Association of Home Appliance Manufacturers (AHAM).

American Society of Heating, Refrigerating, and Air-Conditioning Engineers (ASHRAE). Engineering organization that produces a series of four outstanding handbooks, as well as widely used standards, including Standards 62 (ventilation), 90 (energy), and 189 (high-performing buildings), and relevant guidelines, including Guideline 14 (measurement and verification). Prior versions of many standards are available for browsing online.

Association of Home Appliance Manufacturers (AHAM). Maintains online directories with energy ratings of consumer appliances, including clothes dryers, clothes washers, dehumidifiers, dishwashers, refrigerators, freezers, and room air conditioners. Some of the directories are downloadable spreadsheets; others are searchable online directories.

Building Science Corporation (BSC). A private consulting and research firm that maintains a highly useful web site, and that develops and shares outstanding research on the building envelope (primarily insulation and air sealing), as well as related topics such as humidity control and ventilation.

California Energy Commission (CEC). California's energy agency maintains the state's energy code, called Title 24, which typically leads the nation in energy efficiency requirements. The Title 24 standards, and research to support the standards, are a rich source of data on existing equipment and high-efficiency equipment. The CEC also maintains an outstanding appliance database, which includes the rated energy efficiency of archived appliances, as well as heating/cooling and water heating equipment, stretching back for several decades.

ENERGY STAR. Maintained by the U.S. Environmental Protection Agency (EPA), ENERGY STAR is a voluntary program with better-than-code energy certifications for a wide variety of equipment and products, as well as whole buildings. Online resources include databases of efficient products and equipment, and some energy calculators (typically downloadable spreadsheets).

Food Service Technology Center (FSTC). Operated by the Fisher-Nickel Inc., and funded by Pacific Gas and Electric, FSTC's online resources, calculators,

Figure 23.1 The promise of transforming our existing buildings is one worth pursuing.

and reports cover a wide range of energy topics relating to food service, including ventilation, hot water, cooking equipment, and refrigeration.

Lighting Research Center (LRC). An independent research and education organization, addressing all areas of lighting, based at Rensselaer Polytechnic Institute, in Troy, New York. LRC's web site contains a wide variety of best practices, research reports, lighting patterns for specific applications, and educational materials related to lighting.

Natural Resources Canada (NRCAN). NRCAN is the Canadian government agency responsible for energy. NRCAN maintains a variety of resources, including online product and equipment directories. NRCAN is also the home of CanmetENERGY, an energy science and organization that disseminates information on energy efficiency and renewable energy and maintains a suite of software for modeling energy savings.

Whole Building Design Guide (WBDG). A web-based portal with online resources and calculators, such as a useful mechanical insulation calculator. While more oriented to new building design, the site nonetheless has useful material for the improvement of existing buildings. A program of the National Institute of Building Sciences.

U.S. Department of Energy (DOE). In collaboration with its national labs, DOE supports a variety of energy research and demonstration efforts. In collaboration with consultants, DOE supports the development and maintenance of national standards for appliances and other energy equipment.

Appendix A

Building Material R-Values

Material	R-Value for Thickness Shown	R-Value per Inch	Material	R-Value for Thickness Shown	R-Value per Inch
Air film			**Masonry materials**		
Still air, vertical surface, nonreflective	0.68		Brick, fired clay, 120 lb/ft^3 density		0.15–0.18
Moving air, 15 mph wind	0.17		Concrete block, 8″, 26–29 lb, medium weight aggregate	1.28–1.71	
Moving air, 7.5 mph wind	0.25		Stone, lime, or sand (average density)		0.03–0.05
Building board			Concrete, typical (150 lb/ft^3)		0.05–0.10
Gypsum or plaster board, $1/2$″	0.45		**Siding materials (on flat surfaces)**		
Particleboard, medium density		1.06	Shingles, wood, 16″, 7.5″ exposure	0.87	
Plywood (Douglas fir), $1/2$″	0.62		Shingles, wood, double, 16″, 12″ exposure	1.19	
Finish flooring materials					
Carpet and fibrous pad	2.08				
Carpet and rubber pad	1.23		Siding, asbestos-cement, 0.25″, lapped	0.21	
Cork tile, $1/8$″	0.28				
Tile—asphalt, linoleum, vinyl, rubber	0.05		Asphalt roll siding	0.15	
Tile—ceramic		0.19	Hardboard siding, 0.4375″	0.67	
Wood, hardwood finish, $3/4$″	0.68		Plywood, $3/8$″, lapped	0.59	
Roofing			**Woods (12% moisture content)**		
Asphalt roll roofing	0.15		Oak		0.80–0.89
Asphalt shingles	0.44		Birch		0.82–0.87
Built-up roofing	0.33		Southern pine		0.89–1.00
Slate	0.05		Douglas fir—Larch		0.99–1.06
Wood shingles, plain and plastic film faced	0.94		Southern cypress		1.09–1.11
			Hem-Fire, Spruce-Pine-Fir		1.11–1.35
			West coast woods, Cedars		1.11–1.48
			California redwood		1.22–1.35

Source: ASHRAE Handbook of Fundamentals, 2005. The above is a partial list only. For the full list, consult the *ASHRAE Handbook.*

Units:
R-value: h-SF-F/Btu
R-value per inch: h-SF-F/Btu-inch

Appendix B

Window Ratings

Year	Code/Standard	Climate Zone	U-factor (Btu/hr-SF-F)	SHGC	Notes
2000	IECC Commercial	1	0.7	Depends on projection factor	No U-factor requirements below 25% window-to-wall ratio (WWR)
		2	0.7		
		3	0.7		
		4	0.7		
		5	0.7		
		6	0.7		
		7	0.7	"	No requirements below 10% WWR
		8	0.7		
2006	IECC Commercial	1	1.2	"	Requirements are for metal frame windows; lower U-factors required for nonmetal for Zone 4 and higher
		2	0.75		
		3	0.65		
		4 except Marine	0.55		
		5, 6, and Marine 4	0.55		
		7 and 8	0.5		
2009	IECC Commercial	1	1.2	"	"
		2	0.75		
		3	0.65		
		4 except Marine	0.55		
		5, 6, and Marine 4	0.55		
		7 and 8	0.45		
2012	IECC Commercial	1 and 2	0.65	0.25	
		3	0.60	0.25	
		4 and 5	0.45	0.40	
		6	0.43	0.40	
		7 and 8	0.37	0.45	
2015	IECC Commercial	1 and 2	0.65	Depends on projection factor	
		3	0.60		
		4 and 5	0.45		
		6	0.43		
		7 and 8	0.37		

(Continued)

Year	Code/Standard	Climate Zone	U-factor (Btu/hr-SF-F)	SHGC	Notes
1996	CABO Model Energy Code (MEC)	Houston, TX (Zone 2)	0.65	Any	Example
		Kansas City, KS (Zone 4)	0.35	Any	Example
		Madison, WI (Zone 6)	0.35	Any	Example
1998	IECC Residential	<2,000 HDD	0.75	0.40	
		2,000–3,499 HDD	0.50	0.40	
		3,500–3,999 HDD	0.50	Any	
		4,000–5,999 HDD	0.40	Any	
		>6,000 HDD	0.35	Any	
2000	IECC Residential	<2,000 HDD	0.75	0.40	
		2,000–3,499 HDD	0.50	0.40	
		3,500–3,999 HDD	0.50	Any	
		4,000–5,999 HDD	0.40	Any	
		>6,000 HDD	0.35	Any	
2009	IECC Residential	Zone 1	1.20	0.30	
		Zone 2	0.65	0.30	
		Zone 3	0.50	0.30	
		Zones 4–8	0.35	Any	
2012	IECC Residential	Zone 1	Any	0.25	
		Zone 2	0.40	0.25	
		Zone 3	0.35	0.25	
		Zone 4	0.35	0.40	
		Zones 5–8	0.32	Any	
2015	IECC Residential	Same as 2012			
1998	ENERGY STAR Residential	<3,499 HDD	0.75	0.40	
		3,500–5,999 HDD	0.40	0.55	
		>6,000 HDD	0.35	Any	
2003	ENERGY STAR Residential	<3,499 HDD	0.75	0.40	
		3,500–5,999 HDD	0.40	0.55	
		>6,000 HDD	0.35	Any	
2009	ENERGY STAR Residential	Southern	0.60	0.27	
		South-Central	0.35	0.30	
		North-Central	0.32	0.40	
		Northern	0.30	Any	
2015	ENERGY STAR Residential	Southern	0.40	0.25	
		South-Central	0.30	0.25	
		North-Central	0.30	0.40	
2016	ENERGY STAR Residential	Northern	0.27	Any	

Appendix C

Air-Mixing Method of Airflow Measurement

If two airstreams are at different temperatures, the relative airflow rate (fraction of each airflow) can be estimated by examining each airflow's entering temperature, and the mixed air temperature.

Example applications include estimating the fraction of outside air being pulled into an air handler, or the fraction of outside air leaking into a room air conditioner.

If one air stream is at temperature T1, if the other air stream is at temperature T2, and if the mixed temperature is T3, the fraction of total airflow that is at temperature T1 is:

Fraction of air that is at temperature T1 = (T3 – T2) / (T1 – T2)

For example, outside air enters an air handler at 30 F. Return air is 70 F. The mixed airflow is 60 F. The fraction of return air is:

Fraction return air = (60 – 30) / (70 – 30) = 75%

Therefore, the fraction of outside air is 25%.

Appendix D

Recommended Illuminance

Building Area and Task	Average Maintained Illuminance (Foot-Candles - Horizontal)	Range of Maintained Illuminance (Foot-Candles - Horizontal)	Notes
Warehousing and Storage			
Bulky items—large labels	10		
Small items—small labels	30		
Cold storage/warehouse	20	10–30	
Commercial Office			
Office (open or private)	40	30–50	At 30″ above the finished floor.
Conference room	30		
Restroom	18	7.5–30	
Lunch/Break room	15	5–20	
Educational (Schools)			
Classroom	40	30–50	At 30″ above the finished floor.
Gymnasium			
Class I (pro or Div 1 college)	125		
Class II (Div 2 or 3 college)	80		
Class III (high school)	50		
Class IV (elementary)	30		
Auditorium	7.5	3–10	
Corridor	25	10–40	
Industrial/Manufacturing			
Assembly			
Simple (large items)	30	15–60	
Difficult (fine)	100	50–200	
Component manufacturing			
Large	30	15–60	
Medium	50	25–100	

(Continued)

Building Area and Task	Average Maintained Illuminance (Foot-Candles - Horizontal)	Range of Maintained Illuminance (Foot-Candles - Horizontal)	Notes
Exterior			
Parking (covered)	5		1 FC min, 10:1 Max to Min Uniformity
Parking (open)(medium activity)			
Lighting Zone 3 (urban)	1.5	0.75–3	
Lighting Zone 2 (suburban)	1	0.5–2	
Gas station canopy	12.5	10–15	
Safety (building exterior)	1	0.5–2	
Retail			
General retail (ambient)		50	
Department store	40	20–80	
Accent lighting (displays)			3–10 times greater than ambient light levels
Automotive			
Showroom/Service area	50	25–100	
Sales lot (exterior)			
Lighting Zone 3 (urban)	20	10–40	
Lighting Zone 2 (suburban)	15	7.5–30	
Grocery			
Circulation	20	10–40	
General retail	50	25–100	
Banking			
ATM	20	10–40	

Based on IESNA Lighting Handbook, 10th ed., 2011.
Refer to the full handbook for specific project work.

Appendix E

Lighting Power Allowances—Space-by-Space

Common Space Types[a]	LPD (W/SF)	Building Type–Specific Spaces[a]	LPD (W/SF)
Audience seating area, in an auditorium	0.63	Facility for visually impaired[b]	
Audience seating area, in a convention center	0.82	In a chapel (and not used primarily by the staff)	2.21
Audience seating area, in a gymnasium	0.65	In a recreation room (and not used primarily by staff)	2.41
Audience seating area, in a motion picture theater	1.14	Convention center—exhibit space	1.45
Audience seating area, in a penitentiary	0.28	Dormitory—living quarters	0.38
Audience seating area, in a performing arts theater	2.43	Fire station—sleeping quarters	0.22
Audience seating area, in a religious building	1.53	Gymnasium/fitness center, in an exercise area	0.72
Audience seating area, in a sports arena	0.43	Gymnasium/fitness center, in a playing area	1.20
Audience seating area, otherwise	0.43	Healthcare facility	
Banking activity area	1.01	In an exam/treatment room	1.66
Classroom/lecture hall/training room, penitentiary	1.34	In an imaging room	1.51
Classroom/lecture hall/training room, other than penitentiary	1.24	In a medical supply room	0.74
Conference/meeting/multipurpose room	1.23	In a nursery	0.88
Copy/print room	0.72	In a nurse's station	0.71
Corridor, facility for visually impaired (not used primarily by staff)[b]	0.92	In an operating room	2.48
Corridor, in a hospital	0.79	In a patient room	0.62
Corridor, in a manufacturing facility	0.41	In a physical therapy room	0.91
Corridor, all other buildings	0.66	In a recovery room	1.15
Courtroom	1.72	Library, in a reading area	1.06
Computer room	1.71	Library, in the stacks	1.71
Dining area, penitentiary	0.96	Manufacturing facility	
Dining area, facility for visually impaired, not used primarily by staff)	1.90	In a detailed manufacturing area	1.29

(Continued)

Common Space Types[a]	LPD (W/SF)	Building Type–Specific Spaces[a]	LPD (W/SF)
Dining area, in bar/lounge or leisure dining	1.07	In an equipment room	0.74
Dining area, in cafeteria or fast food dining	0.65	In an extra high bay area (>50′ height)	1.05
Dining area, in family dining	0.89	In a high bay area (25–50′ height)	1.23
Dining area, otherwise	0.65	In a low bay area (<25′ floor-to ceiling height)	1.19
Electrical/mechanical room	0.95	Museum: In a general exhibition area	1.05
Emergency vehicle garage	0.56	Museum: In a restoration room	1.02
Food preparation area	1.21	Performing arts theater—dressing room	0.61
Guest room	0.47	Post office—sorting area	0.94
Laboratory, in or as a classroom	1.43	Religious buildings	
Laboratory, otherwise	1.81	In a fellowship hall	0.64
Laundry/washing area	0.60	In a worship/pulpit/choir area	1.53
Loading dock, interior	0.47	Retail facilities—In a dressing/fitting room	0.71
Lobby, in facility for visually impaired (not used primarily by staff)[b]	1.80	Retail facilities—In a mall concourse	1.10
Lobby, for an elevator	0.64	Sports arena—playing area: For a Class I facility	3.68
Lobby, in a hotel	1.06	Sports arena—playing area: For a Class II facility	2.40
Lobby, in a motion picture theater	0.59	Sports arena—playing area: For a Class III facility	1.80
Lobby, in a performing arts theater	2.00	Sports arena—playing area: For a Class IV facility	1.20
Lobby, otherwise	0.90	Transportation facility—in a baggage/carousel area	0.53
Locker room	0.75	Transportation facility—in an airport concourse	0.36
Lounge/break, except health care (health care lounge/break: 0.92)	0.73	At a terminal ticket counter	0.80
Office, enclosed	1.11	Warehouse—storage area—for medium to bulky, palletized items	0.58
Office, open plan	0.98	Warehouse—storage area—for smaller, hand-carried items	0.95
Parking area, interior	0.19		
Pharmacy area	1.68		
Restroom, in facility for visually impaired, not used primarily by staff[b]	1.21		
Restroom, otherwise	0.98		
Sales area	1.59		
Seating area, general	0.54		
Stairwell	0.69		
Storage room	0.63		
Vehicular maintenance area	0.67		
Workshop	1.59		

[a]In cases where both a common space type and a building area specific space type are listed, the building area specific space type shall apply.
[b]A "Facility for the Visually Impaired" is a facility that is licensed or will be licensed by local or state authorities for senior long-term care, adult day care, senior support or people with special visual needs.
Note: Partial table. For full table, consult IECC 2015.

Appendix F
HID Lighting Designations

Standard Mercury Vapor				Metal Halide			
ANSI	Nominal Power Rating (watts)	Actual Power Consumption (watts)	Notes	ANSI	Nominal Power Rating (watts)	Actual Power Consumption (watts)	Notes
H33	400	454		M47	1,000	1,080	Probe start
H36	1,000	1,075		M48	1,500	1,610	Probe start
H37	250	285		M57	175	210	Probe start
H38	100	125		M58	250	295	Probe start
H39	175	205		M59	400	460	Probe start
H43	75	95		M80	250	295	Probe start—double ended
H46	50	75		M81	150	185	Probe start—double ended
				M85	70	93	Probe start—double ended
Low-Pressure Sodium (SOX/SOX-E)				M90	100	129	Probe start/pulse start
L69	18	30		M91	100	129	Probe start—double ended
L70	35	60		M98	70	93	Probe start
L71	55	80		M102	150	185	Probe start/pulse start
L72	90	125		M107	150		Energy saving probe start
L73	135	180		M110	50	70	Probe start/pulse start
L74	180	208		M128	400	452	Pulse start
				M130	35/39	52	
High-Pressure Sodium				M131	350	400	Pulse start
S50	250	295	Standard	M132	32	38	Pulse start
S51	400	465	Standard	M133	1,500	1,610	Pulse start—double ended
S52	1,000	1,100	Standard	M134	2,000	2,140	Pulse start—double ended
S54	100	128	Standard	M135	400	452	Pulse start
S55	150	188	55 volt	M137	175	208	Pulse start
S62	70	91	Standard	M138	250	290	Pulse start
S66	200	241	Standard	M139	70	90	Ceramic—double ended
S68	50	64	Standard	M140	100	128	Ceramic—double ended
S76	35	45	Standard	M142	150	190	Ceramic—double ended
S104	50	64	White	M155	400	452	Pulse start
S105	100	128	White				
S106	600	665	Standard				
S111	750	840	Standard				

Appendix G
Lighting Software

Lighting Software	Vendor	Online/Download	Notes
Agi32	Lighting Analysts Inc.	Download	
Autolux	Independent Testing Laboratories (ITL)	Download	AutoCAD add-in
Calczone	Philips Lightolier	Online	
Genesys III	Philips	Download	
LitePro DLX	Hubbell	Download	
Lumen Micro	Lighting Technologies	Download	No longer supported
Luxicon	Cooper Lighting	Download	
Parking Lot Luminare Calculator	Lighting Research Center	Online	
Simply Indoor	Lighting Technologies	Download	No longer supported
Visual	Lithonia	Download/Online	

Appendix H
Lighting Reflectances

Paints:		Carpet:	
Highly reflective white	90%	Low maintenance (dark)	2–5%
Typical white	70–80%	Moderate maintenance	5–9%
Light cream	70–80%	Higher maintenance	9–13%
White	70–80%	Very high maintenance	13+%
Light green*	53%	Linoleum	
Kelly green*	49%	White	54–59%
Medium blue*	49%	Black	0–9%
Medium yellow*	47%	Concrete:	
Medium orange*	42%	Black polished	0%
Medium green*	41%	Gray polished	20%
Medium red*	20%	Light polished	60%
Medium brown*	16%	Reflective concrete floor coatings	66–93%
Dark blue-gray*	16%	Dark paneling walls	10%
Dark brown*	12%	Burlap	10%
Woods:		Plywood	30%
Maple	54%	Bulletin boards	10%
Poplar	52%	Gray fabric partitions	51%
White pine	51%	Gray plastic-coated steel desk	63%
Red pine	49%	Ceiling tile	
Oregon pine	38%	Typical reflectance	76–80%
Birch	35%	High reflectance	90%
Beech	26%		
Oak	23%		
Cherry	20%		

*Estimated for flat paints. For gloss paints, add 5 to 10%.

Appendix I
Room Air Conditioner Efficiency Requirements

Type	Requirements	Capacity (Btu/hr)							
		< 6000	6000–7999	8000–10999	11000–13999	14000–19999	20000–24999	25000–27999	> = 28000
Without reverse cycle, with louvered sides	Federal 1/1/1990	8.0	8.5	9.0	9.0	8.8	8.2	8.2	8.2
"	Federal 10/1/2000	9.7	9.7	9.8	9.8	9.7	8.5	8.5	8.5
"	Federal 6/1/2014	11.0	11.0	10.9	10.9	10.7	9.4	9.0	9.0
"	ENERGY STAR 11/16/2005	10.7	10.7	10.8	10.8	10.7	9.4	9.4	9.4
"	ENERGY STAR 10/1/2013	11.2	11.2	11.3	11.3	11.2	9.8	9.8	9.8
Without reverse cycle, without louvered sides	Federal 1/1/1990	8.0	8.5	8.5	8.5	8.5	8.5	8.5	8.2
"	Federal 10/1/2000	9.0	9.0	8.5	8.5	8.5	8.5	8.5	8.5
"	Federal 6/1/2014	10.0	10.0	9.6	9.5	9.3	9.4	9.4	9.4
"	ENERGY STAR 11/16/2005	9.9	9.9	9.4	9.4	9.4	9.4	9.4	9.4
"	ENERGY STAR 10/1/2013	10.4	10.4	9.8	9.8	9.8	9.8	9.8	9.8
With reverse cycle, with louvered sides	Federal 1/1/1990	8.5	8.5	8.5	8.5	8.5	8.5	8.5	8.5
"	Federal 10/1/2000	9.0	9.0	9.0	9.0	9.0	8.5	8.5	8.5
"	Federal 6/1/2014	9.8	9.8	9.8	9.8	9.8	9.3	9.3	9.3
"	ENERGY STAR 11/16/2005	9.9	9.9	9.9	9.9	9.9	9.4	9.4	9.4
"	ENERGY STAR 10/1/2013	10.4	10.4	10.4	10.4	10.4	9.8	9.8	9.8
With reverse cycle, without louvered sides	Federal 1/1/1990	8.0	8.0	8.0	8.0	8.0	8.0	8.0	8.0
"	Federal 10/1/2000	8.5	8.5	8.5	8.5	8.0	8.0	8.0	8.0
"	Federal 6/1/2014	9.3	9.3	9.3	9.3	8.7	8.7	8.7	8.7
"	ENERGY STAR 11/16/2005	9.4	9.4	9.4	9.4	8.8	8.8	8.8	8.8
"	ENERGY STAR 10/1/2013	9.8	9.8	9.8	9.8	9.2	9.2	9.2	9.2

(Continued)

Type	Requirements	Capacity (Btu/hr)							
		< 6000	6000–7999	8000–10999	11000–13999	14000–19999	20000–24999	25000–27999	> = 28000
Casement-only	Federal 1/1/1990	NA	NA	NA	NA	NA	NA	NA	NA
"	Federal 10/1/2000	8.7	8.7	8.7	8.7	8.7	8.7	8.7	8.7
"	Federal 6/1/2014	9.5	9.5	9.5	9.5	9.5	9.5	9.5	9.5
"	ENERGY STAR 11/16/2005	9.6	9.6	9.6	9.6	9.6	9.6	9.6	9.6
"	ENERGY STAR 10/1/2013	10.0	10.0	10.0	10.0	10.0	10.0	10.0	10.0
Casement-slider	Federal 1/1/1990	NA	NA	NA	NA	NA	NA	NA	NA
"	Federal 10/1/2000	9.5	9.5	9.5	9.5	9.5	9.5	9.5	9.5
"	Federal 6/1/2014	10.4	10.4	10.4	10.4	10.4	10.4	10.4	10.4
"	ENERGY STAR 11/16/2005	10.5	10.5	10.5	10.5	10.5	10.5	10.5	10.5
"	ENERGY STAR 10/1/2013	10.9	10.9	10.9	10.9	10.9	10.9	10.9	10.9

Sources:

1. Gregory Rosenquist, "Window-Type Room Air Conditioners," *ASHRAE Journal*, January 1999, p. 35.

2. ENERGY STAR Program Requirements for Room Air Conditioners—Partner Commitments, Version 3.1

3. http://www1.eere.energy.gov/buildings/appliance_standards/product.aspx/productid/41. Accessed April 7, 2015.

4. Southern California Edison Company, Design & Engineering Services, Work Paper WPSCREHC0001, Revision 1, Energy Star Room Air Conditioners, October 16, 2007.

5. ENERGY STAR Program Requirements for Room Air Conditioners—Partner Commitments, Version 2.1 draft.

6. As of June 1, 2014, the federal requirement is a combined energy efficiency ratio (CEER).

Appendix J

Chiller Efficiency Requirements

	Air Cooled (EER)		Positive Displacement (kW/ton) (Reciprocating/Rotary/Screw/Scroll)					Centrifugal (kW/ton)					Source
	<150 tons	≥150 tons	<75 tons	<150 tons	150–299 tons	300–599 tons	>600 tons	<150 tons	150–299 tons	300–399 tons	400–599 tons	>600 tons	
ASHRAE 90–1975, effective 1/1/1977	7.2	7.2	1.101	1.101	1.101	1.101	1.101	0.930	0.930	0.930	0.930	0.930	9
ASHRAE 90–1975, effective 1/1/1980	7.5	7.5	1.034	1.034	1.034	1.034	1.034	0.882	0.882	0.882	0.882	0.882	9
ASHRAE 90.1–1989, Full Load	9.2	8.5	0.926	0.926	0.837	0.676	0.676	0.926	0.837	0.676	0.676	0.676	10
ASHRAE 90.1–1989, IPLV	9.6	8.5	0.902	0.902	0.782	0.664	0.664	0.902	0.782	0.664	0.664	0.664	10
ASHRAE 90.1–2001 and 2004 Full Load	9.6	8.5	0.790	0.790	0.718	0.639	0.639	0.703	0.634	0.576	0.576	0.576	3
ASHRAE 90.1–2001 and 2004 Part Load	9.6	8.5	0.676	0.676	0.628	0.572	0.572	0.670	0.596	0.549	0.549	0.549	3
IECC 2003 Full Load	9.6	8.5	0.790	0.790	0.720	0.640	0.640	0.700	0.630	0.580	0.580	0.580	1
IECC 2003 Part Load	9.6	8.5	0.780	0.780	0.710	0.630	0.630	0.700	0.630	0.580	0.580	0.580	1
FEMP, Path A, Full Load	10.4	10.4	0.750	0.710	0.680	0.580	0.580	0.620	0.590	0.560	0.560	0.550	2, 6
FEMP, Path A, Part Load	12.5	12.75	0.630	0.610	0.580	0.540	0.540	0.600	0.600	0.550	0.550	0.400	2, 6
FEMP, Path B, Full Load	9.56	9.56	0.800	0.790	0.720	0.640	0.640	0.640	0.640	0.600	0.600	0.570	2, 6
FEMP, Path B, Part Load	15.39	15.07	0.600	0.510	0.500	0.480	0.480	0.360	0.350	0.360	0.360	0.350	2, 6
IECC 2009, Path A, Full Load	9.562	9.562	0.780	0.775	0.680	0.620	0.620	0.634	0.634	0.576	0.576	0.570	5
IECC 2009, Path A, Part Load	12.5	12.75	0.630	0.615	0.580	0.540	0.540	0.596	0.596	0.549	0.549	0.539	5
IECC 2009, Path B, Full Load	NA	NA	0.800	0.790	0.718	0.639	0.639	0.639	0.639	0.600	0.600	0.590	5
IECC 2009, Path B, Part Load	NA	NA	0.600	0.586	0.540	0.490	0.490	0.450	0.450	0.400	0.400	0.400	5
ASHRAE 189.1–2009, Path A, Full Load	10	10	0.780	0.775	0.680	0.620	0.620	0.634	0.634	0.576	0.576	0.570	4
ASHRAE 189.1–2009, Path A, Part Load	12.5	12.75	0.630	0.615	0.580	0.540	0.540	0.596	0.596	0.549	0.549	0.539	4
ASHRAE 189.1–2009, Path B, Full Load	NA	NA	0.800	0.790	0.718	0.639	0.639	0.639	0.639	0.600	0.600	0.590	4
ASHRAE 189.1–2009, Path B, Part Load	NA	NA	0.600	0.586	0.540	0.490	0.490	0.450	0.450	0.400	0.400	0.400	4
IECC 2012, Path A, Full Load	10	10	0.780	0.775	0.680	0.620	0.620	0.634	0.634	0.576	0.576	0.570	7
IECC 2012, Path A, Part Load	12.5	12.75	0.630	0.615	0.580	0.540	0.540	0.596	0.596	0.549	0.549	0.539	7
IECC 2012, Path B, Full Load	NA	NA	0.800	0.790	0.718	0.639	0.639	0.639	0.639	0.600	0.600	0.590	7
IECC 2012, Path B, Part Load	NA	NA	0.600	0.586	0.540	0.490	0.490	0.450	0.450	0.400	0.400	0.400	7
ASHRAE 189.1–2014, Path A, Full Load	10.1	10.1	0.750	0.720	0.660	0.610	0.560	0.610	0.610	0.560	0.560	0.560	8
ASHRAE 189.1–2014, Path A, Part Load	13.7	13.7	0.600	0.560	0.540	0.520	0.500	0.550	0.550	0.520	0.500	0.500	8
ASHRAE 189.1–2014, Path B, Full Load	9.7	9.7	0.780	0.750	0.680	0.625	0.585	0.695	0.635	0.595	0.585	0.585	8
ASHRAE 189.1–2014, Path B, Part Load	15.8	16.1	0.500	0.490	0.440	0.410	0.380	0.440	0.400	0.390	0.380	0.380	8

Sources:

1. PSO Chiller Energy Use Estimating Tool v3.
2. http://energy.gov/eere/femp/covered-product-category-water-cooled-electric-chillers. Accessed April 12, 2015.
3. U.S. Environmental Protection Agency, ENERGY STAR Building Upgrade Manual, Chapter 9, "Heating and Cooling." Revised January 2008, p. 4.
4. ASHRAE 189.1–2009.
5. IECC 2009.
6. http://energy.gov/eere/femp/covered-product-category-air-cooled-electric-chillers. Accessed April 12, 2015.
7. http://publiccodes.cyberregs.com/icod/iecc/2012/icod_iecc_2012_ce4_sec006.htm?bu2=undefined (IECC 2012). Accessed April 12, 2015.
8. ASHRAE 189.1–2014.
9. ASHRAE 90–1975.
10. ASHRAE 90.1–1989.

Appendix K

Existing Exhaust Schedule

	Fan 1	Fan 2	Fan 3	Fan 4	Fan 5	Fan 6	Notes
Location							
Roof	——	——	——	——	——	——	
Exterior wall	——	——	——	——	——	——	
Above ceiling	——	——	——	——	——	——	
Mechanical room	——	——	——	——	——	——	
Other	——	——	——	——	——	——	
Serves							
Bathrooms	——	——	——	——	——	——	
Kitchens	——	——	——	——	——	——	
Other	——	——	——	——	——	——	
Schedule (hour of day)	**Exhaust Airflow (CFM)**						
1							
2							
3							
4							
5							
6							
7							
8							
9							
10							
11							
12							
13							
14							
15							
16							
17							
18							
19							
20							
21							
22							
23							
24							

Appendix L
Existing Outdoor Air Schedule

	AHU 1	AHU 2	AHU 3	AHU 4	AHU 5	AHU 6	Notes
Location							
Rooftop unit	——	——	——	——	——	——	
Mechanical room air handler	——	——	——	——	——	——	
Other	——	——	——	——	——	——	
Serves	——	——	——	——	——	——	
Schedule (hour of day)	**Outside Airflow (CFM)**						
1							
2							
3							
4							
5							
6							
7							
8							
9							
10							
11							
12							
13							
14							
15							
16							
17							
18							
19							
20							
21							
22							
23							
24							

Appendix M

Proposed Outside Air Schedule

Location					
Rooftop unit		_____			
Mechanical room air handler		_____			
Other					
Serves					
	Weekday		Weekend		
	Occupancy (people)	Proposed Outside Air (CFM)	Occupancy (people)	Proposed Outside Air (CFM)	Notes
Schedule (hour of day)					
1					
2					
3					
4					
5					
6					
7					
8					
9					
10					
11					
12					
13					
14					
15					
16					
17					
18					
19					
20					
21					
22					
23					
24					

Appendix N

Simplified Model of a Building Entering or Recovering from Setback

A lumped capacitance heat transfer analysis approach can be used to model a building's change in temperature during setback, and during its recovery from setback. Assuming a uniform indoor temperature (air and building materials), and neglecting radiation effects, it can be shown that:

$$T = To + (Ti - To) \times [exp(-hAt/M) + (Q/hA) \times (1 - exp(-hAt/M)) / (Ti - To)]$$

Where:

T = indoor temperature, F
To = outdoor temperature, F
Ti = initial indoor temperature, F
hA = heat loss rate of the building, Btu/hr/F
t = time, hours
M = thermal mass, Btu/F
Q = heat delivered to the building by its heating system, Btu/hr

As a building goes into setback, the heating system is off ($Q = 0$), and the model simplifies to:

$$T = To + (Ti - To) \times [exp(-hAt/M)]$$

The time (t) that the building takes to cool from an initial temperature (Ti) to a second indoor temperature (T), at an average outdoor temperature (To) can be used to solve for the thermal mass of the building (M), using the building's heat loss rate (hA − obtained from the building's design load divided by its design temperature difference; the design load might be estimated as slightly lower than the heating capacity Q). It might be noted that the inverse of hA / M, in other words M / hA, is also known as the time constant of the equation.

With the thermal mass (M) calculated, the first equation can be solved, to obtain a relationship of the building temperature as it recovers from setback with the heating system (Q) on. In this case, the initial indoor temperature (Ti) is set equal to the setback temperature. It might also be noted that the term Q / hA represents the ratio of the heating system capacity to the heat loss rate of the building.

From: http://wwwme.nchu.edu.tw/Enter/html/lab/lab516/Incropera%20-%20PDF/5.pdf. Accessed June 6, 2015.

Appendix O

Gas Pilot Sizes and Gas Use

Pilot lights are used on a variety of older appliances and heating equipment, including boilers, furnaces, water heaters, clothes dryers, stoves, ovens, gas fireplaces, and more. On most of these appliances, the energy use of the pilot contributes little or no useful energy, and so is a waste. On some of the appliances, such as storage water heaters, some portion of the pilot energy is useful energy. Newer equipment typically has an electronic ignition that only uses energy during ignition. Typical pilot light use might be assumed to be 1,000 Btu/hr as a rule of thumb, but it can vary widely depending on the pilot orifice size and gas pressure. The following table provides energy consumption for different typical pilot orifice sizes and gas pressures. Low-pressure natural gas delivery pressures are typically 7″ water column, at the meter, but appliances frequently reduce this pressure to 3.5 to 4″, and propane (LP gas) pressure is typically 11″. Natural gas pilot orifice sizes typically vary from approximately 0.014 to 0.026 (inches diameter). Examples include 0.026″ for a commercial cooking appliance pilot (1,835 Btu/hr at 4″ pressure), and 0.018″ for a natural gas water heater pilot (879 Btu/hr at 4″ pressure). Propane (LP) orifice sizes typically vary from 0.010 to 0.016″ diameter, but 0.010″ appears common, for which energy consumption is 548 Btu/hr at a typical 11″ pressure. In some cases, pilot lights for old natural appliances have been found to consume as much as 4,000 Btu/hr, so if the pilot light appears large, or for instances where a building has many gas appliances (e.g., a multifamily building), it may be worth measuring the orifice size, obtaining the gas pressure, and obtaining the actual pilot energy use from the table below, or measuring gas consumption at the gas meter while all other gas loads are shut off.

Note that the orifice sizes are the same as the drill bit sizes used to drill the orifice holes. So by stocking drill bits 71 to 80 (plus 87 for the 0.0100 orifice) in an energy audit toolkit, the size of an orifice can be measured.

Gas Pilot Light Energy Usage (Btu/hr)

		Pressure (inches water column)						
		Natural Gas				LP		
Orifice No.	Orifice Size (inches)	3	3.5	4	4.5	7	11	Notes
	0.0100						548	Estimated.
80	*0.0135*	428	463	495	525	654	1273	
79	*0.0145*	494	534	571	605	755	1469	
1/64	*0.0156*	572	618	661	701	874	1700	
78	*0.0160*	602	650	695	737	919	1788	
77	*0.0180*	762	823	879	933	1163	2263	
76	*0.0200*	940	1016	1086	1152	1436	2794	
75	*0.0210*	1037	1120	1197	1270	1583	3081	
74	*0.0225*	1190	1285	1374	1457	1818	3536	
73	*0.0240*	1354	1462	1563	1658	2068	4024	
72	*0.0250*	1469	1587	1696	1799	2244	4366	
71	*0.0260*	1589	1716	1835	1946	2427	4722	

Source: http://www.hvacredu.net/gas-codes/module2/Gas%20Orifice%20Capacity%20Chart.pdf

Appendix P

Estimated Existing Motor Efficiencies, Pre-1992

RPM	HP	Open Drip-Proof (ODP)				Totally Enclosed Fan-Cooled (TEFC)			
		Load				Load			
		100%	75%	50%	25%	100%	75%	50%	25%
900	10	85.3	86.1	84.8	78.3	85.5	85.8	84.8	77.3
	15	86.3	87.5	86.6	79.6	86.4	87.2	86.4	79.0
	20	87.6	88.3	87.3	81.8	87.9	88.9	88.2	84.4
	25	88.3	88.8	88.1	83.0	87.9	88.4	86.8	78.6
	30	88.1	89.1	88.5	84.5	88.6	89.2	88.6	85.2
	40	87.5	87.6	87.1	84.5	89.0	88.8	87.0	82.5
	50	89.3	90.2	89.6	87.1	89.8	89.7	88.5	82.5
	60	89.9	90.5	89.9	86.4	90.6	91.1	90.3	86.9
	75	90.9	91.4	90.8	85.8	90.6	90.8	89.9	83.6
	100	91.3	91.7	91.2	86.8	91.1	91.6	91.0	87.9
	125	91.6	92.1	91.6	89.5	91.5	91.4	90.5	87.5
	150	91.9	92.6	92.2	89.7	91.5	91.7	91.0	88.0
	200	92.6	93.5	93.1	90.2	93.0	93.8	93.1	90.1
1,200	1	74.5	75.8	75.3	70.2	73.4	74.4	73.1	65.1
	1.5	77.6	78.9	78.4	73.3	77.9	78.9	77.6	69.6
	2	79.9	81.2	80.7	75.6	78.3	79.3	78.0	70.0
	3	81.7	83.0	82.5	77.4	80.4	81.4	80.1	72.1
	5	83.6	84.9	84.4	79.3	83.1	84.1	82.8	74.8
	7.5	85.5	86.8	86.3	81.2	84.4	85.4	84.1	76.1
	10	86.1	87.4	86.9	81.8	85.7	86.7	85.4	77.4
	15	87.8	88.8	88.3	82.0	86.6	87.5	86.0	74.9
	20	88.3	89.4	88.7	83.3	88.5	89.3	88.5	82.9
	25	88.9	90.0	89.8	86.6	89.3	89.9	88.8	82.1
	30	88.9	90.6	90.5	87.4	89.6	90.2	89.2	83.6
	40	90.0	90.3	89.4	85.7	90.2	90.4	89.2	81.6
	50	90.7	91.2	90.4	87.4	91.3	91.6	90.9	84.1
	60	91.3	91.7	90.8	86.8	91.8	91.9	90.8	85.2
	75	91.9	92.3	91.6	87.7	91.7	92.0	90.9	85.1
	100	92.1	92.7	92.2	89.2	92.3	92.2	91.2	86.2
	125	92.2	92.8	92.2	88.2	92.2	91.6	90.5	84.0
	150	92.8	93.1	92.4	88.6	93.0	92.8	91.5	86.3
	200	93.0	93.4	93.0	90.3	93.5	93.3	92.0	86.3

(Continued)

| RPM | HP | Open Drip-Proof (ODP) | | | | Totally Enclosed Fan-Cooled (TEFC) | | | |
| | | Load | | | | Load | | | |
		100%	75%	50%	25%	100%	75%	50%	25%
1,800	1	77.6	78.9	78.4	73.3	76.7	77.7	76.4	68.4
	1.5	79.3	80.6	80.1	75.0	79.1	80.1	78.8	70.8
	2	80.5	81.8	81.3	76.2	80.8	81.8	80.5	72.5
	3	82.4	83.7	83.2	78.1	81.5	82.5	81.2	73.2
	5	83.8	85.1	84.6	79.5	83.3	84.3	83.0	75.0
	7.5	85.2	86.5	86.0	80.9	85.5	86.5	85.2	77.2
	10	86.1	87.4	86.9	81.8	85.7	86.7	85.4	77.4
	15	87.8	88.8	88.3	82.0	86.6	87.5	86.0	74.9
	20	88.3	89.4	88.7	83.3	88.5	89.3	88.5	82.9
	25	88.9	90.0	89.8	86.6	89.3	89.9	88.8	82.1
	30	88.9	90.6	90.5	87.4	89.6	90.2	89.2	83.6
	40	90.0	90.3	89.4	85.7	90.2	90.4	89.2	81.6
	50	90.7	91.2	90.4	87.4	91.3	91.6	90.9	84.1
	60	91.3	91.7	90.8	86.8	91.8	91.9	90.8	85.2
	75	91.9	92.3	91.6	87.7	91.7	92.0	90.9	85.1
	100	92.1	92.7	92.2	89.2	92.3	92.2	91.2	86.2
	125	92.2	92.8	92.2	88.2	92.2	91.6	90.5	84.0
	150	92.8	93.1	92.4	88.6	93.0	92.8	91.5	86.3
	200	93.0	93.4	93.0	90.3	93.5	93.3	92.0	86.3
3,600	1	76.2	77.3	76.2	69.9	73.0	72.7	70.7	62.1
	1.5	77.3	78.4	77.3	71.0	75.2	74.9	72.9	64.3
	2	79.6	80.7	79.6	73.3	78.9	78.6	76.6	68.0
	3	79.1	80.2	79.1	72.8	79.6	79.3	77.3	68.7
	5	82.6	83.7	82.6	76.3	82.4	82.1	80.1	71.5
	7.5	82.9	84.0	82.9	76.6	82.6	82.3	80.3	71.7
	10	85.0	86.1	85.0	78.7	85.0	84.7	82.7	74.1
	15	86.6	87.7	86.8	80.5	85.7	85.8	83.6	73.2
	20	88.1	88.9	88.8	85.4	86.6	87.7	86.1	76.7
	25	88.4	89.2	88.7	83.7	87.5	87.4	85.3	75.2
	30	87.7	88.9	88.8	84.7	87.7	87.0	84.7	75.4
	40	88.6	89.7	89.9	86.9	88.5	88.0	85.8	75.2
	50	89.1	90.1	89.8	88.4	89.0	88.7	86.7	77.8
	60	90.4	90.9	90.9	87.8	89.4	88.4	85.8	76.6
	75	90.4	90.6	90.1	85.7	90.6	89.9	88.0	78.9
	100	90.5	91.2	91.0	89.0	90.9	90.3	88.7	81.9
	125	91.2	91.9	91.4	90.3	90.9	90.1	87.9	77.4
	150	91.7	91.8	91.9	90.1	91.5	90.9	88.4	81.7
	200	91.5	91.7	90.9	83.6	92.7	92.0	90.1	83.5

Appendix Q
Equipment Expected Useful Life

Category	Improvement	Expected Useful Life (Years)	Source	Data Quality
Envelope				
	Air sealing	15	e	○
	Air sealing	15	i	○
	Brick or stone veneer	50+	h	○
	Brownstone	40	h	○
	Curtain wall, metal/glass	30	h	○
	Door, exterior common, aluminum and glass	30	h	○
	Door, exterior common, solid core wood or metal clad	25	h	○
	Door, storm/screen	7	h	○
	Doors, storm	20	i	○
	Exterior insulation finishing systems (EIFS)	20	h	○
	Glass block	40	h	○
	Insulation: blanket	24	a	●
	Insulation: molded	20	a	●
	Insulation	25	e	○
	Insulation, wall	50+	h	○
	Insulation (attic, wall, floor, band joist, basement, crawlspace)	30	i	○
	Painting, exterior	5–10	h	○
	Reflective window film/window treatment	10	e	○
	Roofing section, built-up, asphalt shingles	42	c	*
	Roof, asphalt shingle (three-tab)	20	h	○
	Roof, built-up (ethylene propylene diene terpolymer [EPDM]/thermoplastic polyolefin [TPO])	20	h	○
	Roof, metal	40	h	○
	Roof, slate	40	h	○
	Siding, aluminum	40	h	○
	Siding, cement-board/cementitious	45	h	○
	Siding, wood	20	h	○
	Siding, vinyl	25	h	○
	Shingles, wood (cedar shake)	25	h	○

(Continued)

Category	Improvement	Expected Useful Life (Years)	Source	Data Quality
	Skylight, roof	30	h	○
	Stucco	50+	h	○
	Waterproofing (foundations)	50	h	○
	Window covering	5	h	○
	Windows (frames and glazing), vinyl or aluminum	30	h	○
	Windows—low solar heat gain coefficient (SHGC), or high performance	25	e	○
	Windows, replacement	20	i	○
	Window insulation, moveable	10	e	○
	Window insulation, moveable	20	i	○
	Windows, storm/screen	10	h	○
	Windows, storm	20	i	○
Heating, Ventilating, and Air Conditioning				
	Air conditioners, window unit	10	a	•
	Air conditioning unit, window or wall mounted, 5,000–29,000 Btu/hr	27	c	○
	Air conditioners, room	9	d	○
	Air conditioners, residential single or split package	15	a	•
	Air conditioners, commercial through-the-wall	15	a	•
	Air conditioners, water-cooled package	15	a	•
	Air conditioning unit, self-contained, 3–60 ton	24	c	*
	Air conditioning unit, self-contained variable air volume, 1.5–200 ton	30	c	*
	Air handling units	52	b	*
	Air handling unit, constant volume	40	r	*
	Air handling unit, variable air volume	26	r	*
	Air handling unit, variable volume/temperature (zone damper system)	12	r	*
	Air terminals: diffusers, grilles, and registers	27	a	•
	Air terminals: induction and fan-coil units	20	a	•
	Air terminals: variable air volume and double-duct boxes	20	a	•
	Bathroom vent/exhaust	10	h	○
	Boiler tuneup	2	f	○
	Boiler—condensing	25	i	○
	Boilers, hot water (steam), steel water-tube	24 (30)	a	•
	Boilers, hot water (steam), steel fire-tube	25 (25)	a	•
	Boilers, steel, gas-fired	22	b	*
	Boilers, hot water (steam), cast iron	35 (30)	a	•
	Boilers, electric	15	a	•
	Boilers, electric	20	h	○
	Boilers, oil-fired, sectional	22	h	○
	Burners	21	a	•
	Burner replacement	20	e	○
	Chillers, package: absorption	23	a	•
	Chillers, centrifugal	25	r	*
	Chillers, centrifugal	25	b	*

Category	Improvement	Expected Useful Life (Years)	Source	Data Quality
	Chillers, package: centrifugal	23	a	•
	Chillers, package: reciprocating	20	a	•
	Coils: DX, water, or steam	20	a	•
	Coils: Electric	15	a	•
	Combustion air, motor louver and duct	25	h	○
	Compressors, reciprocating	20	a	•
	Condenser coil cleaning	10	m	*
	Condensers, air-cooled	20	a	•
	Condensers, evaporative	20	a	•
	Controls, electric	16	a	•
	Controls, electronic	15	a	•
	Control systems, electronic/DDC	7	b	*
	Controls, pneumatic	20	a	•
	Controls, pneumatic lines and	30	h	○
	Control systems, pneumatic/hybrid	18	b	*
	Cooling towers: galvanized metal	20	a	•
	Cooling towers, metal	22	b	*
	Cooling towers: wood	20	a	•
	Cooling towers: ceramic	34	a	•
	Cooling tower	25	h	○
	Dampers	20	a	•
	Damper, stack	12	f	○
	Duct insulation	20	e	○
	Duct sealing	20	e	○
	Duct sealing	20	i	○
	Ductwork	30	a	•
	Economizer, enthalpy	7	e	○
	Energy management system	10	e	○
	Engines, reciprocating	20	a	•
	Fan coil	36.5	r	*
	Fan coil unit, electric	20	h	○
	Fan coil unit, hydronic	30	h	○
	Fan hood exhaust, 150–34,000 CFM	39	c	*
	Fans: axial	20	a	•
	Fans: ceiling	10	f	○
	Fans: centrifugal	25	a	•
	Fans: propeller	15	a	•
	Fans: ventilating, roof mounted	20	a	•
	Furnaces, gas or oil fired	18	a	•
	Furnace (gas heat with A/C)	20	h	○
	Furnace—natural gas	20	i	○
	Gas distribution	50+	h	○
	Ground source heat pump: interior components	20	g	○
	Ground source heat pump: ground loop	100	g	○
	Heater, electric baseboard	25	h	○
	Heat exchanger	35	h	○

(Continued)

Category	Improvement	Expected Useful Life (Years)	Source	Data Quality
	Heater, wall mounted electric or gas	20	h	○
	Heat exchangers: shell-and-tube	24	a	•
	Heat pump condensing component	20	h	○
	Heat pumps: residential air-to-air	15	a	•
	Heat pumps: commercial air-to-air	15	a	•
	Heat pump, self-contained, 1.5–50 ton	28	c	*
	Heat pumps: commercial water-to-air	19	a	•
	Heat pumps: water to air, geothermal	25	r	*
	Heat pump, air source	20	i	○
	Heat pump, air to air	17	r	*
	Hydronic heat/electric A/C	20	h	○
	Outdoor air reset	20	f	○
	Packaged terminal air conditioner (PTAC)	15	h	○
	Packaged terminal air conditioner/heat pump (PTAC/PTHP)			
	Pump, circulator, < 1–25 HP	42	c	*
	Pumps: base mounted	20	a	•
	Pumps: condensate	15	a	•
	Pumps: pipe mounted	10	a	•
	Pumps: sump and well	10	a	•
	Radiant heaters, electric	10	a	•
	Radiant heaters, water or steam	25	a	•
	Range hood, kitchen	10	h	○
	Rooftop units, packaged DX	22	r	*
	Rooftop units (packaged HVAC)	20	h	○
	Rooftop air conditioners, single-zone	15	a	•
	Rooftop air conditioners, multizone	15	a	•
	Steam traps	5	f	○
	Stove, wood or solid fuel pellet	20	i	○
	Thermostat, programmable	8	e	○
	Thermostat, programmable	11	i	○
	Turbines, steam	30	a	•
	Tuneup, central air conditioner	10	f	○
	Unit heaters, gas or electric	13	a	•
	Unit heaters, hot water or steam	20	a	•
	Valve actuators: hydraulic	15	a	•
	Valve actuators: pneumatic	20	a	•
	Valve actuators: self-contained	10	a	•
	Valves, motorized	12	h	○
	Water distribution, chilled	50+	h	○
	Water heaters, electric potable	21	b	*
	Water heaters, electric	13	d	○
	Water heaters, gas, commercial	15	f	○
	Water heaters, gas	11	d	○
Lighting				
	Daylight dimming	9	e	○
	Exterior lighting, building mounted	10	h	○
	Exterior lighting, pole mounted	25	h	○
	Fluorescent fixture	13	e	○

Category	Improvement	Expected Useful Life (Years)	Source	Data Quality
	Hardwired compact fluorescent (CFL)	13	e	○
	High-intensity discharge (HID) (exterior and interior)	13	e	○
	LED exit signs	13	e	○
	Lighting controls	18	f	○
	Light fixture	20	f	○
	Lighting power density	15	e	○
	More efficient lighting design	20	e	○
	Occupancy sensors	9	e	○
	Photocell (with time clock)	8	e	○
Plug				
	Clothes dryers, electric	16	q	*
	Clothes dryers, electric	12	d	○
	Clothes dryers, gas	12	d	○
	Clothes washers	11	d	○
	Clothes washers	15	i	○
	Computers, portable	3	d	○
	Dehumidifier	12	e	○
	Dehumidifier	15	i	○
	Dishwasher, common area	15	h	○
	Dishwasher, in dwelling unit kitchen	5–10	h	○
	Dishwashers	12	f	○
	Dishwashers	15	i	○
	Dishwashers	10–15	j	○
	Fax machines	4	d	○
	Freezers	11	d	○
	Freezers	15	i	○
	Microwave ovens	9	d	○
	Power supplies, computer	4	f	○
	Range, in dwelling unit kitchen	15	h	○
	Ranges, electric	16	d	○
	Refrigerators (standard sizes only)	12	d	○
	Refrigerator	10	h	○
	Refrigerator	17	i	○
	Room air conditioners	9	d	○
	Vending machine occupancy controls	10	e	○
Other Electrical				
	Elevator, cab	15	h	○
	Elevator, controller, dispatcher	15	h	○
	Elevator, shaftway hoist rails, cables, traveling	25	h	○
	Elevator, shaftway hydraulic piston and leveling	25	h	○
	Elevator, machinery	30	h	○
	Elevators, electrical switchgear	50+	h	○
	Food service equipment, electric	11	f	○
	Motors, electric	18	a	●
	Motor starters	17	a	●
	Motors	15	e	○

(Continued)

Category	Improvement	Expected Useful Life (Years)	Source	Data Quality
	Panel, electrical, 120–600 V, 15–4,000 amp	47	c	*
	Transformers, Energy Star	30	e	○
	Transformers, electric	30	a	●
	Transformer	30	h	○
	Transformer, low-voltage dry-type	50	p	*
	Variable-frequency drive, nonprocess	15	e	○
Other Gas				
	Food service equipment, gas	11	f	○
	Ranges, gas	17	d	○
Refrigeration				
	Refrigeration condensing unit, 1.25–30 ton	34	c	*
	Refrigeration case, strip curtains	4	e	○
	Refrigeration case, continuous cover	5	e	○
	Refrigeration controls	10	e	○
Renewables				
	Biomass heat	20–30	g	○
	Photovoltaics	25–40	g	○
	Solar vent preheat	30–40	g	○
	Solar water heating	20	e	○
	Solar water heat	10–25	g	○
	Solar hot water	20–25	h	○
	Wind	20	g	○
Water				
	Clothes washers	11	d	○
	Clothes washers	15	i	○
	Clothes washers	20	o	○
	Clothes washers (front loading)	15–20	j	○
	Clothes washers, commercial, single-load—multifamily	11	n	○
	Clothes washers, commercial, single-load—coin-op	7	n	○
	Clothes washers, commercial, multiload washer/washer extractor	15	n	○
	Clothes washers, commercial, tunnel washer	7–15	n	○
	Dishwasher, common area	15	h	○
	Dishwasher, in dwelling unit kitchen	5–10	h	○
	Dishwashers	12	f	○
	Dishwashers	15	i	○
	Dishwashers	10–15	j	○
	Dishwashers, commercial	20–25	k	○
	Drain heat recovery	40	l	○
	Faucet aerators	5	f	○
	Faucet aerators	10	i	○
	Fixtures/faucets, in dwelling unit bathrooms	15–20	h	○
	Fixtures, public bathroom	15	h	○
	Hot and cold water distribution	50	h	○
	Hot water setback	5	f	○

Category	Improvement	Expected Useful Life (Years)	Source	Data Quality
	Low-flow showerhead	10	e	○
	Low-flow showerhead	10	i	○
	Low-flow showerhead	7	f	○
	Pipe wrap	15	e	○
	Pump, commercial sump	15	h	○
	Tank temperature turndown	4	e	○
	Toilet, in dwelling unit bathroom	50+	h	○
	Toilet tank components, in dwelling unit bathroom	5	h	○
	Toilet	20+	j	○
	Urinals	20+	j	○
	Urinals (waterless)	20+	j	○
	Water heaters, electric	13	d	○
	Water heaters, gas	11	d	○
	Water heaters, heat pump	10	i	○

Data Quality:

*High-quality data: recent, independent, statistically significant.

○Data quality unknown.

•Data are old, not independent or not statistically significant.

Sources:

a. Data obtained from a survey of the United States by ASHRAE Technical Committee TC 1.8 (Akalin, 1978). Some updates in 1986.

b. Abrahamson, 2005. ASHRAE/ORNL.

c. Lawrence Livermore National Lab and Whitestone Research, Santa Barbara, CA, for the National Nuclear Safety Administration, U.S. Department of Energy, 2006.

d. Buildings Energy Data Book: 5.7 Appliances, March 2012, Table 5.7.15, Major Residential and Small Commercial Appliance Lifetimes, Ages, and Replacement Picture.

e. Measure Life Report: Residential and Commercial/Industrial Lighting and HVAC Measures, Prepared for The New England State Program Working Group (SPWG), June 2007, by GDS Associates, Inc., Table 2: Commercial and Industrial Measures.

f. http://mn.gov/commerce/energy/topics/conservation/Design-Resources/Deemed-Savings.jsp.

g. www.nrel.gov/analysis/tech_footprint.html. Accessed November 30, 2014.

h. Fannie Mae, Instructions for Performing a Multifamily Property Conditions Assessment (Version 2.0), Appendix F, Estimated Useful Life Tables, 2014. Data shown is for "multifamily/coop," which is typically between values provided for seniors (typically longer life) and students (typically shorter life).

i. NY Home Performance with ENERGY STAR Effective Useful Life of Energy Efficient Measures, NYSERDA, August 2012.

j. R. Bruce Billings and Clive Vaughan Jones, *Forecasting Urban Water Demand*, 2nd ed. American Water Works Association, 2008.

k. http://www.allianceforwaterefficiency.org/commercial_dishwash_intro.aspx. Accessed December 25, 2014.

l. Estimate based on similar materials. For example, a drain heat recovery device is typically made of copper water piping.

m. John Proctor, P.E., Rob deKieffer, Mary O'Drain, and Amalia Klinger, "Commercial High Efficiency Air Conditioners—Savings Persistence," 1999 International Energy Program Evaluation Conference.

n. Robert Zogg, William Goetzler, Christopher Ahlfeldt, et al. "Energy Savings Potential and RD&D Opportunities for Commercial Building Appliances, Final Report." Submitted to U.S. Department of Energy, Energy Efficiency and Renewable Energy, Building Technologies Program, Navigant Consulting, Inc., December 21, 2009.

o. From Richard Bole, "Life-Cycle Optimization of Residential Clothes Washer Replacement." Report No. CSS06–03. Center for Sustainable Systems, University of Michigan, April 21, 2006.

p. National Grid, "Transformer Replacement Program for Low-Voltage Dry-Type Transformers, Implementation Manual, Version 2013.1," April 4, 2013, p. 12.

q. U.S. Environmental Protection Agency. "ENERGY STAR Market & Industry Scoping Report Residential Clothes Dryers," November 2011.

r. ASHRAE Owning and Operating Cost Database. http://xp20.ashrae.org/publicdatabase/. Accessed August 15, 2015.

Appendix R

Request for Proposal for Energy Audits

General/Goals

A comprehensive energy audit is sought for <building/address/description/square feet>. Short-term goals within one to two years are <%> energy reduction, and long-term goals by <year> are <%> energy reduction. Honest energy savings estimates are sought; overestimated savings are discouraged.

Deliverables

Deliverables include a site visit, draft and final reports, the energy model software data file, and a closeout meeting.

Site Visit

A site visit shall be complete within <quantity> weeks of the signed contract. The building will be accessible during <days/hours>. Advance notice shall be given of the site visit, by <quantity> days.

Inspect all areas, including but not limited to attics, basements, occupied areas, above dropped ceilings, exterior, roof.

An interior and exterior space-by-space lighting survey shall be conducted, including lighting power density, light level below and between fixtures, fixture description, proposed lighting power density, existing and proposed lighting control and hours/year, and savings based on comparison to best-practice lighting power density for each space type.

Testing Required:

__ Blower door test (__ whole building, __ guarded test on the following areas: _____)

____ Infrared scans:

100% infrared scans of all walls, windows, doors, and ceiling/roof. Annotate each image clearly with the location.

__ Duct leakage test

__ Combustion efficiency

__ Water delivery flow rates: showers, faucets

____ Other: _____

Comprehensive energy audit including evaluation of the following improvements

___ High-efficiency heating/cooling

___ Ventilation
  ___ Reduced overventilation (e.g., exhaust)
  ___ Demand-controlled ventilation

___ Insulation improvements

___ Air sealing/infiltration reduction

___ Premium-efficiency motors

___ Adjustable speed drives

___ Lighting
  ___ High efficiency
  ___ Right-lighting (reduced overlighting)

___ Controls
Multilevel controls for spaces such as offices, bilevel lighting or other motion control for stairwells and corridors, photocell plus motion control for exterior lighting unless otherwise discussed, motion control for parking, motion control for restrooms.

___ Hot and cold energy and water conservation

___ Renewables
  ___ Photovoltaic
  ___ Solar hot water
  ___ Wind
  ___ Other: _____

___ Other: _____

The energy manager will provide two years of energy bills within one week of contract signing. The energy manager will provide occupancy schedules (number of people), by room, by hour for a typical day, for purposes of ventilation, lighting, and temperature control calculations by the energy auditor. The energy manager will provide building drawings as available, and other documents as available, such as equipment manuals. The energy manager will inventory heated and cooled spaces that could possibly be removed from the heating and cooling systems, such as stairwells, basement spaces, mechanical and electrical utility spaces, and the like. The energy manager will establish existing control sequences for lighting, including off-delay settings for motion sensors, and timer settings. The energy manager will provide an inventory of make/model for all refrigerators, vending machines, cooking equipment, laundry machines, and other major equipment, prior to the site visit. The energy manager will do 24-hour energy consumption (kWh) monitoring on plug loads, and provide results to the energy auditor.

Report

Building and energy systems description, including insulation R-values, description of the thermal boundary and deficiencies, description of unconditioned spaces and deficiencies, window U-factors, heating and cooling plant efficiency, heating and cooling distribution system type and condition, domestic hot water heater and distribution/delivery system, lighting survey, other electric loads (e.g., transformers on the building side of the electric meter), plug load survey, other gas loads (e.g., clothes dryers, ovens, etc.).

A list of energy improvements observed that have previously been made to the building.

Billing analysis, including a summary of a minimum two years of bills, benchmarking against similar buildings using EPA Portfolio Manager or approved equivalent, calculation of energy rates to be used in energy savings calculations, and estimated fraction of end loads by pie chart (heating, cooling, lighting, domestic hot water, plug loads, other).

A building energy model shall be built and trued-up against existing energy bills, within +0/–5 percent. In other words, the model cannot use more energy than the existing energy bills, but can use up to 5 percent less than the existing energy bills. The building model shall be done with eQuest, or approved equivalent. The model data file shall be delivered to the Owner on completion of the project, and shall become the property of the Owner.

Improvements, including estimated installed cost and source/basis, estimated annual energy savings including assumptions and calculation methods, estimated annual energy cost savings, a description of noncost trade-offs (comfort, health and safety, etc.), estimated useful life, net present value, carbon emissions reduction, and a written scope of work to guide implementation. The scope of work is not expected to be a detailed design document, but shall at a minimum provide location and quantity, required/rated efficiency, features that impact energy use, and testing required. Improvement energy savings shall be interactively calculated and listed in order from high to low <choose one: net present value; carbon emissions; energy savings; etc.>. In other words, the entire package of improvements shall be interactively calculated and presented first in the list. Then, the improvement that, when removed, makes the package next most attractive shall be removed and the package results provided, and so on for other improvements.

Deliver draft report by e-mail electronically, as a PDF, to <e-mail address>, by <date/time>. No printed/bound copies are required.

Deliver revised final report within <quantity> weeks of receiving comments on the draft report.

Closeout meeting. Attend a closeout meeting at <address> to answer questions about the report findings.

Quality Control. The energy audit report will be subject to internal quality control review by a senior staff member.

Qualifications

Energy auditor must have at least <quantity> years experience. Energy auditor must have completed at least <quantity> energy audits at similar facilities. Provide qualifications. Provide minimum <quantity> references from clients where you have completed energy audits.

Fee: _____

The Owner is not obligated to accept the lowest or any proposal, and reserves the right to reject any or all proposals.

Appendix S

Energy Audit Review Checklist

___ Title page: Author, building name, date, name of portfolio program (if any)

___ Have all requirements of the energy audit RFP/contract been met? Use the RFP/contract as its own checklist to ensure that the work is complete.

___ Review the building description for completeness:
   ___ R-values of walls, roof and foundation
   ___ U-factors for windows and doors
   ___ Description of infiltration sites and thermal bridging. Is window and door infiltration described? Is attic and basement/crawlspace infiltration described? Other infiltration sites?
   ___ Description of stack effect pathways: vertical pathways, entrance and exit to vertical pathways, horitontal pathways
   ___ Results of tests such as blower door and infrared imagery
   ___ Heating and cooling plant: type, age, efficiency, condition
   ___ Heating and cooling distribution system
   ___ Heating and cooling controls, including temperature control set points and schedule
   ___ Ventilation system and operation schedule
   ___ Domestic hot water (heater, distribution, fixtures)
   ___ Cold water (fixtures)
   ___ Occupancy schedule
   ___ Space-by-space lighting inventory, lighting power density, control, and schedule
   ___ Motors and motor controls/schedule
   ___ Major appliances
   ___ Plug load inventory and short term monitoring results
   ___ Other major energy loads (refrigeration, commercial cooking, elevators/escalators, etc.)

___ Comparing the building description to the improvements, were any significant potential improvements missed?
   ___ Envelope improvement
      ___ Insulation
      ___ Infiltration reduction
      ___ Windows and doors (U-factor improvement or reduction/elimination)
      ___ Heating/cooling plant improvement (including not only high-efficiency replacement, but fuel-switching or

363

distribution-switching such as elimination of steam heating systems)

___ Heating/cooling distribution improvement (e.g. pipe/duct insulation, duct sealing, removing heating or cooling from zones that do not need it)

___ Ventilation improvements (demand control, ventilation reduction while still complying with code requirements, heat or energy recovery)

   ___ Domestic hot water

      ___ Heater efficiency (or fuel switching)

      ___ Distribution efficiency (insulation, recirculation controls)

      ___ Fixture efficiency

   ___ Cold water fixture efficiency

   ___ Controls improvements

      ___ Adding thermal zones

      ___ Unoccupied temperature control

      ___ Ventilation control

   ___ Motor improvements

      ___ High-efficiency motor replacements

      ___ Variable speed drives

   ___ Lighting improvements

      ___ High-efficiency replacement

      ___ Right-lighting (reduced over-lighting)

      ___ Lighting controls

         ___ Interior lighting controls

            ___ Dividing spaces into multiple areas

            ___ Allowing lighting reduction (dimming, bilevel, or inboard/outboard control)

            ___ Motion control

            ___ Photocontrol (daylighting)

         ___ Exterior lighting controls

            ___ Photocontrol to keep lights off during the day, PLUS either:

            ___ Motion sensor (preferred), or

            ___ Timer

  ___ Other

     ___ Transformers

     ___ Elevators/escalators

     ___ Refrigeration

     ___ Cooking

     ___ Appliances

     ___ Other: _____

___ Are improvements substantive, or are they token?

___ Are total savings substantive and do they meet the owner's goals, or are they token?

Energy savings calculations:

   ___ For whole-building computer models, are reports provided for both inputs and outputs?

   ___ For spreadsheet calculations, are inputs, outputs, and equations provided?

   ___ Check assumptions that are used for each improvement: Are they reasonable? Are they conservative, to avoid overestimating savings? Key assumptions: existing equipment efficiency (check that they are not too low), proposed equipment efficiency (check that they are not too high), temperature set-point changes, hours per day runtime for lighting and appliances (pre- and post-improvement).

__ Check measurements/quantities: light fixtures, insulation square footage, etc.

__ Are savings for each improvement a reasonable fraction of the fuel/electricity being used at the facility? Most importantly, the total savings should not exceed the total energy use of the fuel/electricity being used, or a reasonable fraction of the total energy use.

__ Is existing building energy use "trued-up" to utility bills?

Billing analysis:

__ Are a minimum of one year's bills included?

__ Are all meters included? Are common area meters included? Tenant meters? Are there any meters for outdoor lighting? Remote buildings?

__ Are costs/bills included for both supply and delivery?

__ Are batch fuel deliveries (oil, propane, etc.) included – delivery dates and quantities?

__ Are water bills included?

__ How were energy rates calculated? Are they reasonable?

__ Is benchmarking provided, to compare the building's energy use to the energy us of other buildings of the same type?

__ Are expected equipment lifetimes reasonable?

__ Is life cycle analysis included (such as net present value, savings-to-investment ratio, etc.)?

__ Are installed cost estimates included and reasonable?

__ Do calculation inputs match the building description?

__ Were renewables improvements evaluated, if they were part of the energy audit scope?

__ Are written workscopes clear for each new proposed device or system? Workscopes should include:

L: location and quantity
E: efficiency rating
F: features that affect energy use
T: testing requirements

__ Does the energy audit comply with rules of any portfolio program in which it is partipating?

Appendix T
Energy Preventive Maintenance Schedule

Item	Frequency
Energy audit	5 years
Infrared scans—look for settled insulation and other voids	5 years
Inspect air conditioning fins for damage; comb bent fins	5 years
Clean air filters (air handlers, fan coils, sealed combustion equipment, room air conditioners, dehumidifiers)	Annual
Inspect and clean indoor coils (air handlers, fan coils, room air conditioners, etc.) and outdoor coils (condensing units, air cooled chillers, etc.)	Annual
Utility meter readings	Monthly
Bulk fuel storage readings (fuel oil, wood pellets, etc.)	Monthly
Check boiler makeup water consumption	Annual
Condenser fans	Annual
Vegetation growth around condensing units	Annual
Test efficiency of boilers and furnaces	Annual
Inspect for water leaks—by observation (at water fixtures like faucets, below sinks, ceilings, basements and mechanical rooms, toilets and urinals, water heaters) and at leak indicator on water meter	Annual
Recommissioning	5 years
Plug-in appliance meter readings	3 years
Check refrigerators: temperatures and ice buildup	Annual
Storm windows—check if closed	Annual—September
Update occupancy schedules—compare to temperature schedules, ventilation schedules, lighting schedules	Annual
Check that temperature control settings have not been changed	Annual
Check outdoor reset control settings (boilers, chillers)	Annual
Clean photocell lenses	Annual
Clean lighting lenses and reflectors	5 years
Check exterior lights are not on during the day, cloudy days	Annual
Gas leak measurements at the gas meter, also check pilot lights are lit	Annual
Log changes in occupancy, new additions, major new equipment and other that will affect energy usage	Ongoing
Log comfort complaints and locations—too hot, too cold, drafts	Ongoing
Core sample of flat roof, to assess insulation condition and water penetration	10 years
Check attic insulation depth markers; check that any removed insulation has been re-installed	5 years

(Continued)

Item	Frequency
Inspect windows and doors for air leakage, weather-stripping and caulk condition, and condensation	5 years
Mold survey—basements, attics, kitchens, bathrooms	5 years
Hazardous materials—photo survey. Asbestos pipe and duct insulation, asbestos floor tile, lead paint.	5 years
Inspect roof for ice dams or other evidence of snow melt due to heat loss	Winter, after snow
Check outdoor air dampers: open when the building is occupied, closed when the building is unoccupied	Annual
Check outdoor air dampers for leakage using air-mixing method	5 years
Check combustion air dampers: Open when boiler or furnace is firing, closed when not firing	Annual
Check that air handler and exhaust fans are operational, without vibrations, and are securely fastened to motor shafts	
Check lightbulbs for failure	Annual
Check lighting motion sensors: Do they come on reliably? Do they turn off after a short off-delay?	5 years
Check hot water temperature	5 years
Flush sediment from water heater	Annual
Clean and tune boilers and furnaces	Annual
Clean clothes dryer vents—low use	Annual
Clean clothes dryer vents—high use (coin op)	Monthly
Clean air conditioner condensate pan and drain—systems where water leakage could damage room finishes (e.g., ductless fan coils)	Annual
Clean air conditioner condensate pan and drain—systems where water leakage will not damage room finishes	5 years
Inspect ductwork for unusual leakage (detached connections, etc.) and condition of duct sealant	5 years
Inspect duct and pipe insulation	5 years
Check air conditioners for signs of refrigerant leakage: low capacity (reports of comfort problems), high-supply air temperature (track over time); have checked professionally if there are signs of leakage	Annual
Test carbon monoxide detectors	5 years

Index

actuator, 61, 154, 171, 176–178, 302, 354
adaptive control, 182
adjustable speed drive, 221
adjusted volume, 224
aerator, 198–199, 356
aerosol duct sealing, 140–142
AFUE, 100–103, 109–111
AHAM, 121, 255, 317
AHRI, 102, 114, 121, 125, 132, 161–164, 166, 170, 190–193, 213, 224, 255, 317
AHU, 142, 342
air compressor, 226
air curtains, 59, 65, 113, 280
air door, 113
air handler, 58, 124, 130, 134, 139, 141–144, 154, 156, 160, 175–176, 179, 186–187, 221, 266, 323, 342, 344, 368
air handling unit, 144, 352
air leakage, 19, 37, 43, 45, 52, 56, 58–62, 64–65, 69, 111, 120–121, 130, 137–141, 143, 154–155, 158, 265, 301, 368
air lock, 59
Air Movement and Control Association, 165
air temperature rise, 109–110
air-cooled chiller, 130–132, 134, 167
air-side economizers, 187
air-to-air, 123, 161, 166, 354
air-to-water, 123, 162
Alliance for Water Efficiency, 201
AMCA, 165
American Institute of Architects, 284, 294
angle stop, 211
Annual Fuel Utilization Efficiency, 100, 109
ANSI, 79, 95, 163, 166, 192, 329
aquastat, 103–104
ARI, 170, 224
as-built drawings, 17, 27, 48, 81, 154, 179
asbestos, 15, 29, 140, 146–147, 149, 284, 290, 319, 368
ASD, 220
ASHRAE 62, 156–157, 160, 166, 271
ASHRAE 90, 51, 74, 101, 133, 271, 338
ASHRAE 189, 76, 146, 338
ASHVE, 156
ASME, 190
axial fans, 133, 152

backdraft damper, 63
balcony, 21, 31, 38, 40, 42, 56–57, 59, 69, 241, 274
ballast, 17, 20, 77, 81, 84, 91, 94–95, 98, 255, 310
ballast checker, 17, 20, 77
balometer (flow hood), 18, 154
balsam wool, 29
BAS, 171
base-mounted pump, 147
basement entry doors, 50
bathroom exhaust grille, 154
belt-driven, 152–153
Bilco doors, 50
bilevel lighting, 89–90, 94, 360
bin method, 157

bin modeling, 250
biomass, 99, 356
blower, 8, 18, 21, 30, 52, 61–68, 109, 112, 133–134, 140, 143, 152, 168, 240, 299, 359, 363
blower door, 8, 18, 21, 30, 52, 61–67, 140, 240, 299, 359, 363
blower wheels, 152
BMS, 171
borescope, 18, 28
BPM, 111, 147, 220
brake, 217, 219
branch ducts, 108, 137
breathing zone, 160
breeching, 168
bridge financing, 285
bridging, 27, 30–32, 36–38, 51, 241, 282, 363
brushless permanent magnet motors, 111, 220
building automation systems, 171
building management systems, 171
Building's Current Requirements, 302–304
bulkhead doors, 50
bypass, 65, 110, 112, 120–121, 161, 175–176, 183, 274, 278, 280

cabinet heaters, 105, 112, 148
California, 74, 98, 114, 121, 134–135, 159, 187, 199–200, 206, 211–212, 214–215, 225, 235, 255, 317, 319, 336
California Energy Commission, 98, 121, 206, 225, 235, 255, 317
call for cooling, 174
call for heating, 173–174, 177, 302
capillary tube, 118
carbon dioxide, 18, 155–156, 159, 177, 180, 238
CCT, 74
ceiling fans, 114, 226–227
cellar door, 52
cellulose insulation, 36
center of glass, 44–45, 47
centrifugal fan, 144, 152
CFL, 6, 74, 77–78, 355
CHP, 239
circulator, 146, 354
clerk of the works, 283
close-coupled pumps, 146
closed loop control, 90
closed motors, 217
clothes dryer, 65, 190, 229–231, 368
clothes washer, ix, 204–206, 208–209, 215, 357
CM/GC, 282
CMAR, 282
cobra head, 81
COG, 44–45
cogeneration, 239, 243, 277, 280
collateral effect, 307
color rendering index, 73
combined heat and power, 193, 239, 249, 277, 280
combined motion and photocell control, 92
combustion efficiency, 100–103, 105, 108, 111–112, 184, 359
commissioning, 4, 90–91, 285, 287–288, 302
compact fluorescent lamp, 77
competing investment, 261

compressor, 25, 118–119, 122–125, 128, 131–132, 134, 148–149, 167–168, 191, 217, 223, 226–228, 279, 353
condenser, 118, 122–124, 126–128, 130–134, 146, 167, 178–179, 186–187, 221, 223–224, 226–228, 279, 295, 353, 367
condensing boiler, 104, 184
condensing unit, 123–124, 127, 168, 356
constant volume, 112, 129, 143, 175, 352
construction cost, 258
construction management, 282–283
Construction Management Association of America, 282
construction management at risk, 282
control optimization, 186
control stop, 211
cooling tower, 123, 130–134, 167–168, 178, 186–187, 271, 274, 353
core losses, 222
correlated color temperature, 74
cost control, 13, 262, 281, 284–285, 292, 305, 315
cost-benefit ratio, 260
counterweight, 34, 50
cove lighting, 20, 93, 95
crank timer, 90
crawlspace, 12, 27, 32, 37–40, 60, 69, 137, 139, 259, 296, 351, 363
cream-skimming, 307–308
CRI, 73
cut sheets, 287
CV, 129, 143

daylighting, 49, 81, 84, 90–91, 94, 242, 279, 364
DDC, iv, 171, 353
declining block rate, 257
decorative lighting, 79–80, 94–96, 275, 278
dedicated outdoor air systems, 160, 246, 275
deep energy retrofits, 9
defect detection, 268, 287, 311–312
defect prevention, 268–269, 287, 311–312
dehumidification, 117, 142–143, 158, 186
demand-controlled ventilation, 158–159, 245, 252, 275–277, 279–280, 360
depth marker, 37
design/bid/build, 282–283
design/build, 282–283
destratification, 114–115, 226
desuperheater, 193, 211
dew point temperature, 117
DHW, 107, 189, 193–194, 226, 239, 262, 283
diffuser, 65, 67, 141, 144, 154, 168–169, 352
dimming control, 91–92
direct digital control, 171
direct expansion, 117–118, 181, 295, 301–302
direct-drive, 152
direct-fired, 108
dish machines, 201
dishwasher, 6–7, 14, 24, 197, 201–203, 214, 227, 317, 355–356
diverter, 111, 212–213, 215, 274
DOAS, 160
downblast, 152
drain heat recovery, 207, 210–211, 356–357
Dreyfuss patent, 174
dual fuel heat pumps, 122
dual-flush, 211
dual-technology motion controls, 89
duct furnace, 108
duct leakage, 8, 18, 58, 138–143, 250, 273, 293, 359
duct leakage test, 140–142, 359
duct lining, 138–139
duct-board, 137
ductless systems, 124–125
DX, 117–118, 125–126, 143, 353–354

ECM, 9, 111, 220
ECO, 9
economizer, 104, 128, 162, 165, 187, 275–276, 353
edge of glass, 44–45
Edison, 77, 79, 135, 187, 215, 336
EF, 194–195, 202, 205, 208
effective leakage area, 63
EIFS, 37, 351
ELA, 63
electric baseboard, 112–113, 183, 353
electrodeless lighting, 80
electronic expansion valve, 118
electronically commutated motors, 111, 220
elevator, 21, 38, 53, 55–57, 61, 64, 67, 181, 221–222, 234, 260, 265, 271, 273, 328, 355, 363–364
elevator room, 221
elevator shaft vent, 64, 260
EMS, 171, 187
enclosure, 27, 104, 148
end use, 13, 249–250, 265, 268, 280, 300
end-use metering, 296
energy conservation measures, 9
energy conservation opportunities, 9, 274
energy cost savings, 11, 15, 257–261, 263, 265, 279, 291, 305, 310, 361
energy efficiency ratio, 118, 121, 287, 336
energy factor, 192, 194–195, 202, 204–205, 207–209, 229–230
energy management system, 155, 168, 177, 186, 302, 353
Energy Policy Act, 102, 197, 218
energy recovery ventilator, 154, 161–162
energy service companies, 8, 177, 263, 291
Energy Utilization Index, 13, 299, 303
EnergyGuide, 121, 202, 205–206, 208–209
enthalpy, 158, 162–163, 353
EOG, 44–45
EPA, 21, 29, 68–69, 98, 120, 197–203, 211–212, 214, 294, 299, 304, 310, 317, 361
EPAct, 197, 200, 211–212, 218, 234
EPS, 28–29, 37
EqLA, 63
equipment service life, 261
equivalent leakage area, 63
errors and omissions insurance, 289
ERV, 161, 163
escalator, 221–222, 234, 363–364
ESCO, 291
EUL, 261
evacuation drawing, 20
evaporator, 118, 122–123, 126–128, 131, 148, 227
exhaust fan, 63–64, 90, 138, 152–153, 165, 168
exit light, 96–97
expanded polystyrene, 28
expected life, 175, 211, 244, 260–261, 265
expected useful life, vii, 261–263, 351–357
explosion-proof motors, 217
exterior insulated finishing system, 37
extruded polystyrene, 28, 37
EXV, 118, 126

fan coil, 123–125, 148, 168, 171, 175, 177–178, 353
fan efficiency, 156, 165, 227
fan efficiency grade, 165
faucet, 19, 22, 197–200, 210, 212, 356, 359, 367
faucet aerator, 198–199
faucet insert, 198–199
FC, 72, 81, 85, 93, 326
FEA, 38
Federal Trade Commission, 202

FEG, 165
fiber-optic camera, 28
fiberglass, 28–30, 34, 37, 42, 50, 145, 147, 196
filter housing, 112, 139
fin comb, 122
fin damage, 26, 122
finite element analysis, 38
first cost, 258
first-hour rating, 194
first-year savings, 257
flex duct, 137–138
floor plan, 19
flow hood, 18, 154
flue, 65, 102–104, 109, 111, 129, 184, 190–192, 280
flue gas economizer, 104
fluorescent, 6, 71, 73–79, 84, 87, 90–91, 94–97, 258, 277, 279, 354–355
flush valve, 211–212
Flushometer, 211, 215
folk measurements, 22
folk quantification, 22–23
foot-candle, 10, 72, 74–75, 83, 85–87, 92, 325–326
four-pipe fan coil, 148, 175
four-way valve, 122
fractional horsepower motors, 218
free cooling, 49, 128, 162, 165, 176, 279
free dridership, 307
free ridership, 307
FTC, 121, 202, 205–206, 208
fuel oil, 99, 168, 190, 249–250, 296–297, 300, 367

gaming, 268–269, 312, 314
garbage bag test, 23
gen-set, 221
general conditions, 284
general contractor, 282–283, 289
general liability, 289
generator, 17, 64, 107, 221
geoexchange, 123
Geoffrey Chaucer, 283
geothermal, 123, 194, 240–241, 354
glass block, 49, 51, 351
glass brick, 49
gooseneck, 152, 154
gravity ventilator, 152–153
greenwashing, 293
grid-tied photovoltaic system, 243
grille, 23–24, 62, 119–120, 139, 141–142, 144, 152–154, 158, 167–169, 352
gross floor area, 21
gross savings, 307
ground-source, 123
guarded blower door test, 66

halogen, 71, 73–74, 79–80, 84, 95–97, 260
heat pump, 6–7, 25, 108, 119, 122–124, 127, 129–130, 135, 148, 162, 167–168, 176, 190–195, 208, 229–231, 235, 240, 249, 254, 259, 267, 272, 278, 293, 295, 299, 353–354, 357
heat pump water heater, 191–193, 259
heat recovery ventilation, 161
heat recovery ventilator, 162, 164
heating capacity, 11, 100, 104, 345
HID, vi, 6, 71, 78–79, 82, 84, 95, 329, 355
high-bay, 78, 81, 95, 277, 279
high-efficiency toilet, 211
high-intensity discharge, 6, 71, 78, 355
high-pressure sodium, 71, 73–74, 78–79, 81, 83–84, 329
high-velocity system, 125
holistic, 7
Home Ventilating Institute, 162

horsepower, 10, 18, 156, 180, 217–220, 248
hot water generator, 107
hourly model, 158, 252
HP, 10, 18, 78, 133, 156, 165, 217–220, 248, 349–350, 354
HRV, 161
HUD, 271, 284
humidity, 18, 40, 108, 151, 158, 161, 163, 168, 171, 177, 181–182, 186, 231, 290, 303, 317
HVI, 162, 165–166
hydraulic elevators, 221–222, 271
hydronic, 99, 104–108, 113–115, 129, 146–147, 181, 184, 186, 189, 221, 353–354

I=B=R, 100
IBC, 74
ice dams, 25, 60–61, 368
ICM, 220
IECC, 43, 51, 72, 74, 76, 146, 148, 165, 321–322, 328, 338
IEER, 124–125
IESNA, 74, 85, 326
IG, 45
IGU, 45
illuminance, vi, 72, 74–75, 80–83, 85–87, 90, 92, 279, 314, 325–326
Illuminating Engineering Society, 74, 97, 294, 303
illumination, 72, 96
in-line pumps, 146–147
inboard-outboard lighting, 88
incandescent, 71, 73–75, 77, 79–80, 84, 96–97, 258
incentives pendulum, 314
incremental rate, 257
indirect water heaters, 190
individual metering, 272, 296
indoor air quality, 3, 8, 26, 108–109, 138, 151, 156–157, 160, 166, 277, 290
indoor coil, 120–121, 124
inducer, 109
induction lamp, 80
induction lighting, 71, 80, 84
infrared heater, 114
infrared thermography, 8, 28, 64
inner envelope, 38–41, 61
installation cost, 86, 241, 258, 260, 262, 265, 309
installed cost, 9, 11, 13, 94, 207, 242, 246, 258–259, 262, 282, 361, 365
instantaneous water heater, 192, 200
Institute of Boiler and Radiator Manufacturers, 100
insulated piping, 146
insulated window shades, 49
insulating glass, 45
insulating glass units, 45
insulation depth marker, 37
Integrated Water Factor, 204
interacting improvements, 9, 267
interactive improvements, 9
interior storm, 45–46, 48
interior window panels, 45
International Building Code, 61, 74
International Energy Conservation Code, 72, 141, 146, 165, 181
inversion calculation, 248–249
investment, 15, 67, 146, 194, 239, 244–246, 249, 257–261, 272, 283, 291, 294, 309, 312, 365
isolated, 247–250, 252, 254, 312–313

joist bay return, 137

lab fume hood, 155
latent cooling, 117
latent effectiveness, 162–163
lay-in troffer, 80
leak indicator, 213, 367

LED, 6, 71, 73–77, 79–80, 84, 87, 89–92, 94–98, 241, 258, 260, 277–279, 294, 355
LEFT, 6, 15, 24, 29, 34, 39, 45–46, 58–59, 92, 146, 187, 196, 245, 266, 284, 297
lenses, 26, 74, 80, 86, 97, 367
level 1 energy audit, 13, 245
level 2 energy audit, 13
level 3 energy audit, 13
lid sink, 211–212
life cycle costing, 253
Life Safety Code, 74, 290
light meter, 17, 72, 80, 82, 86, 92
lighting pattern, 87
lighting power allowance, 72, 76
lighting power density, 10, 21, 71, 76, 82, 85, 97, 355, 359, 363
limited portfolio programs, 307–308, 311
line voltage, 173, 175
linear diffuser, 169
liquid chillers, 117
liquid line, 148–149
liquid-filled transformer, 223
louver, 127, 129, 154–155, 168, 353
low-bay, 78, 95
low-cost/no-cost, 8–9, 13, 67, 245, 307
low-e coating, 47
low-flow, 198–200, 357
low-flow faucet insert, 198–199
low-hanging fruit, 307
low-pressure sodium, 71, 73–74, 78–79, 83–84, 329
LPA, 72
LPD, 71–72, 76, 83, 327–328
luminous efficacy, 73
lux, 10, 72

M&V, 250, 312
maintained illuminance, 74, 325–326
major vertical penetrations, 21
makeup air, 54, 109, 128–129, 153–154, 158, 160, 165–166, 273, 276
master meter, 272
measurement and verification, 244, 250, 288, 291, 312, 315, 317
mechanical floor plan, 19
mechanical ventilation, 49, 151, 157, 246
median service life, 261–262
MEF, 204–210
mercury vapor, 71, 73–74, 78–79, 84, 329
metal halide, 71, 73–74, 78–79, 81, 84, 94, 277, 329
metal studs, 31–32
metering, 135, 199–200, 243, 272, 274, 276, 292, 296, 301
metering faucet, 200
MG, 218, 221–222
MG 1 standard, 218
MG set, 221–222
mineral wool, 29
mini-split, 124
mixing box, 154
mixing valve, 196
Modified Energy Factor, 205, 207–209
motion control, 81, 84, 87, 89, 92, 229, 360, 364
motion sensor, 6, 87, 89–90, 233, 266, 294, 364
motor actuator, 178
motorized valve, 177
motors, 17, 20, 25, 111–112, 125–126, 134, 143, 147, 158, 177, 180, 217–222, 234, 239, 275, 277, 280, 355, 360, 363
MR, 95
multi-split, 124
multifaceted reflector, 95
muntin, 44

nameplate, 19–21, 24–25, 100, 102, 110, 119, 121, 133, 143, 167, 169, 190, 217–218, 225, 227, 233–234
National Electrical Manufacturer's Association, 218, 222
National Fenestration Research Council, 43
natural gas, 99, 127, 190, 237–238, 249, 265, 268, 295–297, 347, 353
natural ventilation, 42, 49, 151, 157, 246
natural-draft furnaces, 109
naturally occurring conservation, 306–307
needle flow, 198
NEMA, 218, 222, 234, 266
net floor area, 21, 71
net metering, 243, 296
net present value, 260, 361, 365
net rating, 100
net savings, 200, 307
net-to-gross ratio, 307
net-zero, 2, 9, 112, 232, 243–246, 303
neutral pressure plane, 52–56
New York, ix, 1, 7, 68, 74, 97, 105, 115, 132, 187, 234, 255, 280, 294, 298, 309, 318
NFPA 101, 74
NFRC, 43–44, 47, 50, 69
no-cost/low-cost, 5, 67
no-load losses, 222
nominal, 24, 31, 79, 145–146, 148, 153, 217, 219–220, 227, 329
nonwater urinals, 212
NSF International, 201
NYSERDA, 70, 97, 114–115, 234, 264, 357

OA, 153
octopus furnaces, 108
ODP, 217, 349–350
off-delay, 74, 82, 89, 92, 199, 233, 254, 266, 303, 360, 368
oil, 6, 99–100, 102, 109, 112, 168, 190–193, 196, 248–250, 296–297, 300, 314, 352–353, 365, 367
on-demand heaters, 190
open drip-proof, 217, 349–350
open loop control, 90
open motors, 217
outdoor air damper, 155, 178–179
outdoor reset, 103–104, 179, 182, 184, 186, 266, 367
outer envelope, 38–39, 41, 61
outside air, vi, 60, 109, 123, 129, 141, 143, 151, 153–158, 160, 163, 176, 187, 273, 275, 323, 343–344
overhead doors, 50–51, 54, 56, 59, 280
overlighting, 7, 71, 81, 84–85, 91, 242, 294, 311, 360
overventilation, 151, 158, 160, 179–180, 360

PACE, 22, 309
packaged, 108, 111, 114, 117, 119–120, 128, 130, 142, 148, 152–153, 168, 176, 187, 240–241, 274, 317, 354
packaged air handlers, 142
packaged rooftop unit, 128
packaged terminal air conditioner, 119, 354
packaged terminal heat pumps, 114, 119, 274
packages of improvements, 9
paddle fans, 226
pan, 50, 368
pan-style, 50
panning, 137–139
PAR, 79
parabolic, 79–80
passive infrared, 89
Passivhaus, 63–64, 303
pause valve, 200–201
payback, 9, 11, 201, 245, 258–261, 263, 265–267, 282, 310
PEA, 13, 300
performance contract, 291
perimeter radiation, 104, 147, 168

perlite, 29
persistence of savings, 242
photocell, 6, 12, 81–82, 84, 86, 89–92, 241, 355, 360, 367
photocontrol, 7, 90, 92, 94, 364
photoluminescent exit lighting, 96
photovoltaic, 9, 108, 112, 241–245, 247–249, 258, 272, 291, 296–297, 300, 306–308, 356, 360
PIR, 89
PL lamp, 78
plant, 7, 9, 99, 113, 133, 139–140, 145, 161, 177, 181, 239, 245, 252, 257, 265, 267, 275, 293, 302, 360, 363
plastic vent piping, 103, 109
PM, 295
polyiso, 28, 31
polyisocyanurate, 28–29, 31, 37, 40
polystyrene, 28–29, 37, 50–51, 225
polyurethane, 29, 50–51, 225
Portfolio Manager, 21, 299, 361
portfolio program, 266, 268–269, 299, 305–310, 312, 363, 365
power roof exhausters, 152
PRE, vii, 43, 60, 94, 101–102, 138, 152, 193, 199, 203, 205, 208, 214, 223, 234–235, 240, 250, 313, 349–350, 365
preheating, 143
preliminary energy-use analysis, 13, 300
premium efficiency, 218, 220, 222, 234, 266
prerinse spray valve, 2, 202
preventive maintenance, vii, 295, 367–368
professional liability insurance, 289
project manager, 281, 285–287, 289, 311
propane, 99, 190–191, 193, 296–297, 314, 347, 365
propeller fan, 167
Property Assessed Clean Energy, 309
PTAC, 119–122, 124, 127, 148, 167–168, 176, 274, 317, 354
PTHP, 119–123, 274, 354
public private partnership, 282

quality assurance, 291, 311
quality control, 1, 13–14, 139, 246, 251, 254, 265, 267–269, 281–284, 286–288, 292, 303, 305–306, 308, 310–312, 315, 361
quartz halogen, 79

R-value, vi, 12, 28–36, 42–43, 45, 50–51, 68, 225, 247, 265–266, 319, 360, 363
radiant barrier, 39, 69
radiant floor heating, 104
radiation, 28, 100, 104–106, 113, 147–149, 168, 186, 345
radiator, 99–100, 105–108, 112–113, 145, 168, 172–173, 184, 186, 212
rapid modeling, 254
realization rate, 307
recessed downlight, 6, 81
recirculating hot water, 196–197, 200
recovery efficiency, 194
recovery ventilator, 154, 161–165
reflectance, vi, 85–87, 333
reflectors, 86, 98, 367
refrigerant, 25, 99, 118, 120, 122–127, 130–132, 135, 143, 148–149, 168, 179, 193, 225, 280, 295, 299, 368
refrigerator, 2, 24, 223–225, 227–229, 234–235, 244, 249, 272–276, 278–280, 301, 317, 355, 360, 367
register, 139, 144, 204, 213, 215, 235, 352
reheat, 144, 183
reinvent treadmill, 314
request for proposal, vii, 281, 359–361
reset control, 103, 179–180, 184, 266, 367
reset ratio, 184
restrike, 78–80, 95, 277
retrocommissioning, 7, 302–304
return grille, 62, 168–169
return on investment, 15, 146, 260–261

reverse cycle air conditioners, 122
reversing valve, 122–123, 167
RFP, 281, 291, 363
right-lighting, 71–72, 76, 81, 84–86, 96, 241–242, 250, 258, 262, 275–276, 279, 294, 311, 360, 364
rigid insulation, 31, 37–39, 67, 137, 240, 266
riser, 57, 106–107, 137–138, 145, 153, 158
rock wool, 29
ROI, 260
rolled duct sealants, 140
roof drain, 38
roof insulation, 36, 38, 261, 267
room air conditioner, vi, 35, 60, 63, 65, 118–121, 183, 191, 323, 335–336
run-around, 162
runout, 137, 145
runout ducts, 137

savings-to-investment ratio, 260, 365
schedule control, 13, 281, 283, 288–289, 292
schedule of values, 286
scope control, 13, 281, 283–284, 292
sealed combustion, 56, 58, 109, 367
seasonal energy efficiency ratio, 121
security light, 79
SEER, 121, 124–125, 129–130
sensible cooling, 117
sensible effectiveness, 162–164
service hot water, 99, 189
service life, 261–262
set point, 82, 103–104, 159, 171–175, 179, 181–182, 192, 194, 196–197, 228
set-point, 159, 172–173, 175, 180–181, 226, 228, 364
setback, vi, 9, 14, 110, 112–113, 119, 123, 175, 178, 180–183, 186, 195, 213, 252, 278, 293, 302, 314, 345, 356
setup, 141, 175, 180–181, 314
shared savings, 291
sheet metal, 77, 137, 140, 170
SHGC, 43, 45–46, 321–322, 352
shower diverter, 213
shower pause valve, 200
showerheads, 198–200, 212
side discharge, 127–128
sidearm water heaters, 190
sidelighting, 90
single-pipe, 106, 168
SIR, 260
sleeve, 3, 26, 34, 60, 63, 119
slip, 14, 217, 220
SMACNA, 140–142
smoke-guided infiltration diagnosis, 67
snap-back, 307
software, vi, 63, 69, 85–86, 219, 247, 253–254, 299, 318, 331, 359
solar heat gain coefficient, 43, 352
solar thermal, 146, 193, 240–242, 258
solid-state lighting, 71, 73, 80, 84
spatial control plus output control, 88
specular reflectors, 86, 98
spillover, 307
spiral duct, 137
split incentive, 272, 276, 313
split system, 123–125, 194
spray foam, 28, 34, 36–38, 67
spray valve, 2, 202
squirrel-cage wheels, 152
SSL, 71, 80
stack effect, 7, 39, 51–58, 61, 65, 67–68, 138, 171, 181, 191, 240, 246, 250, 271, 273, 275, 293, 363
Standard 62, 156–157, 159, 166
Standard 90, 72, 76, 102, 132–133, 141, 146, 165, 253

standby loss, 192–194
standing pilot, 111, 191, 193, 230, 237, 276
storage water heater, 197
storm window, 45–46, 48, 50, 64
submetering, 272, 296
submittal, 287
suction line, 148–149
superheat, 148–149
supplemental conditions, 284
supply diffuser, 168
suspended unit heater, 112
synchronous speed, 217

T86, 173
TAB, 77, 154, 351
take-back, 307, 311
tankless heaters, 190
tapered roof insulation, 36, 38
task lighting, 80, 85, 98
TEFC, 217, 349–350
temperature rise, 109–111, 184, 194, 203, 209
TENV, 217
test and balance, 154–155
theater lighting, 97, 278
thermal boundary, 7, 27, 30, 32, 35–36, 38, 40, 50, 59–61, 65, 112, 130, 137, 139–140, 265, 293, 296, 360
thermal bridging, 27, 30–32, 36–38, 51, 241, 363
thermal efficiency, 100–103, 109–112, 192–195, 266
thermal imaging, 28, 30, 68
thermal zoning, 171–172, 175, 183, 240, 245, 278
thermal zoning diagram, 172
thermostat, 113, 119, 122, 128, 168, 171–178, 183, 187, 197, 228, 266, 354
thermostatic expansion valve, 118
thermostatic radiator valve, 172–173
through-wall air conditioners, 118, 172
time delay, 82, 89
timeout, 82, 89, 98
timer, 7, 84, 86, 89–92, 94, 158, 178, 195, 222, 229–230, 233, 241, 278, 360, 364
Title 24, 74, 253, 317
toaster, 232
toilet, 198, 211–212, 215, 275, 357, 367
toilet lid sink, 211–212
tons of cooling, 10, 117
toplighting, 90
total effectiveness, 162–163, 165
totally enclosed fan-cooled, 217, 349–350
totally enclosed nonventilated, 217
TP1, 223
track lighting, 94–95
traction elevators, 221, 271
transformational energy improvements, 1, 7, 312
transformer, 173, 177, 222–223, 234, 265, 356–357, 360, 364
trickle valves, 200
trigeneration, 239
triple-pane window, 45, 47–48
troffer, 80, 82, 87
true-up, 251
tubular, 77
tungsten halogen, 79
turbine ventilator, 154
twinning, 108
two-pipe, 106–107, 148, 168, 177–178
two-pipe fan coil, 148
TXV, 118, 126, 135

U-factor, 33, 42–47, 49, 51, 69, 247, 265–266, 321–322, 360, 363
UF, 29
UFFI, 29
ultrasonic, 89
uncapped wall cavity, 67

unconditioned space, 36, 38–40, 293
uniformity ratio, 74
unit heater, 112–113, 149
unit ventilators, 148
united inches, 43
unvented, 108
upblast, 152, 165, 168
upflow, 108, 127
urea formaldehyde, 29
urinal, 197–198, 211–212, 215, 357, 367
utility data, 314

v-strip, 68, 70
vacancy sensor, 87, 89
vacuum steam, 107
variable air volume, 129, 143, 168, 352
variable frequency drive, 18
variable speed drive, 134
VAV, 129, 143–144, 168, 183
vegetated (green) roof, 39
vent damper, 58, 105, 191
ventilation effectiveness, 158, 160
ventilation zoning, 160
vermiculite, 29, 68
vertical discharge, 124, 127, 129
VFD, 220
vindauga, 42
visible transmittance, 43
void, 27–28, 30, 32–33, 42, 367
VSD, 220
VT, 43
VVT, 175
VVVF, 222

walk-through survey, 13
warewasher, 201
warm edge spacers, 47
warmup time, 78–79
Water Factor, 204–205
water heater, 7, 19, 149, 189–197, 200, 202, 208, 210, 213, 242, 259, 360, 368
water heater wrap, 192
water meter, 24, 189, 194, 211–213, 301, 367
water meter leak indicator, 213
water-cooled chiller, 130–133, 168, 187
water-side economizers, 187
water-source, 123, 194
water-to-air, 123, 354
water-to-water, 123, 210
waterless urinal, 212
WaterSense, 197, 211
weather-strip, 6, 63
WF, 204–205, 207, 209
window air conditioner, 64
window area, 43, 49, 254
window condensation, 46, 51
window film, 50, 351
window rating, 49
window spacer, 49
window surface numbering, 47
wood chips, 99
wood pellets, 99, 296, 367
workscope, 265–266, 268, 365

XPS, 28–29, 37

ZEB, 244
zero energy building, 244
zone damper system, 176, 183
zoning, 112, 160, 171–172, 175, 183, 187, 240, 245, 278
Zonolite, 29